GeoPlanet: Earth and Planetary Sciences

For further volumes:
http://www.springer.com/series/8821

Paweł Rowiński
Editor

Experimental Methods in Hydraulic Research

Springer

Editor
Paweł Rowiński
Polish Academy of Sciences
Institute of Geophysics
Ks. Janusza 64
01-452 Warsaw
Poland
p.rowinski@igf.edu.pl

The GeoPlanet: Earth and Planetary Sciences Book Series is in part a continuation of Monographic Volumes of Publications of the Institute of Geophysics, Polish Academy of Sciences, the journal published since 1962 (http://pub.igf.edu.pl/index.php).

ISSN 2190-5193 e-ISSN 2190-5207
ISBN 978-3-642-26805-2 ISBN 978-3-642-17475-9 (eBook)
DOI 10.1007/978-3-642-17475-9
Springer Heidelberg Dordrecht London New York

Cover design: deblik, Berlin

Printed on acid-free paper

Springer is part of Springer Science+Business Media (www.springer.com)

Advisory Board

Tilman SPOHN
Institut für Planetenforschung
Deutsches Zentrum für Luft- und Raumfahrt
in der Helmholtz Gemeinschaft
Berlin, Germany

Krzysztof STASIEWICZ
Swedish Institute of Space Physics
Uppsala, Sweden

Roman TEISSEYRE
Earth's Interior Dynamics Lab.
Institute of Geophysics,
Polish Academy of Sciences
Warszawa, Poland

Jacek TRONCZYNSKI
Laboratory of Biogeochemistry
of Organic Contaminants,
IFREMER DCN_BE
Nantes, France

Steve WALLIS
School of the Built Environment,
Heriot-Watt University, Riccarton,
Edinburgh, Scotland, UK

Wacław M. ZUBEREK
Department of Applied Geology
University of Silesia
Sosnowiec, Poland

Preface

Hydraulic research is developing beyond traditional civil engineering to satisfy increasing demands in natural hazards assessment and also environmental research. In such conditions, experimental methods draw from various areas of human activities and research, i.e., from physics, biology, chemistry, aerospace research, oceanic research, etc. Our ability to describe processes in nature rests on the observation and experimental methods as well as on theoretical basics of various disciplines. The current volume, being a result of the meeting that took place in Wiejce Palace in Poland during the 30th International School of Hydraulics, has an ambition of presenting a kind of state-of-the-art as well as ongoing research projects in which experimental methods play a key role. It is obviously the task that may be only partially fulfilled in one volume but definitely provides a valuable material for researchers and students involved in hydraulic studies all over the world. The authors from numerous laboratories in various countries guarantee a representative sample of different studies lying at the frontier of the field.

This book covers problems differing in scale, subject, used methods, and geographical location. Chapters have been prepared by the leading experts in the field as well as by young researchers from 11 countries from four continents. There were seven invited speakers at the meeting (but only six chapters were submitted to this volume), and they prepared relevant contributions based on their lectures. In fact, two comprehensive chapters deal with various aspects of sediment transport in rivers. The chapter of Dey provides an overview of the theoretical basis of sediment transport with a special emphasis put on the problems of entrainment in the streams characterized by loose boundary. Coleman treats the problem of sediment transport from the perspective of the measuring techniques allowing to investigate the dynamics of the riverbeds. Manson describes unique series of experiments in Iceland rivers pertaining to better understanding of stream metabolism. Two contributions deal mostly with laboratory techniques and the detailed physics of particle movement in open channels. Di Cristo provided a state of the art in particle imaging velocimetry and its applications in hydraulics. Ferreira concentrated more on revealing and comparing the applicability of laser Doppler anemometry (LDA) and particle imaging velocimetry (PIV) in the studies of turbulence in open-channel flows over mobile and armored beds and to scour mechanisms. Majewski discussed a case study related to the hydrodynamics of Lake Żarnowiec influenced by nuclear and pumped storage power plant. Other chapters in the book deal with a variety of subjects devoted to hydraulic engineering with special emphasis put on experimental techniques.

The meeting in Wiejce was a jubilee 30th School of Hydraulics, and therefore, a few words on the history of this event are relevant in this place. International School of Hydraulics (till 2003 a national event) has very long tradition – it was initiated in 1981 and took place without interruption each year. Throughout all that long 25-year period, it has been successfully organized by the Institute of Hydro-Engineering of the Polish Academy of Sciences. Starting from the 26th, the organization of International Events was taken over by the Institute of Geophysics, Polish Academy of Sciences. The 26th School took place at Goniądz in Biebrza National Park, the 27th at Hucisko (the Eagle's Nest) situated in the heart of Jura Region in southern Poland, and the 28th one at the Podewils Castle in a small town of Krag, the largest fifteenth-century Knight's Castle in Pomerania. Those Schools have been carried out under the auspices of the Committee for Water Resources Management of the Polish Academy of Sciences and International Association of Hydro-Environment Engineering and Research IAHR, particularly its Committee on Experimental Methods and Instrumentation. All events turned out to be a great success and gathered many researchers from various countries, among them the European leaders in the field of hydraulics. Starting from 2008, the International School of Hydraulics takes place every second year.

The next article, 30 Years of the School of Hydraulics, is an excerpt from the lecture of Professor Majewski, the head of 25 Schools, and it provides an overview of the long history of the School of Hydraulics.

There are a few individuals that have to be acknowledged herein. The 30th International School of Hydraulics, and consequently this book would not be possible without the dedicated work of the staff members Anna Łukanowska, Anna Zdunek, Monika Kalinowska, and Robert Bialik. Anna Dziembowska carefully checked the format of all submitted chapters, made various corrections, and assured good quality of English of all chapters of the book. Special thanks go to the members of International Scientific Committee of the School for their continuous support, namely, to Andreas Dittrich, Ian Guymer, Andrea Marion, Vladimir Nikora, Steve Wallis, and Anders Wörman. All chapters have been reviewed, and I am especially grateful to Włodzimierz Czernuszenko, Janusz Kubrak, Wojciech Majewski, Marek Mitosek, Jarosław Napiórkowski, Michał Szydłowski, and Robert Bialik who assured high professional standard of all contributions.

Institute of Geophysics, Polish Academy of Sciences Paweł M. Rowiński

30 Years of the School of Hydraulics

Introduction

Thirty years passed since the first School of Hydraulics was organized. This 30-year anniversary of School of Hydraulics generates both retrospections and outlooks to the future. In such a situation, two questions may be asked:

- Where are we going? Which means, what will be the future of the School of Hydraulics?
- Where we come from? What was the beginning of the School and how it developed till the present day?

Answer to the first question belongs now mainly to the organizers of the next Schools. I would like, however, to concentrate my presentation on the past of the School of Hydraulics. What was the idea of the School, how it was organized, what topics were discussed, and what can be regarded as achievements and what as failures.

Organization of Schools

In the beginning of 1981, the Committee on Water Resources Management of the Polish Academy of Sciences (CWRM PAS) decided to initiate the School of Hydraulics (SH) under the general heading *Contemporary Hydraulic Problems of Inland Waters*. This idea was to follow, the already operating, School of Hydrology – *Contemporary Problems of Hydrology,* which was also organized under the auspices of the CWRM PAS. The initiator of the School of Hydraulics was Professor Bolesław Kordas, chairman of the CWRM PAS. Wojciech Majewski from the Institute of Hydro-engineering (IH) of the Polish Academy of Sciences, at that time associate professor, was appointed as the scientific chairman and organizer of the IH.

The main aim of the SH was to create a forum for discussion and education in the field of inland water hydraulics including also hydraulic structures. The aim was also to improve the scientific and engineering level of hydraulics, understood in a very wide sense. The School meetings were planned to take place every year in September. It was expected that participants of SH will come from universities, technical and agricultural universities, research institutes, enterprises dealing with hydraulic engineering, and administration for water resources management.

Each School was devoted to a selected topic of inland hydraulics and consisted of lectures given by eminent specialists and presentations of the participants in the form of papers or communications. These included excerpts from doctor dissertations or habilitation thesis. It was envisaged to present during SH reports from international congresses, symposia, or seminars.

Each time, the School was organized in a different place and connected with a visit to interesting hydraulic structure during construction or operation. Assistance for organization of each SH was provided by a given institute or university, which had the opportunity to present its organization and scope of research.

The first SH was organized in Osieczany in September 1981. Professor B. Kordas, who was the initiator of the SH, died tragically several days before the School started.

The Initial Years of SH

The initial years were very difficult because of political unrest in Poland, which resulted in the shortage of funds for science and difficulties in food supply. The situation worsened in 1982 when the Martial Law was introduced. All this produced additional organizational difficulties. Fortunately, due to the involvement of many people, all these problems were overcome and SH was regularly organized every year. Organization and main topics of all 30 Schools are presented in Table 1.

Further information about SH (place of the School and visited hydraulic facility) are provided in Table 2.

School Organization and Proceedings

The only records that remained from the first Schools were short reports. There was no publication of proceedings. For the first time, full proceedings including papers were prepared from the V SH (1985). Next there was a book of abstracts of lectures and papers from the X SH (1990). Since 1991 (XI SH), proceedings including lectures and papers were published regularly as monographs of the Institute of Hydro-engineering. In general, these proceedings were published before each School. All papers were reviewed before acceptance for presentation.

Table 1 Organization and main School topics

Year	School number	National or international	School organizer	Main topic of the School
1981	I	N	IH	Hydraulics of hydraulic structures
1982	II	N	IH	Flow in open channels
1983	III	N	IH	Sediment transport and its modeling
1984	IV	N	IH	Modeling of thermal regime in rivers
1985	V	N	IH	Scale effects in physical models
1986	VI	N	IH	Problems of flow in rivers and conduits
1987	VII	N	IH	Sediment movement and unsteady flow
1988	VIII	N	IH	Contemporary problems of hydraulic research and methods of solution
1989	IX	N	IH	New trends in river training
1990	X	N	IH	Unsteady flow in open channels
1991	XI	N	IH	Modeling of the flow (field and laboratory measurements
1992	XII	N	IH	Influence of hydraulic structures on the environment
1993	XII	N	IH	Hydraulics of open channels
1994	XIV	N	IH	Rivers and water resources management
1995	XV	N	IH	Physical processes in river flow
1996	XVI	N	IH	GIS in hydraulics and water management
1997	XVII	N	IH	Hydraulic aspects of floods
1998	XVIII	N	IH	Extreme phenomena in hydraulics
1999	XIX	N	IH	Flow modeling in open channels
2000	XX	N	IH	Flow modeling in open channels (continued)
2001	XXI	N	IH	Modeling of flow in open channels
2002	XXII	N	IH	Hydraulic structures and flood protection
2003	XXIII	N	IH	Modeling of floods and sediment transport
2004	XXIV	I	IH	Hydraulic problems in environmental engineering
2005	XXV	I	IH	Hydraulic problems in view of WFD
2006	XXVI	I	IG	Environmental hydraulics
2007	XXVII	I	IG	Transport phenomena in hydraulics
2008	XXVIII	I	IG	Hydraulic methods for catastrophes
2009	XXIX	N	AU	Contemporary problems of river channels in view of WFD
2010	XXX	I	IG	Experimental methods and techniques in hydraulic research

N National School in Polish language, *I* International School in English language, *IH* Institute of Hydro-engineering, Gdańsk, Polish Academy of Sciences, *IG* Institute of Geophysics, Warsaw, Polish Academy of Sciences, *AU* Agricultural University, Kraków

In 1995, it was decided to create a logo for the SH. Several proposals of logo were presented by the participants. The best one, prepared by Marian Mokwa from the Agricultural University, was finally accepted. It exists in the form of two letters in a circle SH. They can stand for Polish abbreviation (Szkoła Hydrauliki) but also for the School of Hydraulics in English. This logo exists till the present day.

Upto 2003, XXIII SH, the proceedings were published in Polish. Only some lectures presented by foreign scientists were published in English with Polish abstract.

Table 2 Location of SH and visited hydraulic facility

School number	Location of School	Visited hydraulic facility
I	Osieczany	Dobczyce Reservoir (construction)
II	Stawiska	Hydraulics Laboratory of IH
III	Janowice	Czorsztyn project (under construction)
IV	Augustów	Augustowski Navigation Canal
V	Szklarska Poręba	Tailing reservoir Żelazny Most
VI	Władysławowo	Pumped-storage power plant and nuclear power-plant (under construction)
VII	Błażejewko	Jeziorsko hydraulic project (Warta River)
VIII	Płock	Włocławek hydraulic project and Hydraulics laboratory of Hydroprojekt Włocławek
IX	Straszyn	Hydraulic powerplants on Radunia River
X	Kobyla Góra	Kobyla Góra dam
XI	Kraków – Wola Zręczycka	Dobczyce Reservoir, Czorsztyn hydraulic project (under construction)
XII	Międzyzdroje	Harbour in Świnoujście
XIII	Szczyrk	Cascade of River Soła
XIV	Lesko	Meteorological Station Lesko, Solina Dam
XV	Wrocław Trzebieszowice	Wrocław Water Node
XVI	Grodno near. Międzyzdroje	Szczecin Lagoon
XVII	Sobieszewo	Mouth of Vistula River
XVIII	Zawoja	Reservoir Świnna Poręba (under construction), Czorsztyn hydraulic project
XIX	Frombork	Ostróda-Elbląg Navigation Canal
XX	Kraków-Ustroń Jaszowiec	Upper Vistula Waterway, Wisła-Czarne dam
XXI	Sasino	Lighthouse Stilo, Sand Dunes near Łeba
XXII	Lubniewice	Boat-lift Niederfinow, Germany
XXIII	Tleń	Hydraulic structures on Wda and Brda Rivers
XXIV	Jastrzębia Góra	Pumped-storage power-plant Żarnowiec
XXV	Debrzyno	Hydraulic power-plants on Radunia River
XXVI	Bartlowizna	Biebrzański National Park
XXVII	Orle Gniazdo	Poraj dam, Jura Krakowsko-Częstochowska
XXVIII	Podewil	Hydraulic power-plants on Słupia River
XXIX	Jałowcowa Góra	Waterway of Upper Vistula, Dobczyce Dam
XXX	Wiejce	

In 2002, the Institute of Hydro-engineering won a competition for the Centre of Excellence (CE) of EU. It was decided to include Schools of Hydraulics in 2003 and 2004 into the activity of the CE. This way, two subsequent Schools, XXIV and XXV, were organized as international. Additional funds facilitated the invitation of eminent lecturers from abroad as well as foreign participants. This way, SH got an international flavor with all presentations in English.

In 2004, Prof. W. Majewski completed his position as the director of the IH. The new director of IH was no more interested in organization of SH. It was decided to transfer the organization of SH to the Institute of Geophysics of the Polish Academy of Sciences in Warsaw. Paweł Rowiński, at this time associate professor, assumed the position of scientific chairman of the School. Three subsequent SHs were organized at the international level by the Institute of Geophysics. Proceedings of these Schools were published by Springer (in Acta Geophysica) and Publications of the Institute of Geophysics (Monographic Volumes). This ISH received good international reputation and support of IAHR.

Unfortunately, the small number of participants in the ISH from Poland resulted in a discussion in the CWRM PAS and it has been decided to organize SH one year as national and the next year as international. This way, the XXIX SH was organized at the national level by the Agricultural University in Kraków. Proceedings from this SH are in preparation.

Participants

Participants represented numerous universities of technology, agricultural universities, research institutes, design and consulting offices, engineering enterprises, and water administration boards. The number of participants varied from 36 (1981) to 76 (1995) and 85 (1996). The number of participants depended not only on the topic of the School but also on the financial situation of institutions sending their participants. There were participants who regularly attended SH, while some attended just one or two SHs. Over the first 25 SHs, there were about 150 participants who attended the School.

Schools were considered not only as a scientific forum, but also as a social and very friendly meeting leading to many long-lasting friendships.

Lectures, Papers, Communications, and Discussions

Lectures were longer presentations concerning selected subjects. Papers were submitted and presented by participants. These usually concerned the topic of the School; however, other interesting papers were also admitted. In the course of time, it has been decided to distinguish the presentations of participants as papers or communications. Introduction of communications allowed, especially young scientists, to present some of their studies, which were not yet completed. In order to encourage young scientists to present papers during SH, a competition for the best paper was established. This came into force in 1999. The special evaluating committee established during each SH took into account not only the scientific value of the paper but also the form of its presentation. The best paper was awarded with a diploma.

During the first SH, six lectures and ten papers were presented. In 1995, there were eight lectures and thirty papers. Foreign lecturers were Prof. H. Kobus from Stuttgard and Prof. P. Larsen from Karlsruhe. After each presentation, there was discussion in the form of questions and comments. There were also numerous informal discussions concerning various scientific and engineering problems.

During two SHs, a discussion on teaching of hydraulics was organized. From the experience of those who had lectures on hydraulics, it appeared that one of the most difficult problems for students to understand is hydrostatics.

Since 2004, the number of foreign lecturers considerably increased as well as the number of foreign participants. This way, the idea of the International School of Hydraulics was accomplished.

School Achievements and Drawbacks

One of the achievements of SH was the amount of scientific degrees awarded to the participants. We did not have exact record of the number of doctor degrees awarded to School participants. However, we have a record of doctor habilitatus degrees awarded to School participants during the years 1981–2004. This number is quite impressive and amounts to 41. These participants came from Kraków, Warsaw, Wrocław, Poznań, Szczecin, and Gdańsk. It is worth mentioning that 14 participants out of these 41 became professors. It is not possible to assume that this is the only achievement of SH; however, it can be regarded that SH provided some kind of stimulus.

A serious drawback during this time was the lack of international contacts. Participations in foreign conferences and seminars were very few. It is difficult to find out what was the real reason for this situation: the lack of funds, low scientific or engineering level of proposed presentations, or simply insufficient command of English.

Organization of International Schools in 2003 and 2004 by the Institute of Hydro-engineering was the first step to improve the situation. Continuation in this direction by the Institute of Geophysics with Prof. Rowiński as th chairman of ISH is a very valuable activity. Unfortunately, the number of Polish participants is not sufficient and should increase.

Problems Which Formed the Topics of SH

During consecutive years, the topics were gradually evolving from studies on various types of hydraulic structures to hydraulic model investigations, including measuring techniques and scale effects. Several hydraulic laboratories existed in Poland during the initial years of SH. They performed model investigations of newly designed hydraulic structures in Poland, but there were also hydraulic laboratories mainly for teaching purposes. Some studies were carried out to better understand various physical processes existing in open channel flow as, e.g., dispersion of pollutants.

Flow in open channels was considered not only as water flow but also thermal regime, sediment transport, or influence of ice cover was taken into account. These problems were investigated on hydraulic models and in natural conditions in the form of in situ measurements. Measuring techniques were gradually developing. Some problems began to be solved by means of mathematical models. These became more and more popular. Problems of unsteady flow in open channels were becoming very important, especially because of flood problems.

Problems related to hydraulic structures were rapidly decreasing due to the fact that only few new hydraulic structures were designed and constructed. However, the problems of the influence of hydraulic structures on the environment were still present.

When Poland entered EU in 2004, numerous new problems connected with hydraulics appeared. In 2000, the Water Framework Directive was established and Poland accepted its rules. In general, it is now possible in retrospection to state that SH always tried to follow problems which were important to science and engineering.

Conclusions

Looking back to 30 years of SH, which passed, it is possible to state that this yearly meeting of scientists and specialists in the realm of hydraulics fulfilled the aims, which were assumed when the School was established. In retrospection of 25 years when I had the duty and honor to chair SH, I would like to say that the School could not be regularly organized without involvement of many participants, who prepared papers, lectures, and also served as reviewers, organizers of study tours, or preparing publications. It would be difficult to mention all their names, because the list would be very long. Some of them unfortunately passed away.

I am very glad that the School of Hydraulics has assumed now an international character. I would like to wish Professor Paweł Rowiński and the organizing team from Institute of Geophysics further success.

Committee on Water Resources Management, Wojciech Majewski
Polish Academy of Sciences

Contents

Experimental Investigations of Sandy Riverbed Morphology

Stephen Coleman

1 Introduction

The innate creation of order out of randomness that is evident in nature both entrances and mystifies. In this regard, mobile beds of granular materials almost invariably arrange themselves into highly organised patterns, with trains of regular water-formed sediment waves ubiquitous on riverbeds and often found in geological strata. In a pragmatic sense, understanding of these intriguing structures is significant in terms of analyses and management of near-bed habitats, transport of sediments and attached micro-organisms and chemicals (e.g. nutrients, contaminants), and design of structures in the fluvial environment (e.g. Amsler and García 1997; Coleman and Melville 2001).

In order to provide insight into developments in understanding of fluvial bedforms, the historical progression of observations and measurements of sand-bed morphology is overviewed in the following, culminating in a summary statement of the recognised position at the start of the twenty-first century. Based on outstanding questions of bedform dynamics at that time, the subsequent sections highlight recent investigations of fluvial bedforms, particularly highlighting the development of instruments and experimental methodologies to enable progression in understanding.

1.1 Early Observations and Measurements of Sand-Bed Morphology

With mankind's love and utilisation of natural and formed waterways, there can be little doubt that the transport of sediment, and the intrinsically associated bedforms, has been observed and studied throughout the ages, in the founding societies in

S. Coleman
Department of Civil and Environmental Engineering, The University of Auckland, Private Bag 92019, Auckland, New Zealand
e-mail: s.coleman@auckland.ac.nz

P. Rowinski (ed.), *Experimental Methods in Hydraulic Research*, Geoplanet: Earth and Planetary Sciences, DOI 10.1007/978-3-642-17475-9_1,

Mesopotamia, China, India, Egypt, and the Roman Empire, for example. Graf (1984) attributes the first basic scientific statements about loose-boundary hydraulics to the Italian hydraulician Guglielmini (1655–1710), with the French hydraulician duBuat (1734–1809) providing detailed observations of bedform mechanics, including "furrow" shapes, sizes, propagation speeds, and associated sediment motions.

In deference to the hydraulician's interest in observable bedform dynamics, geologists use bedforms preserved in stratigraphic records to infer details of historical currents and sedimentary environments. In this regard, de la Beche (1851) and Sorby (1859) discuss ripple marks formed by currents, including their orientation, shape, sizes, motion, and transport of sediment. Sorby (1859) notes that such structures "are so common that they cannot have escaped the attention of anyone who has carefully examined stratified rocks." Jukes (1862) also discusses ripple marks formed by currents, including their sizes and motions. He concludes that current ripples indicate "that the strength, velocity and mode of action of moving water in the old geological periods was precisely of the same kind and intensity as those with which we are familiar at the present day."

Efforts regarding observations of bedform dynamics in nature and through controlled experiments accelerated in the late 1800s. Raudkivi (1967) cites an expression for bedform celerity given by Partiot (1871) based on observations in the River Loire. Numerous field observations of ripple marks, including their sizes and their dynamics, are provided by Hunt (1882), with a principal focus on ripples occurring in oscillatory flows. Darwin (1883) also discusses the respective fine-sand ripples formed in uniform currents and oscillating flows. Of particular relevance to the present chapter, he describes early experimental observations of current-ripple sizes, shapes, orientation, growth, and dynamics, noting that Forel (1883) refers to bedwaves formed by continuous currents as dunes in summarising observations and his experimental studies. Deacon (1894) reports further observations of experiments regarding the orientation, shape, propagation speed, transport rate, and associated sediment dynamics for current ripples, and how these vary with increasing flow strength. At the turn of the century, Bertololy (1900) carried out studies of the formation of waves on a flattened creek bed, recording their simultaneous appearance over the whole bed, orientation, downstream migration, lengthening, and lateral uniting to form normal current ripples (Bucher 1919).

In a notable work, Cornish (1901), a student of Kumatology, describes observations of current-generated sand waves that are much larger than ripple marks, i.e. dunes in today's terminology. He comments that Reynolds (1891) noted the presence of these large tidal sand waves below low water in model estuaries, where the existence of such waves in nature had been overlooked, principally as they are out of sight, until the model results encouraged searches for them. Having sought out field sites evidencing these tidal-current sand waves under conditions favourable to examination, Cornish (1901) notes the shape, orientation, and variation with flow of these larger waves and provides measurements of lengths, heights, and migration rates. He also provides observations of sand wave growth, including inferences of bedform elimination (or coalescence as described in more recent works, e.g. Führböter 1983; Raudkivi and Witte 1990; Coleman and Melville 1994). Cornish (1901) goes further to offer a theory on the origin, growth, and decay of sand waves in currents, where these waves can appear suddenly and subsequently grow considerably, and they are mechanistically different to the smaller current ripples. He attributes the origination of sand waves to a form of deposition-scour wave (identified later by Inglis 1949; Raudkivi 1963, 1966; etc.), where the initial waves extend themselves laterally even more quickly than they grow vertically owing to near-bed fluid dynamics. Growth is suggested to be limited by a combination of the water depth and decreasing effectiveness of the lee vortex as the bedform grows. Cornish (1908) further discusses the sizes of current ripples, sand waves, and upstream-moving bedforms (antidunes in today's terminology).

Studies of ripples and larger sand waves to this time were based on observations and sparse manual measurements of sizes and migration speeds. In a novel approach, a sounding lead covered in tallow was used by Siau (1841a, b) to prove the existence of oscillatory-flow ripple marks at depths of up to 617 ft (Johnson 1916). 1910–1920 saw a notable increase in bedform measurements and experimental studies of bedform dynamics, and a consolidation of understanding of current-formed bed waves, notably through the German works of Blasius (1910) and Forchheimer (1914), and the works of Gilbert (1914), Johnson (1916), and Bucher (1919).

The study of Gilbert (1914), carried out with the assistance of Edward Charles Murphy, is recognised as a landmark experimental investigation (e.g. Bucher 1919; Kramer 1935; Kondrat'ev et al. 1959; Raudkivi 1967; Graf 1984). Although the primary aim of the study was to determine the laws governing bedload transport, the tests of sediment diameters of 0.3–7 mm involved bedforms from threshold conditions to antidunes occurring for supercritical flows. Gilbert (1914) provides detailed observations of bedform types, origination, growth, migration and speeds, sizes (and relations to controlling factors such as sediment and flow depth and speed), transport, shape and three-dimensionality, variation with discharge (including sizes, speeds, and the dune-plane bed and plane bed-antidune transitions), interactions (including coalescence, bedform dividing, generation on larger bedforms, and antidune cycling), and associated sediment dynamics (including related actions of turbulent structures). In regard to bedform types, Gilbert (1914) discusses dunes, plane bed, antidunes, and shoals, commenting that antidunes were earlier noted by Cornish (1899), and also Cornish (1908) and Owens (1908b) in the discussion and closure of Owens (1908a).

Bucher (1919) provides a review of the work to that time on the origin of ripples and related sedimentary surface forms. He discusses current ripples (conventionally parallel, but also rhomboid and linguoid, and having notable similarities to aeolian ripples in terms of characteristics and origins), meta ripples and sand waves (deemed a serious menace to navigation), related forms, and transitions between them. He summarises data on bedform sizes and speeds of movement, suggests formation mechanics (including a potential Helmholtz instability for dunes), and describes the orientations, anastamosing hierarchy, shapes and three-dimensionality, migration and speeds, sizes (including lengthening with time, and changes with increasing and decreasing velocity), transport, controlling factors (including flow depth, fluid viscosity and sediment size and density), and associated sediment dynamics (including interactions with near-bed flow structures).

A review of European hydraulic research regarding transportation by traction is given by Kramer (1935), who augments this with additional experimental results. In discussing this paper, Straub (1935) presents the results of recent North American experiments, providing measurements of ripple sizes and observations of ripple formation, shape, movement, and variation with flow velocity. Kondrat'ev et al. (1959) provide a later summary of research on the theory of sediment movement, including discussions and analyses of theoretical, experimental, and field investigations on the movement of sediment as bedforms. Regarding Russian experimental studies of bedforms up to that time, they particularly highlight the early work of Velikanov in 1923 and extensive investigations of Goncharov, Lapshin, Roborovskaya, and Pushkarev in 1935–1936. In terms of experimental methodologies, they note developing large-scale use of photography, high-speed motion pictures, and special emulsion drops in numerous experiments over the period 1936–1946 as turbulent flow structure was investigated in conjunction with bedform and sediment dynamics under the guidance of Velikanov. Kondrat'ev et al.'s summary of Russian experimental investigations finishes with the studies of Kudryashov and Znamenskaya in the 1950s. They conclude that most investigators to that time had ignored the internal flow structure and focused on determining empirical relations between bedforms and structureless-stream variables such as mean velocity, depth, slope, sediment diameter, etc. They also note that the wide variety of proposed relationships revealed the extent to which the nature of the interaction between stream and channel was still obscured. In terms of Russian field investigations of bedforms, Kondrat'ev et al. (1959) note the studies on the rivers Volga, Mologa, Luga, and Kemka over the period 1934–1935 in which bedform arrangements, sizes, speeds, transport rates, and superposition were linked to flow and sediment characteristics. They also put forward a theory in which bedform generation is linked to pressure variations at the bed surface and basic laws of oscillation.

Inglis (1949) reviews work on fluvial ripples and dunes carried out in India over the previous decade using sediments sizes of 0.2–19.4 mm. He provides observations regarding bedform types (principally ripplets, ripples, dunes, and sheet movement) and transitions between them, mechanisms of formation (including scour-deposition waves and viscous-turbulent sediment motions), controlling factors (including flow depth, velocity, discharge, sediment load, and sediment characteristics, with cohesion and grain coarseness limiting or preventing ripple formation), shapes and three-dimensionality, sizes, bedform superposition and elimination, hydraulic resistance, and associated sediment dynamics (including interactions with turbulent structures).

1.2 The Ultrasonic Depthsounder Revolution and Non-intrusive Measurements of Bed Dynamics

In 1956, the US Geological Survey established a significant research project to investigate roughness in alluvial channels. Because photographic recording and manual measurements of exposed bedforms after the draining of a channel could be misleading regarding the characteristics of dynamic bedforms, as could subaqueous measurements of the bed surface by mechanical probing (e.g. using point gauges), a sonic depth sounder was developed to allow the non-intrusive measurement of bed levels in shallow water. Measurements of dynamic bed-surface variations in space and time were readily obtainable using the new depth sounder and its relative the dual-channel stream monitor (Karaki et al. 1961; Richardson et al. 1961), from which dynamic data on bedform types, shapes, dimensions, and migration and transport rates could also be easily determined. This development served to revolutionise the measurement and analysis of dynamic subaqueous bed topography in the following decades, becoming the principal experimental tool of use and enabling landmark advances in understanding and description of bedforms (e.g. ASCE 1966; Guy et al. 1966; Nordin and Algert 1966; Ashida and Tanaka 1967; Crickmore 1967; Jain and Kennedy 1971, 1974; Wang and Shen 1980; Willis and Kennedy 1980; Nakagawa and Tsujimoto 1984; Bridge and Best 1988). Later studies used similar approaches for non-intrusive bed measurements that utilised different instruments, e.g. an infra-red probe (Richards and Robert 1986; Robert and Richards 1988) and a laser displacement meter (e.g. Nairn 1998; Coleman et al. 2003; Tuijnder et al. 2009).

1.3 Outstanding Gaps in Understanding and Recent Research

Despite the highlighted studies and advances up until the 1960s, and the significant efforts over the subsequent years to the start of the twenty-first century, ASCE (2002) observed that even then it remained difficult to give assured answers to basic questions such as how to characterise a dune-covered bed, including bedform shapes, sizes, three-dimensionality, and their statistical natures.

In the following, recent research is presented into conjecture that has been put forward regarding gaps in the understanding of fluvial bedforms. In each case, the postulated understanding is outlined, and then experimental investigations involving the writer that test the hypotheses are described. With this approach, it is intended that the use of developments in instrumentation and experimental methods to gain insight into longstanding research questions will be highlighted, along with resulting progressive advances in understanding.

2 Bedform Generation

The generation of bedforms from plane-bed conditions is typically attributed to one of three phenomena: (a) turbulent fluid motions, (b) instability of the fluid-sediment flow system when perturbed, and (c) granular transport mechanics. Through significant research efforts, there has been a wide spectrum of theories developed, with all of these theories still presenting unresolved inconsistencies. Postulated theories are discussed below along with investigations designed to test the hypotheses and resulting findings.

2.1 *Turbulent Fluid Motions and a Viscous-Fluid Flume*

As turbulence is ubiquitous in natural water flows, it is reasonable to assume that turbulence is closely associated with fluvial bedforms (e.g. Velikanov 1955; Kondrat'ev et al. 1959; Jackson 1976). Yalin (1992) ascribes the formation of alluvial dunes and bars to bursting processes associated with the turbulent nature of the flow. Raudkivi (1997) also proposes that the initiation of ripples can probably be ascribed to turbulent bursting processes that exhibit certain orders. Yalin and da Silva (2001) note that no periodic bedforms, including ripples and dunes, occur in laminar flows, thereby inferring the crucial role of turbulence in bedform origination. Challenges to the concept that bedforms are turbulence generated include that the principally random nature of turbulent events in time and space contrasts with the highly structured nature of bedforms (e.g. Liu 1957). Furthermore, turbulence-based theories do not satisfactorily explain the observed scaling of the sand waves initially formed on the bed principally with sediment size rather than flow characteristics (e.g. Coleman and Melville 1996).

The view that ripples and dunes can form only in turbulent flow is attributed by Yalin (1972) primarily to the experimental work of Tison (1949), who carried out a series of experiments to determine whether ripplemarks can be generated in uniform laminar flow. Yalin comments, however, that the experiments of Tison (1949) cannot be regarded as exhaustive. Johnson (1916) also notes the work of de Candolle (1883), who produced ripple marks artificially by experimenting, not only with sand and various substances in powdered form covered by water, but also with liquids of varying viscosity, covered with water and other liquids. De Candolle (1883) was able to make ripples in sand with a variety of fluids, but with olive oil it was found to be impossible (Darwin 1883).

In order to test whether bedform-generation is driven by flow turbulence, it was decided to investigate whether bedforms could be generated for laminar flows over planar sediment beds. To achieve this goal, a glass-walled tilting viscous-fluid channel was constructed measuring 0.3 m × 0.1 m (wide) × 2.55 m (long) with a header tank at the inlet and a collection bay at the outlet (Fig. 1a). The centrifugal pump circulating the flow was controlled using a variable speed drive, with vertical sluice gates installed at the upstream and downstream ends of the channel to aid control of flow within the flume (Coleman et al. 1998; Coleman and Eling 2000). A recess in the wooden channel base measuring 0.025 m × 0.1 m (wide) × 1.3 m (long) was filled with sediment to create an erodible bed section. Sediment was not recirculated. The fluid used in the experiments was Shell Tellus Grade 32 hydraulic oil, with the bed comprised of respective uniform sediments of median sizes $d = 0.28$–1.6 mm. For the temperature range of the experiments, the fluid density could be taken to be essentially constant at $\rho = 870$ kg/m^3. The kinematic viscosity, ν, of the oil was available in chart form as a function of temperature, with $\nu = 0.8$–1×10^{-4} m^2/s for the experiments undertaken. The particular challenge in the experiments lay in ensuring that the flows would entrain and move the sediments whilst retaining a laminar nature.

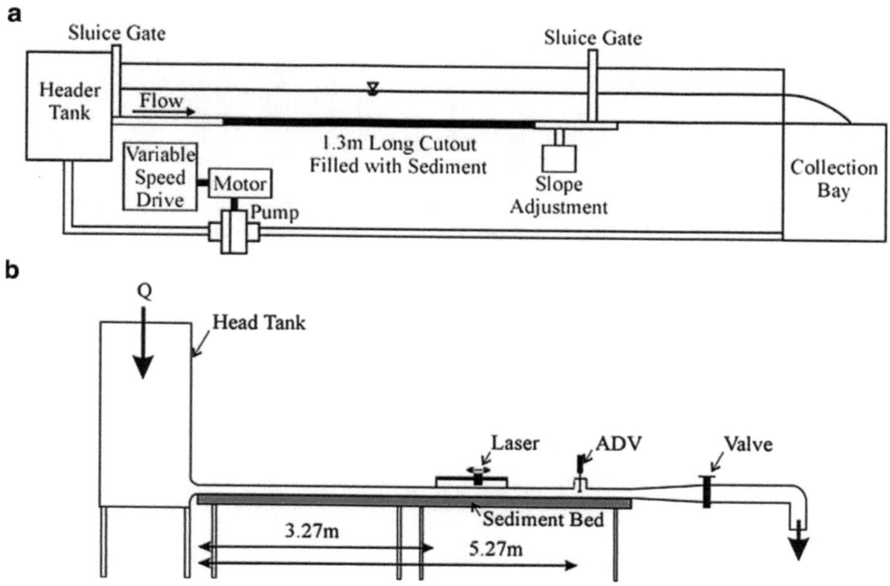

Fig. 1 Schematic drawings of experimental facilities: (**a**) viscous-fluid flume (after Coleman and Eling 2000) and (**b**) water tunnel (after Coleman et al. 2003)

At the start of a bed development run, the bed was smoothed, the settings for the desired flow were established, and then the flow was initiated. For the first runs, bedforms were simply observed. For the later runs, the flow was halted at selected stages of bed development and the centreline bed profile was measured with the channel remaining filled with oil. The bed profiles were measured using a bed-profile measurement system utilising an ultrasonic depthsounder (Coleman 1997), this system being configured to record bed elevation measurements to within ±0.4 mm every 1.2 mm along the flume. The centreline velocity profile midway along the test section was measured using a laser Doppler velocimeter for each run of measured bed profiles.

The results of the tests undertaken indicate that both antidunes and also the seed waves leading to dunes and ripples (Fig. 2) can be generated from plane-bed conditions in open-channel laminar flow, the lengths, shapes, and patterns of generation and growth (Fig. 3) for these sand waves in laminar flows being consistent with observations for alluvial flows (Coleman and Melville 1994; Coleman et al. 1998; Coleman and Eling 2000). In particular, the seed waves are of a preferred wavelength that is relatively insensitive to the characteristics of the applied flow and primarily a function of the size of the sediment. Coleman and Eling (2000) propose that these seed-wave lengths λ for alluvial and laminar open-channel flows over beds of quartz and lightweight sediments of size $d = 0.2$ to $d = 1.6$ mm can be simply described by $\lambda = 175d^{0.75}$, where λ and d are expressed in millimetres. With both ripples and dunes being postulated to subsequently develop from seed

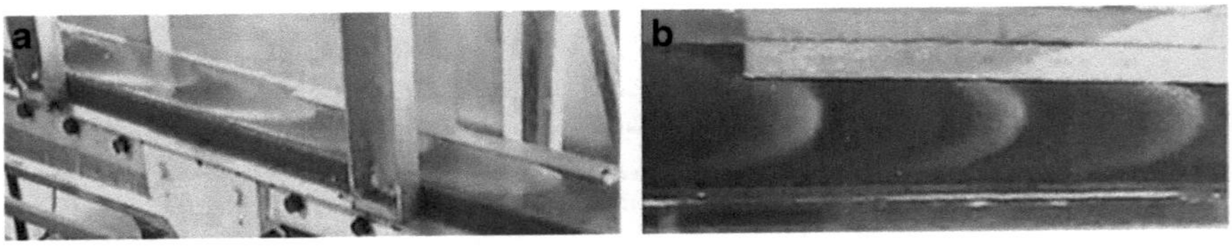

Fig. 2 Oblique (**a**) and plan (**b**) views of seed sand waves generated for a flow of depth-based Reynolds number $R = 169$, Froude number $F = 0.51$, shear velocity ratio $u_*/u_{*c} = 1.3$, flow depth $h = 0.048$ m, and $d = 0.28$ mm, with flow from *left* to *right*

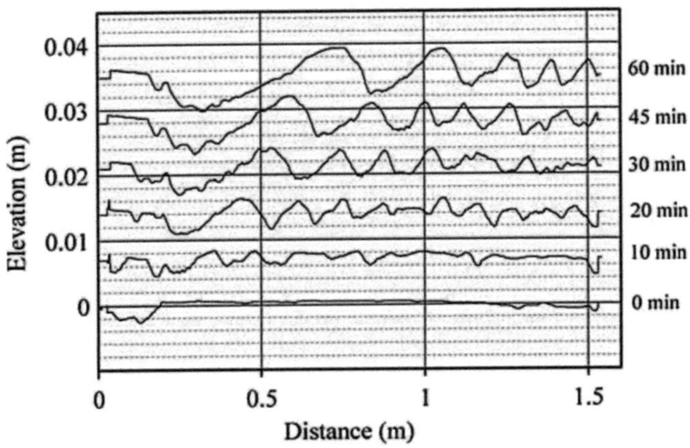

Fig. 3 The development of sand waves for Test 2-600 of $R = 441$, $F = 0.63$, $u_*/u_{*c} = 3.0$, $h = 0.068$ m, and $d = 0.44$ mm. Bed profiles have been offset vertically to aid clarity, with flow from *left* to *right*

waves for alluvial flows (e.g. Coleman and Melville 1996), the laminar-flow bedform data suggest that the generation of ripples and dunes in alluvial flows cannot be attributed to an organised structure of turbulence within the flow. Lajeunesse et al. (2010) provide interesting additional examples of fluvial and submarine morphodynamics that demonstrate the ability of laminar flow to generate morphodynamic features similar to those generally associated with turbulent flow.

2.2 Fluid-Sediment System Instability and Non-intrusive Water Tunnel Measurements

Most researchers have attributed the process of fluvial-bedform generation from plane-bed conditions to instability of the sand–water interface (e.g. Liu 1957;

Smith 1970; Reynolds 1976; Engelund and Fredsøe 1982; de Jong 1983, 1989; Sumer and Bakioglu 1984; McLean 1990). Typically, the continuity equation for the sediment phase is coupled with a suitable flow model to give predictions of unstable regions in the parameter space for the linearised system, together with the most-unstable sinusoidal-perturbation scales that are then predicted to appear in the sediment bed. Flow models that have been adopted generally include a free water surface and are based on potential flow (e.g. Kennedy 1963, 1969; Coleman and Fenton 2000), shallow-water flow (e.g. Gradowczyk 1970) or rotational flow (e.g. Engelund 1970; Fredsøe 1974; Richards 1980; and Colombini 2004).

Challenges to the concept that bedforms are due to instability of the fluid-sediment flow system include the degree to which such potential instabilities are likely to be disrupted by highly turbulent near-bed flow layers. Yalin (1992) further queries why the initial assumed perturbation for these approaches necessarily extends infinitely in the direction of flow without any change in amplitude or period. In general, these approaches also do not satisfactorily explain the observed scaling of initial seed waves principally with sediment size rather than flow characteristics (e.g. Coleman and Melville 1996).

Although instability theories primarily involve a free surface, Engelund and Fredsøe (1982) conjecture that at low flow velocities dunes occur for closed conduits as for open channels. Experimental investigations of bedforms in closed conduits are presented by Ismail (1952), who describes observations of 'dunes' and larger sand 'waves', and Nakagawa and Tsujimoto (1984), who made bed-profile measurements at stages when the flow was stopped and the conduit soffit removed.

In order to investigate whether a free fluid surface acts to control subaqueous bedform generation and growth, tests using a plexiglass mobile-bed water tunnel were undertaken at The University of Illinois. The tunnel (Fig. 1b) has a head tank at one end and a pneumatic butterfly valve at the other end. The head tank maintains pressures on the order of 2 m head along the test bed, while the valve control enables flow to be rapidly changed to a desired rate. Water is supplied to the head tank by laboratory pumps and returned to the laboratory sump via a downstream tank facilitating settlement of transported sediments. The rectangular tunnel cross section measures 0.1 m high and 0.3 m wide, where the upper surface of a 0.13-m-thick sediment bed formed the invert of a 6-m-long test section of the tunnel for the testing described herein. Pressure gradients along the tunnel can be monitored using manometers attached to pressure taps on the tunnel soffit. A specially designed mount on the soffit further enables an acoustic Doppler velocimeter (ADV) to be used to measure the centreline velocity profile.

Twelve experiments were carried out over a range of flows for two uniform quartz sediments of median sizes $d = 0.11$ and 0.87 mm, respectively. For each experiment, the 6 m length of bed was initially screeded to a constant level. The tunnel soffit was then replaced, and the measuring systems were reinstalled prior to the head tank and tunnel being filled with water. The downstream valve was closed at this stage, with inflow diverted over an overflow weir in the head tank. Pressure lines were then bled and correct operation of the measuring systems was confirmed. To commence a run, flow was initiated over a period of seconds using the pneumatic valve control. Closure of the flow control valve signalled the end of the experiment, where runs were typically of 20–60 min duration.

In a significant advance on previous studies of bedforms in closed conduits (e.g. Ismail 1952; Nakagawa and Tsujimoto 1984), bed profiles as these developed from plane-bed conditions for the experiments were automatically and non-intrusively measured through the plexiglass soffit using a high-accuracy high-frequency Keyence LB300 laser moved along the tunnel at a computer-controlled rate by a Velmex BiSlide carriage (Fig. 4a). Bed profiles were measured 0.08–0.10 m from the tunnel side wall over a length of approximately 1 m. For each run, the rate of carriage movement was selected to enable the profile length to be traversed in typically 15–20 s. The frequency of bed elevation measurement was then defined to give horizontal distances between successive elevation measurements of 1–2.5 mm for each run. Because any air bubbles along the soffit during the experiment would prevent local bed-level measurement during a run, prior to each test, a pair of magnets collocated inside and outside of the upper tunnel surface were used to sweep air bubbles from the flooded tunnel soffit along the line of the tunnel to be profiled. Bed elevations, flow velocities, and pressures were measured to resolutions of 0.03 mm, 1 mm/s, and 1 mm, respectively, for each run undertaken.

The experiments undertaken show that analogous to open-channel flows, seed waves on a planar sediment bed of a closed-conduit are instigated by discontinuities in the bed, with seed-wave lengths proposed by Coleman et al. (2003) to be given by $\lambda = 175d^{0.75}$ as for alluvial seed waves (see above). For closed-conduit flows, both ripples and dunes grow from these seed waves (at rates increasing with increasing flow strength and utilising the mechanisms of bed-form speed decreasing with size, coalescence, throughpassing, and interwave generation) to limiting lengths, heights, steepnesses, and bed friction factors that are approximately maintained or possibly decrease thereafter (Coleman et al. 2003). Figure 5 confirms the respective ripple and dune natures of the bed forms for the present finer-sediment and coarser-sediment runs, where the measured data indicate that equilibrium closed-conduit ripple and dune magnitudes can be predicted using relations derived for equivalent open-channel flows. Limitation of free-surface deformation does result in increased rates of bed-wave development for closed-conduit flows in comparison to open-channel flows. The closed-conduit tests undertaken confirm, however, that the interaction of bed waves with associated free surface waves is principally an effect, rather than a cause, of bedform initiation and growth to equilibrium.

2.3 PIV Measurements and Fluid-Sediment System Instability or Turbulent Fluid Motions

In the 1980s, particle image velocimetry (PIV) emerged as a powerful tool for simultaneously measuring velocity time series at multiple points over an extended

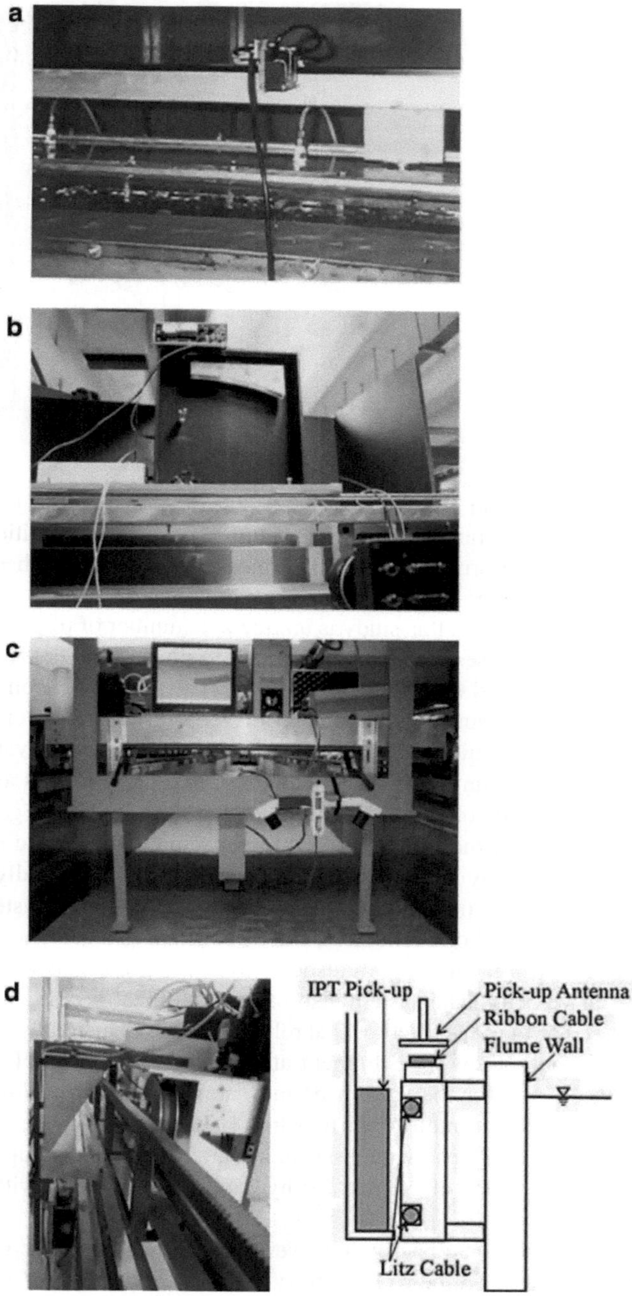

Fig. 4 Experimental equipment: (**a**) Velmex-BiSlide-mounted laser profiler measuring bed-waves; (**b**) galvanometer-driven PIV system (photo courtesy of Stuart Cameron); (**c**) 3D laser scanner measuring a rippled bed surface; and (**d**) IPT carriage for dynamic measurement of large domains

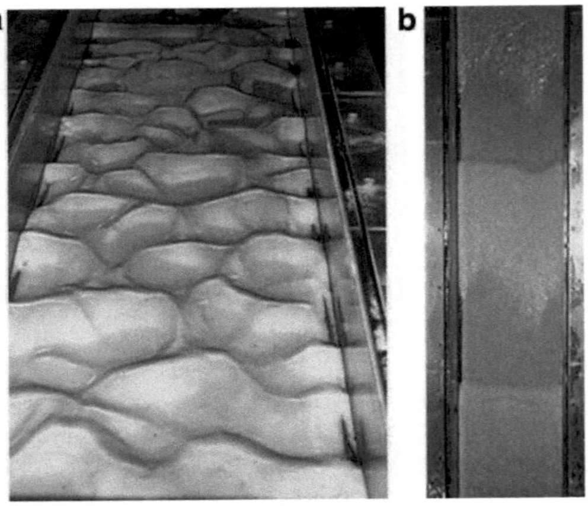

Fig. 5 End-of-run bed surfaces (tunnel soffit removed, flow from *bottom* to *top*): (**a**) Run 9 of $d = 0.11$ mm, $u_*/u_{*c} = 1.27$, and grain Reynolds number $R_* = u * d/\nu = 1.8$ and (**b**) Run 11 of $d = 0.87$ mm, $u_*/u_{*c} = 1.30$, and $R_* = 25$

area of a flow (e.g. Adrian 1986; Hesselink 1988; Gray et al. 1991). In order to further examine the potential roles of fluid-sediment system instabilities or turbulent fluid motions on bedform generation, PIV measurements of near-bed flow were made for bedforms growing from plane-bed conditions.

The PIV system utilised in this study is unique in a number of respects (Schlicke et al. 2007). The system uses a high-speed camera to image flow fields illuminated by a scanning-beam lightsheet, where the lightsheet is generated from a 5W CW 532 nm Nd:YVO4 laser using a galvanometer-driven mirror (computer controlled) together with a parabolic mirror (Fig. 4b). The system has several advantages over rotating-polygon type scanning systems, including: that the mirror is always positioned at the focal point of the parabola, and that the lightsheet generation is extremely versatile, with the lightsheet width (for a single parabolic mirror) and beam scan velocities easily and independently adjustable. Additionally, the beam scan velocity, which is typically non-linear in rotating-polygon systems due to inherent properties of parabolic reflectors, can be constant in a galvanometer-based system (giving a uniform intensity lightsheet) by driving the galvanometer at an unsteady angular velocity. The system also offers additional benefits over equivalent double-pulsed or twin-laser setups that rely on beam expansion by lens systems and that typically only allow measurement at frequencies up to 50 Hz. Integrated synchronisation options for the system permit frame-straddling techniques to be used in order to reduce interframe times to below 1/(camera frame rate). The system as designed is capable of resolving flow fields at frequencies of up to 200 Hz (covering the typical scales of interest for hydraulic researchers), with recording durations of over 8 min.

The bedform-generation experiments were made using a glass-sided tilting recirculating (water and sediment) laboratory flume (measuring 0.44 m × 0.38 m (deep) × 12 m). The tested bed was composed of a uniform quartz sediment of median size $d = 0.8$ mm. The two flows tested were of respective values of $u_*/u_{*c} = 1.8$ and 1.5, and $R_* = 28$ and 23, where the flow depth in each case was

$h = 0.125$ m. Prior to and between tests, flow was maintained in the flume (significantly below threshold for the sediment) in order to keep the 50-μm polyamide ($s = 1.03$) PIV seeding material suspended in the flow. For each test, the bed was initially smoothed using a sharp-edged scraper that was pulled along the flume. With the PIV system in operation, it took at least 9 s for seed waves to be generated after steady flow was established over the flattened bed for a given test. Detailed measurements of granular-bed and flow dynamics were made using the PIV system prior to and during development of waves in the sand bed for each test (Coleman and Nikora 2009).

Figure 6 shows normalised single-sided autospectra of near-bed streamwise velocities measured at selected elevations above the planar mobile bed over 10 s immediately prior to seed-wave generation for Test B1 of the series undertaken (Coleman and Nikora 2009). The wavenumber spectra were calculated from frequency spectra as $S_{uu}(kd)/(u_*^2) = u_e S_{uu}(\omega)/(du_*^2) = f(kd = \omega d/u_e)$, where $k = 2\pi/\lambda$ is angular wavenumber, λ is wavelength, ω is angular frequency, and eddy-convection velocity u_e is used as the normalising flow velocity owing to the near-bed limitation of Taylor's frozen turbulence hypothesis (e.g. Cameron and Nikora 2008; Coleman and Nikora 2009). The waves first generated on the planar bed for this test were of lengths of approximately 100 mm, i.e. of $kd = 0.050$. If bedforms are generated by fluid-sediment system instabilities or turbulent fluid motions, then increased spectral energy in the near-bed flow at about the initial bed wavelength of $kd = 0.050$ could be expected. Of the Fig. 6 spectra, that obtained 4.13 mm above the bed does potentially allow for a possible, albeit subtle, increase in spectral energy at $kd = 0.05$. Overall, however, clear peaks in Fig. 6 near-bed spectra are not apparent at this wavenumber, suggesting that periodic bed-forms are not generated by instabilities or turbulent events.

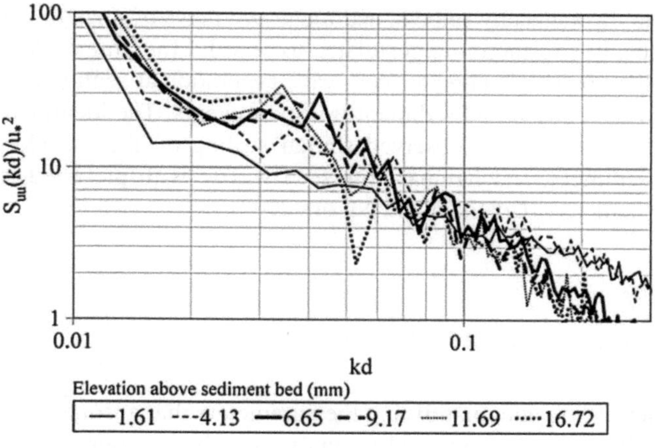

Fig. 6 Normalised autospectra of near-bed streamwise velocities

2.4 PIV Measurements and Granular Transport

Seed-wave generation via a scour-deposition wave arising with boundary-layer redevelopment downstream of a bed perturbation is the most widely considered mechanism of wave generation due to granular motions (e.g. Cornish 1901; Inglis 1949; Raudkivi 1963, 1966; Smith 1970; McLean and Smith 1986; Venditti et al. 2005a). Such an initial bed perturbation could arise through several means, including a random sediment pileup, the impacts of turbulent events on the bed (e.g. Williams and Kemp 1971; Gyr and Schmid 1989; Best 1992), or shear-wave interactions (Gyr and Kinzelbach 2004). The above-described PIV measurements of bedwave formation were able to be used to test this hypothesised generation mechanism, where edge-detection of the bed surface in recorded PIV images can be used to provide high-frequency estimates of bed levels at sub-grain-size accuracy simultaneously across the measurement window (e.g. Schlicke et al. 2005; Coleman et al. 2008b; Younkin and Hill 2009).

The PIV results suggest that the seed waves from which both ripples and dunes develop are generated on planar mobile sediment beds in a two-stage process (Coleman and Nikora 2009). The first stage comprises the motion of random sediment patches. In the second stage (Fig. 7), interactions of the moving patches result in a bed disturbance that exceeds a critical height H_c and interrupts the bed-load layer, with this disturbance then stabilising and growing and generating quasi-regular seed waves successively downstream via a scour-deposition wave. The stage-one sediment patches reflect the passage of sediment-transport events caused by attached eddies. These eddy-transport events propagate at speeds that are proportional to their size and less than overhead eddy convection velocities, but potentially larger than local average fluid and sediment velocities. The second-stage scour-deposition wave arises from the requirement of sediment mass conservation and the sediment-transport and bed-stress (τ_b) distributions downstream of a bed perturbation. This proposed two-stage generation mechanism is potentially valid for fully turbulent, hydraulically smooth, and rough bed flows of small to large sediment transport rates. It is furthermore postulated to be valid for laminar flows, although the critical disturbances leading to seed-wave generation arise through bed discontinuities, and not eddy-based sediment-transport events. The identified generation mechanism can explain the observed similar scaling of alluvial, closed-conduit, lightweight-sediment, and laminar-flow seed waves, with preferred lengths = O (130) grain diameters (Coleman and Nikora 2009).

3 Bed-Surface Characterisation

Although identifying and quantifying the parameters that adequately characterise a field of bedforms is fundamental to analyses of bed topography and its inter-relations with flow and sediment transport, even for the visually regular bedform

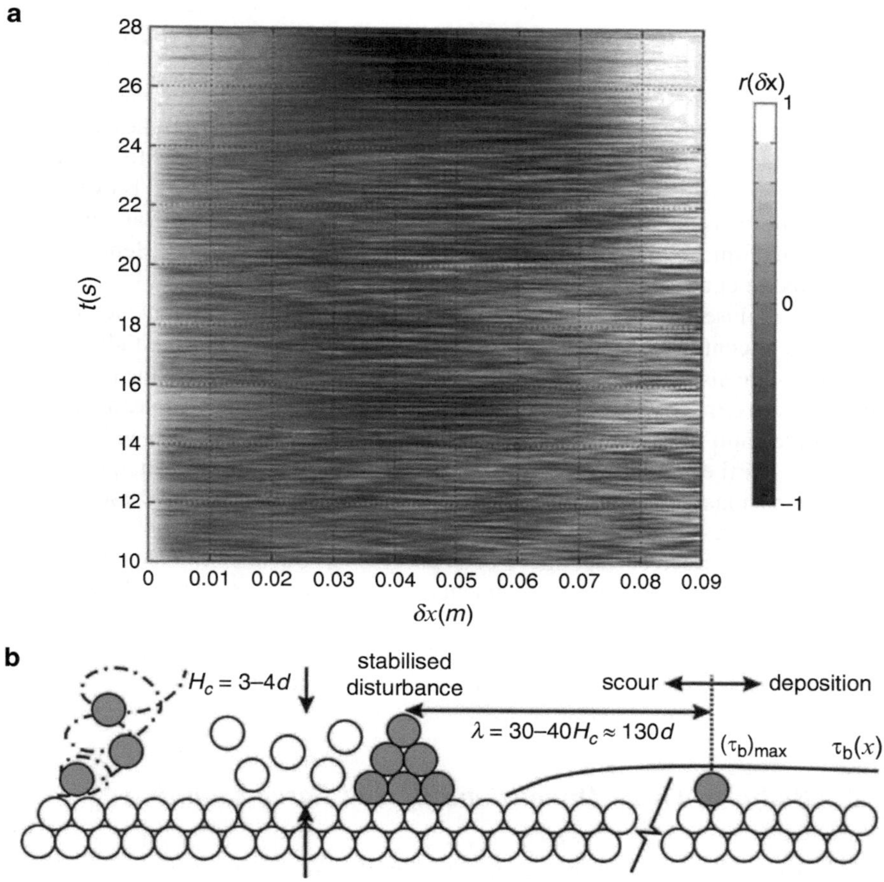

Fig. 7 (**a**) Bed-profile autocorrelation function $r(\delta x)$ for Test A2 generation of seed waves and (**b**) a schematic of the proposed two-stage seed-wave generation process. Test A2 was of $d = 0.8$ mm, $h = 0.125$ m, $U = 0.64$ m/s, and $u_* = 0.037$ m/s. The plot of autocorrelation function shows random patches prior to approximately 15 s, with a regular seed-wave length establishing between 15 s and 24 s for the test

trains of Figs. 2 and 5, it can be difficult to quantify representative bedform characteristics from the measured bed surfaces. Recent research confirms that there remains a real need to determine agreed reliable methodologies for estimating bedform characteristics (e.g. ASCE 2002; van der Mark et al. 2006, 2008; McElroy et al. 2008; Tuijnder et al. 2009; Bartholdy et al. 2010; Coleman and Nikora 2011).

Following the advent of the ultrasonic depth-sounder, a great deal of work has focused on quantification of bedform lengths and heights from measured bed profiles (e.g. Ashida and Tanaka 1967; Crickmore 1970; Nordin 1971; Jain and Kennedy

1974; Wang and Shen 1980; Haque and Mahmood 1985; Moll et al. 1987; Robert and Richards 1988; Coleman 1996; Nikora et al. 1997; McElroy et al. 2008; van der Mark et al. 2008; Tuijnder et al. 2009).

In contrast to analyses of linear sections of bed surfaces, the discussion of this section focuses on recent three- and four-dimensional (3D and 4D) measurements of bed surfaces, where the need to quantify the 3D form is recognised from detailed studies of fixed beds (e.g. Maddux et al. 2003a, b; Venditti 2007) and is even more apparent for considerations of deformable mobile beds of waves that vary markedly in space and time (e.g. Inglis 1949; ASCE 1966; Parsons et al. 2005; Henning et al. 2009; Aberle et al. 2010).

Early analyses of bedform three-dimensionality were based on viewing the crestlines or contours of individual bedforms in plan. Allen (1968) initially quantified three-dimensionality for an individual discrete bedform through the ratio of the streamwise extent of its crestline to its cross-stream crest width. He later adopted an alternative ratio of streamwise crest spacing to cross-stream crest width (Allen 1969). Venditti et al. (2005b) present a further form of this approach, quantifying three dimensionality for an individual bedform as the ratio of curved crest length to cross-stream crest width.

Recently, riverbed surfaces have been analysed as 3D random fields in contrast to combinations of discrete bed elements (e.g. Nikora et al. 1998; Goring et al. 1999; Butler et al. 2001; Nikora and Walsh 2004; Aberle and Nikora 2006). Sand-bed investigations along this line are presented below, along with comments on the approaches, example results, and discussions of implications.

3.1 Bedform Three-Dimensionality and Laser Scanning

Recognising the increasing use of lasers to profile sediment surfaces (e.g. Nairn 1998; Coleman et al. 2003; Aberle and Nikora 2006; Tuijnder et al. 2009), the measurement of bed surfaces using a 3D scanner was investigated. The Polhemus FastSCAN system utilised compiles measurements of 3D surfaces obtained by sweeping the scanning wand over the surface in a manner similar to spray painting. The handheld wand projects a narrow line of laser light onto the surface, with the intersection of the laser line and the surface in 3D space being recorded using the two cameras of the SCORPION wand used in the present tests (Fig. 4c). During scanning, the relative location and orientation of the wand is determined using an electromagnetic tracking system, enabling the 3D surface to be accurately reconstructed from the handheld scanner measurements as the scanner moves throughout space. The system is fast, flexible, portable, and easy to use and has been adopted in biomedical imaging, recording of archaeological artefacts, industrial design, and the production of movies and video games. Of particular relevance to the present study, the FastSCAN system was used by Smart et al. (2004) to analyse the roughness of alluvial beds in the field. Owing to its electromagnetic operation, the system is limited in the presence of large metal objects, which became a significant factor in the present laboratory tests. The system also relies on good reflection from scanned surfaces, inhibiting its use with dark-coloured sediments. The system can include refraction correction to enable scanning through different media, although the flume was drained prior to bed measurement for the present tests.

The flume used for the tests is a glass-sided tilting recirculating (water and sediment) laboratory flume measuring 1.5 m $\times$ 1.2 m (deep) $\times$ 45 m. The flume slope was fixed at 6.74×10^{-4} for the tests. A full-width erodible-bed test section measuring 4.6 m long $\times$ 0.6 m deep was located 25 m downstream of the entrance to the flume. Immediately upstream of this section was an additional full-width length of erodible-sediment bed measuring 1.6 m $\times$ 0.12 m and then 2 m $\times$ 0.02 m that acted to provide sediment input to the test section and also limit any potential scour-hole effects propagating into the upstream end of the test section. For a given experiment, the entire sediment bed was smoothed, the flume was filled to the required water depth, and then steady flow was run for 18–21 h for the four basic tests, or for about 30 min for the additional tests of (a) dunes for high flow velocities, and (b) measurement of a partly developed bed. A side-looking handheld ADV was used to measure a centreline velocity profile immediately prior to the end of each test. The flume was drained after each test and the bed surface measured. All runs utilised a uniform white quartz sand of median size $d = 0.2$ mm and critical shear velocity $u_{*c} = 0.013$ m/s. The flows tested covered depths of 0.15–0.55 m and shear velocities of $u_*/u_{*c} = 1.9$–4.3, producing ripples for the four basic tests and dunes for the high-velocity test.

The scanner setup to measure the exposed bed surface is shown in Fig. 4c. The wooden measuring frame was removed for flow over the bed during each test to prevent any interference with the bed development. Owing to limitations of wand position relative to the magnetic-field transmitter, the wand and tracker (the box immediately above the bed in Fig. 4c) were located approximately 0.5 and 0.2 m above the smoothed sediment bed, respectively. From this elevation, the length of the laser line emitted by the wand was approximately 250 mm in the streamwise direction. The wand was moved along the supporting fixed wooden beam (Fig. 4c) to obtain a 250 mm wide sweep of the sediment bed over a length of about 1 m across the flume. Starting from the downstream end of the test section, the carriage supporting the scanning system was then moved 150 mm in the upstream direction and another transverse sweep of the bed recorded and saved. This continued until a streamwise length of approximately 4–5 m had been measured in successive sweeps. For a given transverse sweep, the bed surface data were output in point-cloud format, with an estimated measurement accuracy of the order of ± 1 mm. The data of each sweep were mapped onto a grid of resolutions of 2 and 5 mm in the streamwise and lateral directions, respectively.

The metal frame and base of the flume introduced distortions in the magnetic field used to locate the wand position for the tests. In order to correct the measured sweeps for this distortion, a set of straight bands installed between fixed locator feet (Fig. 4c) were measured at each carriage position. The corrections required to restore the recorded curves (distorted by the metal flume components) to straight lines at each position were then imposed on the respective measured bed surfaces to

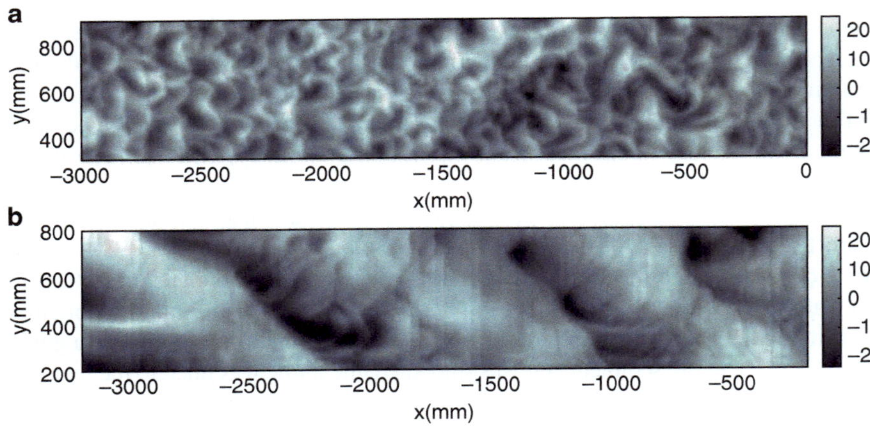

Fig. 8 Bed surfaces measured using a handheld 3D laser scanner: (**a**) rippled bed and (**b**) dune-covered bed. Bed elevations are in mm, with flow from *left* to *right*

correct these surfaces for any magnetic field distortions. The corrected sweeps were subsequently stitched together to give a final measured bed surface for each test, where the relative inter-sweep offsets required were determined based on cross-correlations of adjacent sweeps.

Figure 8 shows two surfaces obtained for a flow depth of 0.15 m. The first surface is a rippled bed obtained after 19 h for Froude number $F = 0.37$, transport stage parameter (van Rijn 1984) $T = 2.46$, stream power $\Omega = 0.79$ W/m^2, and $u_*/u_{*c} = 3.28$. The dunes of the second surface were obtained after 30 min for $F = 0.55$, $T = 6.65$, $\Omega = 2.04$ W/m^2, and $u_*/u_{*c} = 4.33$. As reflected by the results of Fig. 8, bed-surface details over a large domain can be accurately (and quickly) captured by the 3D scanning system, where as discussed in the following section, the geometry of the 2D autocorrelation function (or the closely related second-order structure function) can provide an effective means of assessing the three-dimensionality of the waves of the measured bed surfaces.

3.2 Bedform Four-Dimensionality and Flying Probes

In an attempt to address the noted absence of investigation of flow and bed development for sediment-transporting flows and naturally mobile and variable bed waves (e.g. Best 2005), the central focus of the SWAT.nz programme (Coleman et al. 2008a) was measurement and analysis of the 3D development of sand waves and associated flow from plane-bed conditions to equilibrium bed-form magnitudes. The 96 erodible-bed experiments carried out were undertaken in two glass-sided tilting recirculating (water and sediment) laboratory flumes, measuring

0.44 m × 0.38 m (deep) × 12 m and 1.5 m × 1.2 m (deep) × 45 m, respectively. The tests utilised two uniform quartz sediments of respective median sizes $d = 0.24$ and 0.8 mm.

For each test, 4D bed development from plane-bed conditions to equilibrium sand waves was recorded using a "flying-probe" configuration of a 31-transducer 5 MHz ultrasonic ranging system manufactured by Seatek. Individual transducers were polled at 5 Hz, with one of the transducers being used to map water-surface elevation, and the remainder being arranged to measure the 3D bed morphology. Bed- and free-surface elevations were measured to within ±1 mm. In the narrow flume, a chain-and-sprocket driven programmable-speed motorized carriage enabled bed surfaces measuring 6.25 m long by 0.175 m wide to be recorded every 60 s. Bed surfaces measuring 18.48 m long by 1.35 m wide were recorded every 120 s using a rack-and-pinion driven programmable-speed motorized carriage for the wide flume tests. For each flume, carriage speeds were sufficiently large compared to bed-development rates that each recorded bed surface can be assumed to be representative of a static snapshot of the dynamically changing bed.

For the wide-flume tests, the bed-sensing transducers were arranged in three rows (ten probes along each row) across the flume (Fig. 9), the carriage motion facilitating recording of ten continuous transects parallel to the flume centreline

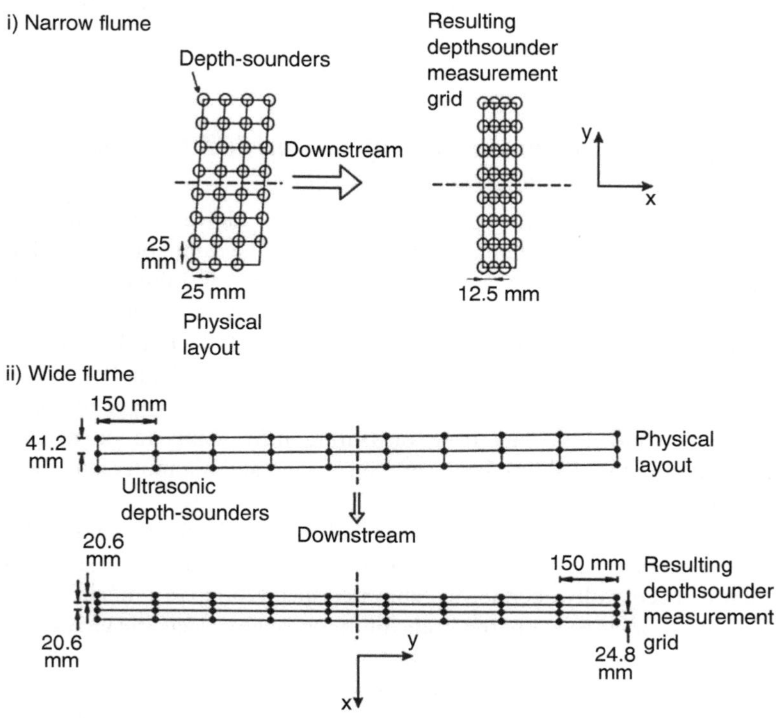

Fig. 9 Flying-probe sensor arrangements

($\Delta x = 20.6$–24.8 mm, $\Delta y = 150$ mm) for each "flight" along the flume length. The narrow-flume flying-probe bed-sensing transducers were arranged in four rows (seven-eight probes along each row) across the flume (Fig. 9), giving eight continuous transects parallel to the flume centreline, with $\Delta x = 12.5$ mm and $\Delta y = 25$ mm. For each flume, the probe positions along the flume were staggered so that regular measurement grids would be obtained when the carriage was moving and the Seatek probes were firing in sequence (Fig. 9). For each recorded bed surface, the collected data were filtered and cleaned to give a final digital elevation map (DEM) from which the bed morphology could be quantified and analysed.

Analyses of the collected flying-probe data confirm earlier indications (e.g. Nordin 1971; Goring et al. 1999; Butler et al. 2001) that the geometry of the 2D autocorrelation function (or the closely related second-order structure function) when applied to sand-bed elevation fields can provide an effective means of assessing the three-dimensionality of sand waves (e.g. Coleman et al. 2008a; Coleman and Nikora 2011). Application of 2D autocorrelation or structure function analyses to measured bed morphologies can potentially aid quantitative assessment of (a) the bed configuration per se, (b) the role of mobile-bed sand-wave form on flow resistance, (c) the effects on sand-wave form of channel side-walls and the channel aspect-ratio, and (d) the role of bedform shape on overhead turbulence and sediment-transport characteristics (e.g. Best 2005; Schindler and Robert 2005; Fernandez et al. 2006).

3.3 Bedform Four-Dimensionality and an IPT Carriage

While undertaking the flying-probe tests of the above SWAT.nz research programme, it was realised that an improved carriage system could be utilised for tests of moving instrumentation. A separate project was initiated to provide a moving platform on which transducers and sensors may be mounted, with power for the platform and the sensors supplied via an IPT (Induction Power Transfer) system. With The University of Auckland being the world leader in IPT technology, the system was also designed to enable both commands for carriage motor control and also wide-bandwidth data from measuring instruments to be dynamically transmitted across the IPT link between the carriage and a remote computer away from the flume.

The IPT carriage facility (Fig. 4d) was installed in 2005, with power supplied and data transferred across an air gap by the IPT system (Saxena 2005; Mu 2005). The carriage was successfully operated at controlled speeds of up to 2 m/s in initial tests, although the use of larger support wheels for the carriage will enable greater carriage speeds. The IPT system means that any potential hazard of accidental electric shock is essentially completely removed, and there are no trailing wires and no sliding contacts (Fig. 4d). The needs met by the IPT carriage include: safely powering a carriage moving above a water-filled flume over distances of up to 50 m; control of carriage motion remotely (direction and speed, traverse and rest periods, including operation over the Internet); remote recording of real-time high-frequency data as the carriage moves along the flume (including high-volume video records); maximisation of installation aesthetics (including minimization of system noise and cabling); and maximisation of safety (including safety of power supply in the peopled and water-filled environment, having carriage control and data-logging removed from the moving carriage, and removing the potential for snagging or damage of power and data cables as conventionally strung to moving carriages). The contactless system means that the carriage simply sits on top of flume rails, with power (for instruments and the carriage drive) and data transferred to and from the carriage in the absence of any fixed physical connections between the carriage and non-carriage objects.

With continuing development of the IPT carriage (e.g. El Sammak 2006; Ibrahim 2006), real-time data that can now be transmitted, analysed, displayed, and saved as the carriage moves include fluid velocities (via ADVs or PIV) and evolving sediment-bed morphology (e.g. using videos or acoustic or laser-based distance-measurement systems). As part of the development, a Java-based software application "E-Flume" has been developed to combine all of the functionality of existing software applications for the carriage motor and identified flume instruments (e.g. El Sammak 2006; Ibrahim 2006). Remote control of the carriage and instruments, e.g. via the Internet, has also been established utilising the IPT link.

4 Conclusions

Leonardo da Vinci's recognition of the value of experimentation is quoted by Levi (1995): 'First I shall try test by experiment before I proceed further, because my intention is to consult experience first, and then with reasoning show why such experience is bound to operate in such a way. And this is the true rule by which those who analyse the effects of nature must proceed; and though nature begins with the cause and ends with the experience, we must follow the opposite course, namely, begin with experience and by means of it investigate the cause.' Through contemporary advances in measurement and analysis frameworks and methodologies, the observations, ideas, and visions of preceding researchers regarding fluvial flows, sediment fluxes, and morphologies are now able to be tested by today's students at increasingly finer temporal and spatial resolutions. This chapter provides a review of recent progressive investigations of fluvial bed morphologies involving the writer, where the review is intended to highlight the use of developments in instrumentation and experimental methods, along with resulting progressive advances in understanding. Investigations of bedform generation using viscous-fluid, water tunnel, open-channel, and custom-built PIV facilities are presented leading to the conjecture that seed sand waves are formed from plane-bed conditions in a two-stage process that is not intrinsically due to turbulent flow structures, nor a fluid-sediment flow-system instability. Three- and four-dimensional measurements of bed-surface development using a 3D laser scanner, a flying-probe system, and an IPT carriage are also presented, where the geometry of the 2D autocorrelation function (or the closely related second-order structure function) is found to provide an effective means of assessing the three-dimensionality of waves in measured bed surfaces.

Acknowledgments I am very grateful to the organising committee of the 30th International School of Hydraulics for their invitation to present this chapter as a lecture. I also acknowledge the support of this committee and the University of Auckland Cross-faculty Research Initiatives Fund to enable my attendance at the school. I wish to thank the colleagues and students who have shared the ideas and adventures leading to the studies described herein, including Bruce Melville, Burkhard Eling, Graeme Twose, Marcelo Garcia, Juan Fedele, Wylie Wong, Rodrigo Musalem, Kirsty Coleman, Vladimir Nikora, Derek Goring, Dougal Clunie, Heide Friedrich, Ted Schlicke, Stuart Cameron, Andries Paarlberg, Joost Lansink, Mark Trevethan, John Cater, Azin Kusari, John Boys, and Grant Covic. In particular, experimental studies are indebted to the technicians who can turn ideas and fancy into reality. In this regard, the present studies have particularly benefitted from the technical skills of Jim Bickner, Ray Hoffman, Gary Carr, Mark Twiname, Jim Luo, and Geoff "Mythbuster" Kirby. The writer's research discussed in this paper was partly funded by the Marsden Fund (Grant UOA220) administered by the Royal Society of New Zealand.

References

Aberle J, Nikora V (2006) Statistical properties of armored gravel bed surfaces. Water Resour Res 42:W11414. doi:10.1029/2005WR004674

Aberle J, Nikora V, Henning M, Ettmer B, Hentschel B (2010) Statistical characterization of bed roughness due to bed forms: a field study in the Elbe River at Aken, Germany. Water Resour Res 46:W03521. doi:10.1029/2008WR007406

Adrian RJ (1986) Multi-point optical measurements of simultaneous vectors in unsteady flow – a review. Int J Heat Fluid Flow 7:127–145

Allen JRL (1968) Current ripples: their relation to patterns of water and sediment motion. Elsevier, New York

Allen JRL (1969) Some recent advances in the physics of sedimentation. Proc Geol Assoc 80:1–42

Amsler ML, García MH (1997) Discussion of 'Sand-dune geometry of large rivers during floods' by Julien PY and Klaassen GJ. J Hydraul Eng ASCE 123(6):582–584

ASCE Task Committee on Flow and Transport over Dunes (2002) Flow and transport over dunes. J Hydraul Eng ASCE 128(8):726–728

ASCE Task Force on Bed Forms in Alluvial Channels of the Committee on Sedimentation (1966) Nomenclature for bed forms in alluvial channels. J Hydraul Div ASCE 92(HY3):51–64

Ashida K, Tanaka Y (1967) A statistical study of sand waves. Proc XII Congr Int Assoc Hydraul Res 2:103–110

Bartholdy J, Flemming BW, Ernstsen VB, Winter C, Bartholomä A (2010) Hydraulic roughness over simple subaqueous dunes. Geo-Mar Lett 30(1):63–76

Bertololy E (1900) Kräuselungsmarken und Dünen. Münchener Geogr. Studien: 9tes Stück (in German)

Best JL (1992) On the entrainment of sediment and initiation of bed defects: insights from recent development within turbulent boundary layer research. Sedimentology 39:797–811

Best J (2005) The fluid dynamics of river dunes: a review and some future research directions. J Geophys Res 110:F04S02. doi:10.1029/2004JF000218

Blasius H (1910) Über die Abhängigkeit der Formen der Riffeln und Geschiebebänke vom Gefläle. Zeitschrift für Bauwesen 60:466–472 (in German)

Bridge JS, Best JL (1988) Flow, sediment transport and bedform dynamics over the transition from dunes to upper-stage plane beds: implications for the formation of planar laminae. Sedimentology 35:753–764

Bucher WH (1919) On ripples and related sedimentary surface forms and their paleogeographic interpretation. Am J Sci 279:149–210, Fourth Series XLVII

Butler JB, Lane SN, Chandler JH (2001) Characterization of the structure of river-bed gravels using two-dimensional fractal analysis. Math Geol 33(3):301–330

Cameron SM, Nikora VI (2008) Eddy convection velocity for smooth- and rough-bed open-channel flows: particle image velocimetry study. In: Proceedings River Flow 2008, Izmir Turkey, 3–5 Sept 2008, pp 143–150

Coleman SE (1996) Wave generation and development on a sandy river bed. Discussion of 'The stability of a sandy river bed' by J Fredsøe. In: Nakato T, Ettema R (eds) Issues and directions in hydraulics. A. A. Balkema, Rotterdam, pp 145–155

Coleman SE (1997) Ultrasonic measurement of sediment bed profiles. In: Proceedings of the 27th congress of the international association for hydraulic research, San Francisco, Aug, pp B221–B226

Coleman SE, Eling B (2000) Sand wavelets in laminar open-channel flows. J Hydraul Res IAHR 38(5):331–338

Coleman SE, Eling B, Twose G (1998) Sand-wave formation in laminar open-channel flow. In: Proceedings of the seventh international symposium on river sedimentation, Hong Kong, China, 16–18 Dec 1998, pp 73–78

Coleman SE, Fenton JD (2000) Potential-flow instability theory and alluvial stream bed forms. J Fluid Mech 418:101–117

Coleman SE, Melville BW (1994) Bed-form development. J Hydraul Eng ASCE 120(4): 544–560

Coleman SE, Melville BW (1996) Initiation of bed forms on a flat sand bed. J Hydraul Eng ASCE 122(6):301–310

Coleman SE, Melville BW (2001) Case study: New Zealand bridge scour experiences. J Hydraul Eng ASCE 127(7):535–546

Coleman SE, Nikora VI (2009) Bed and flow dynamics leading to sediment-wave initiation. Water Resour Res 45:W04402. doi:10.1029/2007WR006741

Coleman SE, Nikora V (2011) Fluvial dunes: initiation, characterisation, flow structure. Earth Surf Process Land 36(1):39–57

Coleman SE, Fedele JJ, García MH (2003) Closed-conduit bedform initiation and development. J Hydraul Eng ASCE 129(12):956–965

Coleman SE, Nikora VI, Melville BW, Goring DG, Clunie TM, Friedrich H (2008a) SWAT.nz: New-Zealand-based 'Sand waves and turbulence' experimental programme. Acta Geophys 56(2):417–439

Coleman SE, Nikora VI, Schlicke T (2008b) Spatially-averaged oscillatory flow over a rough bed. Acta Geophys 56(3):698–733

Colombini M (2004) Revisiting the linear theory of sand dune formation. J Fluid Mech 502:1–16

Cornish V (1899) On Kumatology (the study of the waves and wave-structures of the atmosphere, hydrosphere, and lithosphere). Geogr J 13(6):624–626

Cornish V (1901) On sand-waves in tidal currents. Geogr J 18(2):170–200

Cornish V (1908) Discussion of 'Experiments on the transporting power of sea currents' by JS Owens. Geogr J 31(4):421–423

Crickmore MJ (1967) Measurement of sand transport in rivers with special reference to tracer methods. Sedimentology 8:175–228

Crickmore MJ (1970) Effect of flume width on bed-form characteristics. J Hydraul Div ASCE 96 (HY2):473–496

Darwin GH (1883) On the formation of ripple-mark in sand. Proc R Soc Lon 36:18–43

de Candolle C (1883) Rides formées à la surface du sable déposé au fond de l'eau et autres phénomènes analogues. Archives des Sciences Physiques et Naturelles, Geneve No. 3, vol IX, pp 241–278 (in French)

de Jong B (1983) The formation of dunes in open channel flow on an initially flattened erodible bed. In: Proceedings of the Euromech 156 – Mechanics of sediment transport, Istanbul, 12–14 July 1982, pp 119–126
de Jong B (1989) Bed waves generated by internal waves in alluvial channels. J Hydraul Eng ASCE 115(6):801–817
de la Beche HT (1851) The geological observer. Longman, Brown, Green and Longmans, London
Deacon GF (1894) Discussion of paper on 'Estuaries' by HL Partiot. Proc Inst Civil Eng CXVIII:47–189
El Sammak R (2006) Interactive instrumentation for civil engineering flume. In: Year 4 research projects. Department of Electrical and Computer Engineering, The University of Auckland, Auckland, New Zealand, 10 pp
Engelund FA (1970) Instability of erodible beds. J Fluid Mech 42:225–244
Engelund F, Fredsøe J (1982) Sediment ripples and dunes. Annu Rev Fluid Mech 14:13–37
Fernandez R, Best J, López F (2006) Mean flow, turbulence structure, and bed form superimposition across the ripple-dune transition. Water Resour Res 42:W05406. doi:10.1029/2005 WR004330
Forchheimer P (1914) Hydraulik. Teubner Verlagsgesellschaft, Leipzig, Berlin (in German)
Forel FA (1883) Les rides de fond étudiés dans le Lac Leman. Arch Sci Phys Nat 10(3):39–72
Fredsøe J (1974) On the development of dunes in erodible channels. J Fluid Mech 64(1):1–16
Führböter A (1983) Zur Bildung von makroskopischen Ordnungsstrukturen (Strömungsriffel und Dünen) aus sehr kleinen Zufallsstörungen. Mitteilungen des Leichtweiss-Instituts, vol 79. Technical University of Braunschweig Braunschweig, Germany, pp 1–51 (in German)
Gilbert GK (1914) The transportation of débris by running water. US Geological Survey Professional Paper 86, Government Printing Office, Washington, DC
Goring D, Nikora V, McEwan I (1999) Analysis of the texture of gravel beds using 2-D structure functions. In: River, coastal, and estuarine morphodynamics, Proceedings of the IAHR symposium 2, Geneva, Italy, pp 111–120
Gradowczyk MH (1970) Wave propagation and boundary instability in erodible-bed channels. J Fluid Mech 33:93–112
Graf WH (1984) Hydraulics of sediment transport. Water Resources Publications, Littleton
Gray C, Greated CA, McCluskey DR, Easson WJ (1991) An analysis of the scanning beam PIV illumination system. Meas Sci Technol 2:717–724
Guy HP, Simons DB, Richardson EV (1966) Summary of alluvial channel data from flume experiments, 1956–61. U.S. Geological Survey Professional Paper 462–1, Government Printing Office, Washington, DC
Gyr A, Kinzelbach W (2004) Bed forms in turbulent channel flow. Appl Mech Rev 57(1):77–93
Gyr A, Schmid A (1989) The different ripple formation mechanism. J Hydraul Res 27(1):61–74
Haque MI, Mahmood K (1985) Analytical study on steepness of ripples and dunes. J Hydraul Eng ASCE 112(3):220–236
Henning M, Hentschel B, Hüsener T (2009) Photogrammetric system for measurement and analysis of dune movement. In: Proceedings of the 33rd IAHR congress, Vancouver, Canada, 8 pp
Hesselink L (1988) Digital image processing in flow visualization. Annu Rev Fluid Mech 20:421–485
Hunt AR (1882) On the formation of ripplemark. Proc R Soc Lon 34:1–18
Ibrahim H (2006) Interactive instrumentation for a civil engineering flume. In: Year 4 research projects. Department of Electrical and Computer Engineering, The University of Auckland, Auckland, New Zealand, 10 pp
Inglis CC (1949) The behaviour and control of rivers and canals (with the aid of models). Research Publication No. 13, Central Waterpower Irrigation and Navigation Research Station, Poona, India
Ismail HM (1952) Turbulent transfer mechanism and suspended sediment in closed channels. Trans ASCE 117:409–434
Jackson RG (1976) Sedimentological and fluid-dynamic implications of the turbulent bursting phenomenon in geophysical flows. J Fluid Mech 77:531–560

Jain SC, Kennedy JF (1971) The growth of sand waves. In: Chiu C-L (ed) Stochastic hydraulics. Proceedings of the 1st international symposium on stochastic hydraulics, University of Pittsburgh, Pittsburgh, pp 449–471

Jain SC, Kennedy JF (1974) The spectral evolution of sedimentary bed forms. J Fluid Mech 63:301–314

Johnson DW (1916) Contributions to the study of ripple marks. J Geol 24(8):809–819

Jukes JB (1862) The student's manual of geology. Adam and Charles Black, Edinburgh, UK

Karaki SS, Gray EE, Collins J (1961) Dual channel stream monitor. J Hydraul Div ASCE 87(HY6):1–16

Kennedy JF (1963) The mechanics of dunes and antidunes in erodible bed channels. J Fluid Mech 16:521–544

Kennedy JF (1969) The formation of sediment ripples, dunes and antidunes. Annu Rev Fluid Mech 1:147–168

Kondrat'ev NE, Lyapin AN, Popov IV, Pin'kovskii SI, Fedorov NN, Yakunin II (1959) Channel processes. Gidrometeoizdat, Leningrad

Kramer H (1935) Sand mixtures and sand movement in fluvial models. Trans ASCE 100:798–838

Lajeunesse E, Malverti L, Lancien P, Armstrong L, Metivier F, Coleman S, Smith CE, Davies T, Cantelli A, Parker G (2010) Fluvial and subaqueous morphodynamics of laminar and near-laminar flows: a synthesis. Sedimentology 57:1–26

Levi E (1995) The science of water: the foundation of modern hydraulics. ASCE Press, New York

Liu H-K (1957) Mechanics of sediment-ripple formation. J Hydraul Div 2:1–23, Paper 1197

Maddux TB, Nelson JM, McLean SR (2003a) Turbulent flow over three-dimensional dunes: 1. Free surface and flow response. J Geophys Res 108(F1). doi:10.1029/2003/JF000017

Maddux TB, McLean SR, Nelson JM (2003b) Turbulent flow over three-dimensional dunes: 2. Fluid and bed stresses. J Geophys Res 108(F1). doi:10.1029/2003/JF000018

McElroy B, Mohrig D, Blom A (2008) Determining characteristic scales for the dynamics and geometry of sandy bedforms. In: Proceedings of the marine and river dune dynamics, Leeds, UK, 1–3 Apr 2008, pp 219–225

McLean SR (1990) The stability of ripples and dunes. Earth Sci Rev 29:131–144

McLean SR, Smith JD (1986) A model for flow over two-dimensional bed forms. J Hydraul Eng ASCE 112:300–317

Moll JR, Schilperoort T, de Leeuw AJ (1987) Stochastic analysis of bedform dimensions. J Hydraul Res IAHR 25(4):465–479

Mu YM (2005) IPT system for a civil engineering flume. In: Proceedings of the 2005 year 4 research projects, Department of Electrical and Computer Engineering, vol 2. The University of Auckland, Auckland, New Zealand, pp 60–67

Nairn BJ (1998) Incipient transport of silt-sized sediments. Rep. KH-R-59, WM Keck Laboratory of Hydraulics and Water Resources. California Institute of Technology, Pasadena, California

Nakagawa H, Tsujimoto T (1984) Spectral analysis of sand bed instability. J Hydraul Eng ASCE 110(4):467–483

Nikora V, Walsh J (2004) Water-worked gravel surfaces: high-order structure functions at the particle scale. Water Resour Res 40:W12601. doi:10.1029/2004WR003346

Nikora VI, Sukhodolov AN, Rowinski PM (1997) Statistical sand wave dynamics in one-directional water flows. J Fluid Mech 351:17–39

Nikora VI, Goring DG, Biggs BJF (1998) On gravel-bed roughness characterization. Water Resour Res 34(3):517–527

Nordin CF (1971) Statistical properties of dune profiles. US Geological Survey Professional Paper 562F, Government Printing Office, Washington, DC

Nordin CF, Algert JH (1966) Spectral analysis of sand waves. J Hydraul Div ASCE 92 (HY5):95–114

Owens JS (1908a) Experiments on the transporting power of sea currents. Geogr J 31(4): 415–420

Owens JS (1908b) Discussion of 'Experiments on the transporting power of sea currents' by JS Owens. Geogr J 31(4):424–425

Parsons DR, Best JL, Orfeo O, Hardy RJ, Kostaschuk R, Lane SN (2005) Morphology and flow fields of three-dimensional dunes, Río Paraná, Argentina: results from simultaneous multi-beam echo sounding and acoustic Doppler current profiling. J Geophys Res 110:F04S03. doi:10.1029/2004JF000231

Partiot HL (1871) Mémoire sur les sables de la Loire. Ann Ponts Chaussees 5(1):233–292

Raudkivi AJ (1963) Study of sediment ripple formation. J Hydraul Div ASCE 89(HY6):15–33

Raudkivi AJ (1966) Bed forms in alluvial channels. J Fluid Mech 26(3):507–514

Raudkivi AJ (1967) Loose boundary hydraulics, 1st edn. Pergamon, New York

Raudkivi AJ (1997) Ripples on stream bed. J Hydraul Eng ASCE 123(1):58–64

Raudkivi A, Witte H-H (1990) Development of bed features. J Hydraul Eng ASCE 116(9): 1063–1079

Reynolds O (1891) Third report of the committee appointed to investigate the action of waves and currents on the beds and foreshores of estuaries by means of working models. British association report

Reynolds AJ (1976) A decade's investigation of the stability of erodible stream beds. Nord Hydrol 7:161–180

Richards KJ (1980) The formation of ripples and dunes on an erodible bed. J Fluid Mech 99(3):597–618

Richards KS, Robert A (1986) Laboratory experiments with the HR multipurpose profile follower on a rippled sand bed, Department of Geography Working Paper, University of Cambridge, Cambridge, 22 pp

Richardson EV, Simons DB, Posakony GJ (1961) Sonic depth sounder for laboratory and field use, US Geological Survey Circular 450, Washington DC

Robert A, Richards KS (1988) On the modelling of sand bedforms using the semivariogram. Earth Surf Process Land 13:459–473

Saxena M (2005) An inductive power transfer system for flume in Civil Engineering. In: Proceedings of the 2005 year 4 research projects, Department of Electrical and Computer Engineering. vol 2. The University of Auckland, Auckland, New Zealand, pp 51–59

Schindler RJ, Robert A (2005) Flow and turbulence structure across the ripple-dune transition: an experiment under mobile bed conditions. Sedimentology 52:627–649

Schlicke E, Coleman SE, Nikora VI (2005) A PIV investigation into the interaction between wave motion and sediment ripples. In: 4th IAHR symposium on river, coastal and estuarine morphodynamics, Urbana, 4–7 Oct 2005, pp 981–989

Schlicke T, Cameron SM, Coleman SE (2007) Galvanometer-based PIV for liquid flows. Flow Meas Instrum 18:27–36

Siau M (1841a) Observations diverses faites en 1839 et 1840, pendant un voyage a l'Ile Bourbon. Comptes rendus de l'academie des sciences XII:774–775

Siau M (1841b) Action des vagues à de grandes profondeurs. Annales de chimie et de physique 2(3):118

Smart G, Aberle J, Duncan M, Walsh J (2004) Measurement and analysis of alluvial bed roughness. J Hydraul Res IAHR 42(3):227–237

Smith JD (1970) Stability of a sand bed subjected to a shear flow of low Froude number. J Geophys Res 75(30):5928–5940

Sorby HC (1859) On the structures produced by the currents present during the deposition of stratified rocks. Geologist 2:137–147

Straub L (1935) Discussion of 'Sand mixtures and sand movement in fluvial models' by H Kramer. Trans ASCE 100:867–873

Sumer BM, Bakioglu M (1984) On the formation of ripples on an erodible bed. J Fluid Mech 144:177–190

Tison LJ (1949) Origine des ondes de sable et des bancs de sable sous l'action des courants. 3rd congress IAHR, Grenoble, France, Report II-13

Tuijnder AP, Ribberink JS, Hulscher SJMH (2009) An experimental study into the geometry of supply-limited dunes. Sedimentology 56(6):1713–1727

van der Mark CF, Blom A, Hulscher SJMH, Leclair SF, Mohrig D (2006) On modeling the variability of bedform dimensions. In: 4th IAHR symposium on river, coastal and estuarine morphodynamics, Urbana, 4–7 Oct 2006, pp 831–841

van der Mark CF, Blom A, Hulscher SJMH (2008) Quantification of variability in bedform geometry. J Geophys Res 113(10.1029/2007JF000940):F03020. doi:10.1029/2007JF000940

van Rijn LC (1984) Sediment transport. Part III: Bed forms and alluvial roughness. J Hydraul Eng ASCE 110(12):1733–1754

Velikanov MA (1955) Dynamics of alluvial streams, vol II, Sediment and bed flow. State Publishing House for Theoretical and Technical Literature, Moscow

Venditti JG (2007) Turbulent flow and drag over fixed two- and three-dimensional dunes. J Geophys Res 112:F04008. doi:10.1029/2006JF000650

Venditti JG, Church MA, Bennett SJ (2005a) Bed form initiation from a flat sand bed. J Geophys Res 110:F01009. doi:10.1029/2004JF000149

Venditti JG, Church M, Bennett SJ (2005b) On the transition between 2D and 3D dunes. Sedimentology 52:1343–1359

Wang WC, Shen HW (1980) Statistical properties of alluvial bed forms. In: Proceedings of the third international symposium on stochastic hydraulics, Tokyo, Japan, 5–7 Aug 1980, pp 371–389

Williams PB, Kemp PH (1971) Initiation of ripples on flat sediment beds. J Hydraul Div ASCE 97(4):505–522

Willis JC, Kennedy JF (1980) Sediment transport in migrating bedforms. In: Shen HW, Kikkawa H (eds) Application of stochastic processes in sediment transport. Water Resources Publications, Littleton, pp 6b/1–6b/32

Yalin MS (1972) Mechanics of sediment transport. Pergamon, New York

Yalin MS (1992) River mechanics. Pergamon Press, New York

Yalin MS, da Silva AMF (2001) Fluvial processes. IAHR, Delft, The Netherlands

Younkin BD, Hill DF (2009) Rapid profiling of an evolving bed form using planar laser sheet illumination. J Hydraul Eng ASCE 135(10):852–856

Entrainment Threshold of Loose Boundary Streams

Subhasish Dey

1 Introduction

When a turbulent flow interacts with a loose boundary composed of noncohesive sediments, hydrodynamic forces are exerted on the sediment particles forming the boundary (henceforth bed). With an increase in flow velocity, the sediment particles on the bed surface are intermittently entrained at a random rate if the magnitude of the induced hydrodynamic forces acting on the sediment particles exceeds a certain threshold value. The condition that is just adequate to initiate sediment motion is termed *threshold* or *critical condition of sediment entrainment*. Importantly, the induced boundary shear stress of the stream flow in excess of that of the stream flow in threshold condition governs the sediment transport rate.

The doctoral research study by A.F. Shields (1936) on sediment movement conducted in the Technischen Hochschule Berlin was a phenomenal contribution (Kennedy 1995). His major finding was his diagram, well known as Shields diagram, that represents the variation of nondimensional threshold boundary shear stress (or threshold Shields parameter) with shear Reynolds number corresponding to the threshold of sediment entrainment. It is considered to be the reference of any sediment transport research. His pioneering work which is widely applied to the fields has inspired numerous researchers conducting further studies. However, not many attempts were made before Shields (1936), but they were mostly empirical with limited applicability. Despite the fact that the Shields diagram is widely applied, even as of today, researchers have expressed some dissatisfactions (Mantz 1977; Miller et al. 1977; Yalin and Karahan 1979; Buffington 1999), since the diagram less complies with the experimental data plots in the smooth and rough-flow regimes (Yalin and Karahan 1979). Thus, further attempts have so far been made to modify the Shields diagram, conducting additional experiments and analyzing the problem theoretically based on deterministic

S. Dey
Department of Civil Engineering, Indian Institute of Technology, Kharagpur, West Bengal 721302, India
e-mail: sdey@iitkgp.ac.in

P. Rowinski (ed.), *Experimental Methods in Hydraulic Research*,
Geoplanet: Earth and Planetary Sciences, DOI 10.1007/978-3-642-17475-9_2,

and probabilistic approaches. Miller et al. (1977), Buffington and Montgomery (1997), Paphitis (2001), and Dey and Papanicolaou (2008) presented a survey on this topic. However, after the discovery of the bursting phenomenon in turbulent flows (Kline et al. 1967), it has created a new look to further explore the sediment entrainment problem. The turbulence is so far introduced as an average like Reynolds shear stress. The conditional statistics towards the bursting events can be the obvious treatment of the sediment entrainment problem, as the most important turbulent events remain implicit with an averaging process. Therefore, the merger of turbulence with a sediment entrainment theory demands its way in between a deterministic and a probabilistic approach. It leads to an open question that to what extent the micromechanical process can be studied in a deterministic framework and when the results can be determined by a probabilistic approach.

A brief perspective review of the important laboratory experimental and theoretical studies on entrainment threshold of sediments under steady stream flows is presented, highlighting the empirical formulations and semitheoretical analyses. Special attention is given towards the role of the turbulent bursting on sediment entrainment.

2 Definition of Entrainment Threshold of Sediments

It is always difficult to set a clear definition of the threshold of sediment entrainment. First type of definition corresponds to the sediment flux. Shields (1936) suggested that the boundary shear stress has a value for which the extrapolated sediment flux vanishes. On the other hand, USWES (1936) put forward that the tractive force is such that produces a *general motion* of bed particles. For the median diameter of sediment particles less than 0.6 mm, this concept was found to be invalid. Thus, the general motion was redefined that the sediment in motion should reasonably be represented by all sizes of bed particles, such that the sediment flux should be greater than 4.1×10^{-4} kg/sm. Paintal (1971) suggested from stochastic viewpoint that due to the fluctuating mode of the instantaneous velocity, there is no mean boundary shear stress below which there is no flux. With this consideration, the threshold condition was defined as the boundary shear stress that produces a certain minimal amount of sediment flux.

Second type of definition corresponds to the bed particle motion. Kramer (1935) defined four types of boundary shear stress conditions for which: (1) no particles are in motion, termed *no transport*; (2) a small number of smallest particles are in motion at isolated zones, termed *weak transport*; (3) many particles of mean size are in motion, termed *medium transport*; and (4) particles of all sizes are in motion at all points and at all times, termed *general transport*. However, Kramer (1935) expressed the difficulty in setting a clear demarcation between these regimes, but defined threshold boundary shear stress to be the stress that initiates a *general transport*. Vanoni (1964) proposed that the sediment threshold is the condition of particle motion in every 2 s at any location of a bed. Different threshold definitions that were in use in various studies leading discrepancies in the data sets and introducing difficulties in making comparisons (Paintal 1971; Buffington and Montgomery 1997).

3 Competent Velocity Concept

A competent velocity is a velocity at the particle level (boundary velocity) or the depth-averaged velocity, which is just adequate to start the particle movement for a given size. Goncharov (1964) used the competent velocity as detachment velocity U_n. It was defined as the lowest average velocity at which individual particles continually detach from the bed. He gave an equation of U_n as:

$$U_n = \log(8.8h/d)(0.57\Delta gd)^{0.5} \tag{1}$$

where h = flow depth; d = representative particle diameter, that is median particle diameter; g = acceleration due to gravity; $\Delta = s - 1$; s = relative density of sediment, that is ρ_s/ρ; ρ_s = mass density of sediment; and ρ = mass density of fluid.

Carstens (1966) proposed an equation of competent velocity u_{cr} at the particle level by analyzing a large number of experimental data. It is:

$$u_{cr}^2/\Delta gd \approx 3.61(\tan\varphi\cos\theta - \sin\theta) \tag{2}$$

where φ = angle of repose of sediment; and θ = angle made by the streamwise sloping bed with the horizontal.

Neill (1968) proposed a design curve for the initial movement of coarse uniform gravels in terms of average-velocity U_{cr} as a competent velocity. It is:

$$U_{cr}^2/\Delta gd = 2(h/d)^{1/3} \tag{3}$$

The forces acting on a spherical sediment particle resting on the bed of an open channel were analyzed by Yang (1973) to propose the equations for both smooth and rough boundaries as follows:

$$\frac{U_{cr}}{w_{ss}} = \frac{2.5}{\log R_* - 0.06} + 0.66 \quad \text{for } 0 < R_* < 70 \tag{4a}$$

$$U_{cr}/w_{ss} = 2.05 \quad \text{for } R_* \geqslant 70 \tag{4b}$$

where w_{ss} = terminal fall velocity; R_* = shear Reynolds number, that is $u_* d/\upsilon$; u_* = shear velocity; and υ = kinematic viscosity of fluid.

Zanke (1977) recommended the following equation:

$$U_{cr} = 2.8(\Delta gd)^{0.5} + 14.7c_1(\upsilon/d) \tag{5}$$

where c_1 = coefficient varying from 1 for noncohesive to 0.1 for cohesive sediments. Many researchers have categorically disapproved the concept of competent velocity. The unanswered question is as to what is meant by the competent velocity at particle level u_{cr} or the competent average-velocity U_{cr}. This confusion has insisted the researchers to seek a more acceptable standard quantity like the threshold boundary shear stress. Nevertheless, Yang's (1973) analysis for the estimation of U_{cr} seems to be reasonable.

4 Lift Force Concept

Einstein (1950), Velikanov (1955), Yalin (1963), Gessler (1966), and Ling (1995) thought that the sediment is entrained solely by the lift force. The lift force can primarily be induced for the following reasons: (1) Sediment particles on the bed surface experience maximum velocity gradient, and thus a lift acts on the particles due to considerable pressure difference; (2) sediment particles may experience lift due to the instantaneous vertical velocity fluctuations in the vicinity of the bed; and (3) the spinning motion of sediment particles may result in lift due to Magnus effect (Dey 1999). Note that if the lift force equals the submerged weight of the particle, then drag force is adequate to entrain the bed particles.

Jeffreys (1929) assumed a potential flow over a circular cylinder having its axis perpendicular to the flow arguing that the lift is prevalent if $(3 + \pi^2)U^2 > 9\Delta g r_1$, where r_1 = radius of the cylinder. To apply this result, modification factors should be accounted for, as the two-dimensional model behaves in a different way than a three-dimension spherical particle in a fluid flow. The drawback of the analysis was that the drag force was ignored. Reitz (1936) discussed a similar idea to express the sediment entrainment with a lift model, where circulation and viscosity were important parameters of his analysis. Lane and Kalinske (1939) emphasized on turbulence for the quantification of lift and assumed that (a) the particles experience lift when their terminal fall velocity is smaller than the instantaneous vertical velocity fluctuations in the vicinity of the bed, (b) the variation of velocity fluctuations follows a normal-error law, and (c) a correlation exists between the velocity fluctuations and shear velocities. Einstein and El-Samni (1949) measured the lift force directly as a pressure difference and proposed the lift force per unit area $f_L = 0.5C_L\rho(u_{0.35d})^2$, where C_L = lift coefficient assumed as 0.178; and $u_{0.35d}$ = flow velocity at an elevation $0.35d$ from the theoretical bed. They also studied the turbulent fluctuations on the lift. The experiments revealed a constant average lift force with superimposed random fluctuations that follow the normal-error law. Their results were used by the Task Committee (1966) estimating $f_L/\tau_c \approx 2.5$; where τ_c = threshold boundary shear stress. It suggests that the lift force is an important mechanism of the threshold of sediment entrainment. However, Chepil (1961) pointed out that once the particle moves, the lift and drag tend to diminish and increase, respectively. Chepil (1961) measured that the lift to drag ratio is about 0.85 for $47 < UD/\upsilon < 5 \times 10^3$, in a wind stream on hemispherical roughness having diameter D, while Brayshaw et al. (1983) measured the ratio as 1.8 for the same roughness at $R_* = 5.2 \times 10^4$. Aksoy (1973) and Bagnold (1974) found the lift to drag ratio on a sphere of about 0.1 and 0.5 at $R_* = 300$ and 800, respectively. Apperley (1968) studied a sphere laid on gravels and found lift to drag ratio as 0.5 at $R_* = 70$. Watters and Rao (1971) observed negative (downwards) lift force on a sphere for $20 < R_* < 100$. Davies and Samad (1978) also reported that the lift force on a sphere adjacent to the bed becomes negative if significant underflow takes place beneath the sphere and the flow condition is $R_* < 5$. However, the lift is positive for $R_* \geq 5$, although the negative lift force could not be clearly explained.

While the lift forces obviously contribute to the sediment entrainment, the occurrence of lift on a sediment particle is still unclear. Insufficient experimental results are available to determine quantitative relationships; as such a critical lift criterion has so far not been obtained which could have been a ready reference for the determination of sediment entrainment. The occurrence of negative lift at low R_* has been well established, but its cause and magnitude remain uncertain. It was understood that besides the lift, the drag is always prevalent to contribute towards the sediment entrainment. For higher R_*, the correlation between lift and drag is another uncertain issue, although the lift is definitely positive.

5 Threshold Shear Stress Concept

5.1 Empirical Equations of Threshold Shear Stress

Attempts have been made to correlate the threshold boundary shear stress τ_c with sediment properties for experimental and field data. Kramer (1935) proposed:

$$\tau_c = 29\sqrt{(\rho_s - \rho)gd/M} \tag{6}$$

where τ_c is in g/m^2; M is the uniformity coefficient of Kramer; and d is in mm. Equation (6) is applicable for $0.24 \leq d \leq 6.52$ mm and $0.265 \leq M \leq 1$.

USWES (1936) recommended the following formula:

$$\tau_c = 0.285\sqrt{\Delta d/M} \tag{7}$$

where τ_c is in Pa; and d is in mm. Equation (7) is valid for $0.205 \leq d \leq 4.077$ mm and $0.28 \leq M \leq 0.643$.

A simple equation of τ_c is given by Leliavsky (1966) as:

$$\tau_c = 166d \tag{8}$$

where τ_c is in g/m^2; and d is in mm. None of the equations take into account the effect of fluid viscosity. Further, each of these equations produces results that differ from each other. However, the empirical equations estimate the approximate values of τ_c, and their use cannot be recommended for the precise estimations.

5.2 Semitheoretical Analyses

Shields (1936) was the pioneer to propose a semitheoretical theory for the entrainment threshold of sediments. The threshold of particle motion is governed by balancing the driving force (as drag force) and the stabilizing resistance. At the threshold condition, when the sediment particle is about to move, $u_* \rightarrow u_{*c}$ (i.e., the threshold shear velocity) that has led to the following functional representation:

$$\Theta_c = f(R_*) \tag{9}$$

where Θ_c = threshold Shields parameter, $u_{*c}^2/\Delta gd$ or $\tau_c/\Delta\rho gd$. Figure 1 that shows the Shields' experimental results, which relate Θ_c and R_*, is known as *Shields diagram*. Figure 1 depicts three distinct flow zones: (1) Smooth flow for $R_* \leq 2$: in this case, d is much smaller than the thickness of viscous sublayer, and it is linearly varying that $\Theta_c = 0.1/R_*$. (2) Rough flow for $R_* \geq 500$: the viscous sublayer does not exist. The threshold Shields parameter Θ_c is invariant of the fluid viscosity and has a constant value of 0.056. (3) Transitional flow for $2 \leq R_* \leq 500$: sediment particles are of the order of the thickness of viscous sublayer. There is a minimum value of $\Theta_c = 0.032$ corresponding to $R_* = 10$.

The shortcoming of the Shields diagram is that the viscous parameter does not have any effect for $R_* \geq 70$, but Θ_c still varies with R_* in this range. Furthermore, the τ_c and u_{*c} that are interchangeable are shown as dependent and independent

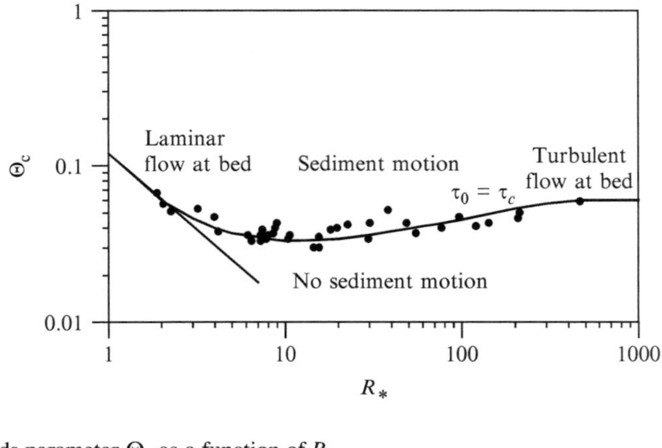

Fig. 1 Shields parameter Θ_c as a function of R_*

Table 1 Explicit empirical equations for the Shields diagram

Researchers	Equation
Brownlie (1981)	$\Theta_c = 0.22R_b^{-0.6} + 0.06\exp(-17.77R_b^{-0.6})$ where $R_d = d(\Delta gd)^{0.5}/\upsilon$
van Rijn (1984)	$\Theta_c(D_* \leqslant 4) = 0.24/D_*$
	$\Theta_c(4 < D_* \leqslant 10) = 0.14/D_*^{0.64}$
	$\Theta_c(10 < D_* \leqslant 20) = 0.04/D_*^{0.1}$
	$\Theta_c(20 < D_* \leqslant 150) = 0.013D_*^{0.29}$
	$\Theta_c(D_* > 150) = 0.055$
	where D_* = particle parameter, that is $d(\Delta g/\upsilon^2)^{1/3}$
Soulsby and Whitehouse (1997)	$\Theta_c = \frac{0.24}{D_*} + 0.055[1 - \exp(-0.02D_*)]$
Paphitis (2001)	$\Theta_c(10^{-2} < R_* < 10^4) = \frac{0.273}{1+1.2D_*} + 0.046[1 - 0.57\exp(-0.02D_*)]$ It is the mean curve of Paphitis (2001)

variables in the diagram. Consequently, τ_c or u_{*c} remains implicit. Thus, attempts are made to derive explicit equations for the Shields diagram (Table 1).

In another study, White (1940) classified a high-speed case ($R_* \geq 3.5$) and a low-speed case ($R_* < 3.5$). High flow velocity is capable of moving larger particles. Therefore, the drag due to skin friction is insignificant as compared to the drag due to pressure difference. The packing coefficient p_f was defined by Nd^2, where N is the number of particles per unit area. The shear drag per particle (i.e., τ_c/N) is $\tau_c d^2/p_f$. At the threshold condition, the shear drag equals the product of the submerged weight of the particle and the frictional coefficient tan φ. Introducing a factor, termed *turbulence factor* T_f, he obtained:

$$\Theta_c = \frac{\pi}{6} p_f T_f \tan\varphi \quad \text{for } R_* \geqslant 3.5 \tag{10}$$

He proposed $p_f = 0.4$ and $T_f = 4$ for fully developed turbulent flow. On the other hand, low flow velocity is capable of moving smaller particles. In this case, the drag due to pressure difference acting on the particle is insignificant as compared to the viscous force. However, the upper portion of the particle is exposed to the shear drag that acts above the center of gravity of the particle. This effect is taken into account introducing a coefficient α_f. He proposed:

$$\Theta_c = \frac{\pi}{6} p_f\, \alpha_f \tan\varphi \quad \text{for } R_* < 3.5 \tag{11}$$

He suggested $p_f\, \alpha_f = 0.34$ as an average value.

Kurihara (1948) extended the work of White (1940) obtaining an expression for T_f in terms of R_*, turbulence intensity and the probability of boundary shear stress increment. The theoretical equations were quite complex. So he proposed simpler empirical equations of threshold boundary shear stress as

$$\Theta_c(X_2 \leqslant 0.1) = (0.047 \log X_2 - 0.023)\beta_2$$
$$\Theta_c(0.1 < X_2 \leqslant 0.25) = (0.01 \log X_2 + 0.034)\beta_2 \qquad (12)$$
$$\Theta_c(X_2 > 0.25) = (0.0517 \log X_2 + 0.057)\beta_2$$

where $X_2 \approx 4.67 \times 10^{-3}[\Delta g/(\upsilon^2\beta_2)]^{1/3}d$; and β_2 $(0.265 \leq M \leq 1) = (M + 2)/(1 + 2M)$.

Egiazaroff (1965) gave yet another derivation for $\Theta_c(R_*)$. He assumed that the velocity at an elevation of 0.63d (above the bottom of particle) equals the fall velocity w_{ss} of particle. His equation is

$$\Theta_c = \frac{1.33}{C_D[a_r + 5.75\log(0.63)]} \qquad (13)$$

where $a_r = 8.5$; and C_D = drag coefficient = 0.4 for large R_*. Both a_r and C_D increase for low R_*. His results do not correspond with the Shields diagram.

Mantz (1977) proposed the extended Shields diagram to obtain the condition of maximum stability (Fig. 2). Yalin and Karahan (1979) presented a curve of Θ_c versus R_*, using a large volume of data collected from literature (Fig. 2). Their curve is regarded as a superior curve to the commonly used Shields curve.

Cao et al. (2006) derived the explicit equation for the curve of Yalin and Karahan (1979). It is:

$$\Theta_c(R_d \leqslant 6.61) = 0.1414/R_d^{0.23}$$
$$\Theta_c(6.61 < R_d \leqslant 282.84) = \frac{[1 + (0.0223R_b)^{2.84}]^{0.35}}{3.09R_d^{0.68}} \qquad (14)$$
$$\Theta_c(R_d \geqslant 282.84) = 0.045$$

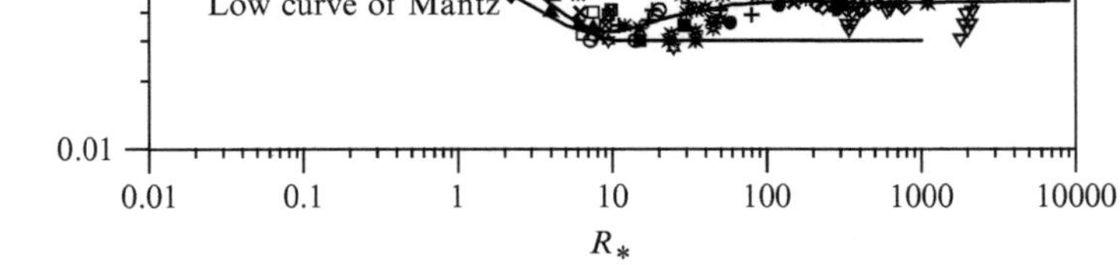

Fig. 2 Curves (Θ_c vs. R_*) of Mantz (1977) and Yalin and Karahan (1979)

Iwagaki (1956) analyzed the equilibrium of a single spherical particle, placed on a rough surface, and found the conditions necessary for the equilibrium of a particle in different conditions of viscous sublayer. The theoretical equation given by him is of the form:

$$\Theta_c = \frac{\tan\varphi}{\varepsilon_s \Psi_s R_*} \tag{15}$$

where ε_s = empirical coefficient for the sheltering effect; and Ψ_s = function of R_*.

The analysis of Ikeda (1982) that is based on Iwagaki (1956) and Coleman (1967) could approximately derive the Shields diagram. The analysis was based on forces acting on a solitary particle placed on a sediment bed. He obtained an equation as follows:

$$\Theta_c = \frac{4}{3} \cdot \frac{\tan\varphi}{(C_D + \tan\varphi C_L)} \cdot \left\{ 10.08 R_*^{-10/3} + \left[\kappa^{-1} \ln\left(1 + \frac{4.5 R_*}{1 + 0.3 R_*} \right) \right]^{-10/3} \right\}^{0.6} \tag{16}$$

where κ = von Kármán constant.

On a horizontal bed, the expression for the force balance given by Wiberg and Smith (1987) leads to:

$$\Theta_c = \frac{2}{C_D \alpha_0} \cdot \frac{1}{f^2(z/z_0)} \cdot \frac{\tan\varphi}{1 + (F_L/F_D)_c \tan\varphi} \tag{17}$$

where $\alpha_0 = A_x d/V$; A_x = frontal area of the particle; V = volume of the particle; z = elevation from the bed; z_0 = zero-velocity level; $F_L/F_D = (C_L/C_D) f^2(z/z_0) / [f^2(z_T/z_0) - f^2(z_B/z_0)]$; z_T = elevation of the top point of the particle from the bed; and z_B = elevation of the bottom point of the particle from the bed.

Dey (1999) and Dey and Papanicolaou (2008) analyzed the hydrodynamic forces acting on a solitary particle resting over a horizontal sediment bed in a three-dimensional configuration (Fig. 3), including the effect of turbulent fluctuations.

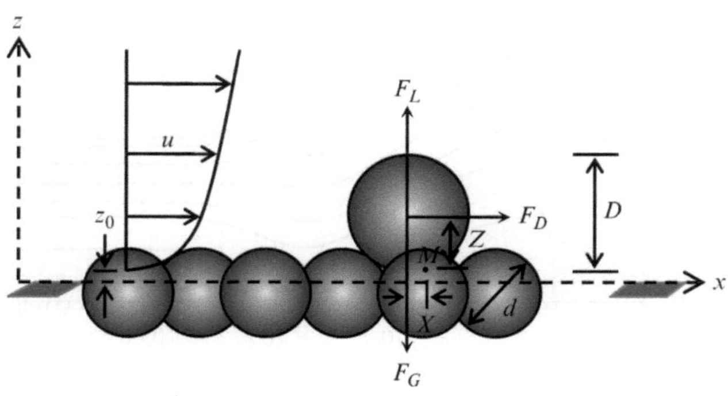

Fig. 3 Forces acting on a solitary particle in a three-dimensional configuration

Considering spherical particles, particle submerged weight $F_G = (\pi/6)\Delta\rho g D^3$ and hydrodynamic forces like drag $F_D = C_D(\pi/8)\rho {u_m}^2 D^2$ and lift (shear and Magnus lift) $F_L = C_L \rho u_m D^2(\partial u/\partial z)^{0.5}[\upsilon^{0.5} + 0.5f(R_*)D(\partial u/\partial z)^{0.5}]$ were taken into account. Here, D = diameter of solitary particle resting over a horizontal bed formed by the sediments of size d; u = flow velocity at elevation z; u_m = mean flow velocity received by the frontal area of solitary particle; $f(R_*) = 1$ for $R_* \geq 3$; and $f(R_*) = 0$ for $R_* < 3$. The lever-arms are $X = 0.433Dd/(D + d)$ and $Z = 0.289D(3D^2 + 6Dd - d^2)^{0.5}/(D + d)$. Taking moment at the pivot M, the equation for threshold condition was given by Dey as:

$$\Theta_c = \frac{2\pi\hat{d}/(1 + p\sqrt{\alpha - 1}\cos\psi)^2}{\pi C_D \hat{u}_m^2(3 + 6\hat{d} - \hat{d}^2)^{0.5} + 6C_L\hat{d}\hat{u}_m(\partial\hat{u}/\partial\hat{z})\{2[(R_*/\hat{d})\partial\hat{u}/\partial\hat{z}]^{-0.5} + f(R_*)\}} \quad (18)$$

where $\hat{d} = d/D$; $\hat{u} = u/u_{*c}$; $\hat{u}_m = u_m/u_{*c}$; $\hat{z} = z/D$; p = probability of occurring sweep event; ψ = sweep angle; $\alpha = \tau_t/\tau_c$; and τ_t = instantaneous shear stress. Using u for different flow regimes, Dey put forward a diagram for entrainment threshold as Θ_c versus D_* for different φ (Fig. 4). Unlike Shields diagram, it can be used directly for the determination of τ_c or u_{*c}.

Besides, James (1990) presented a generalized model of the threshold of sediment entrainment based on the analysis of forces acting on a particle, taking into account the particle geometry, packing arrangements and variations of near-bed flow velocity, drag, and lift. Ling (1995) studied the equilibrium of a solitary particle on a sediment bed, considering spinning motion of particles. He proposed two modes for limiting equilibrium, namely, rolling and lifting. McEwan and Heald (2001) analyzed the stability of randomly deposited bed particles using a discrete particle model. The threshold boundary shear stress could be adequately represented by a distribution of values. A Shields parameter of 0.06 for gravels found to correspond to the distribution for which 1.4% (by weight) of particles is on motion.

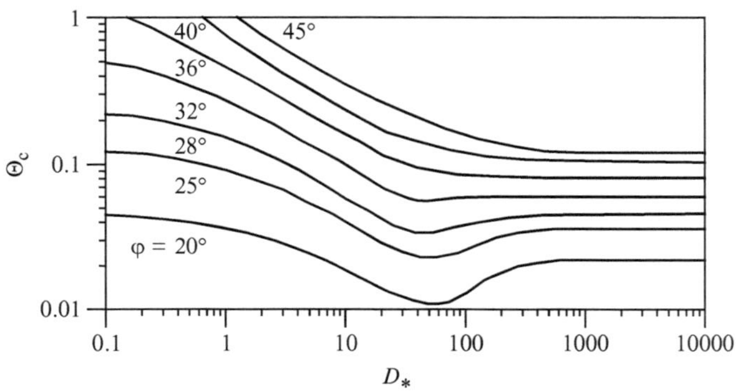

Fig. 4 Dependency of Θ_c on D_* for different φ

An analysis on sheltering of particles revealed that remote sheltering induced by the prominent upstream particles has an effect to increase the apparent threshold boundary shear stress of exposed particles.

6 Probabilistic Concept

The sediment entrainment is probabilistic in nature. It depends primarily on the turbulence characteristics in conjunction with the location of a specific particle relative to the surrounding particles of various sizes and their orientations. Gessler (1970) estimated the probability that particles of a specific size remain stationary. It was revealed that the probability of a given particle to remain stationary depends strongly on the Shields parameter and feebly on the shear Reynolds number. Grass (1970) proposed to use a probabilistic description of the stresses acting on a single particle to achieve motion. He identified two probability distributions: one for the boundary shear stress τ_w induced by the fluid and other for the boundary shear stress τ_{wc} required to put the particle in motion. When these two distributions start overlapping (Fig. 5), the particles that have the lowest threshold boundary shear stress start to move. The representative magnitudes of the probability distributions are their standard deviations being used to describe the distance of the two mean boundary shear stresses as $\bar{\tau}_{wc} - \bar{\tau}_w = n(\sigma_c - \sigma_\tau)$. He experimentally obtained the relationships as $\sigma_\tau = 0.4\bar{\tau}_w$ and $\sigma_c = 0.3\bar{\tau}_{wc}$, which leads to $\bar{\tau}_w = \bar{\tau}_{wc} \times (1 - 0.3n)/(1 + 0.4n)$. For $n = 0.625$, the result collapses with that of Shields.

Mingmin and Qiwei (1982), who developed a stochastic model, expressed the statistical parameters using the velocity of bottom flow and particle size. The probability of incipient motion, life distribution of stationary particles, number of distributions of particles in incipient motion, and intensity of incipient motion were derived. Wu and Chou (2003) studied the rolling and lifting probabilities for sediment entrainment by introducing the probabilistic features of the turbulent fluctuations and particle shape. These probabilities were linked to the two separate criteria for incipient motion to study the threshold entrainment probabilities.

The discovery of the turbulent bursting phenomena has encouraged the researchers in further studying the role of turbulence on sediment entrainment. In an attempt

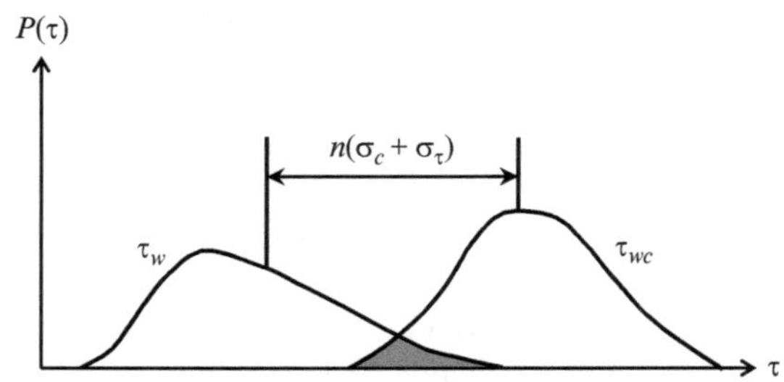

Fig. 5 Probabilities of boundary shear stress τ_w due to flow and threshold boundary shear stress τ_{wc} corresponding to the motion of individual particles

to link the characteristics of turbulent events with the threshold of sediment entrainment, Clifford et al. (1991) and Nelson et al. (1995) suggested that the Reynolds stress component is not the most relevant component to the sediment entrainment. However, the quadrant analysis by Papanicolaou et al. (2001) showed that the ratio of the Reynolds stress to the turbulence intensity is smaller in the beds with low-density particles than the densely packed ones. Hence, sediment entrainment criterion based solely on time-averaged boundary shear stress may underpredict the transport, especially in low-density packing cases. Based on these, Papanicolaou et al. (2002) developed a stochastic sediment threshold model that considered the role of near-bed turbulent structures and bed micro-topography upon the initiation of unisized particle motion. The model was based on the hypothesis that the probability of occurrence of exceeding the minimum moment required to initiate rolling motion equals the probability of first displacement of a particle. The theoretical derivation was complemented by the experimental measurements of the probability and near-bed turbulence for different packing regimes. They found it reasonable to consider that on average (temporal and spatial) for a sufficient large number of data, the probability of the occurrence of intermittent turbulent events equals the sediment entrainment probability. In another attempt, Dancey et al. (2002) proposed a criterion, which might be interpreted as the probability of individual particle motion, considering the statistical nature of sediment motion in turbulent flow and the time-scale of flow. The sediment threshold was specified by a constant value of the probability. However, a threshold criterion based upon the probability of particle motion could yield relatively active sediment beds, where the mechanism is strongly dependent upon the sediment packing density.

7 Role of Turbulence on Threshold of Sediment Entrainment

Cao (1997) proposed a model for the sediment entrainment based on the characteristics of the bursting structures (with time and spatial scaling) that are inherent in wall turbulent flows. He argued that the sediment entrainment is strongly dependent on the shear velocity. In another attempt, Zanke (2003) developed a model for the sediment threshold considering the influence of turbulence. He recognized two important effects as (a) the effective boundary shear stress acting on a particle increases above the time-averaged boundary shear stress owing to turbulent stress peaks and (b) the particles exposed to the flow become effectively lighter due to lift forces. Both the turbulence induced effects are randomly distributed. Dey and Raikar (2007) measured and analyzed the vertical distributions of time-averaged velocity and turbulence intensities in the flow on the near-threshold gravel beds. In the inner-layer, the law of the wall for the time-averaged velocity holds with $\kappa = 0.35$ and a constant of integration 7.8; while in the outer-layer, the law of the wake defines the velocity profiles with an average value of the Coles' wake parameter as 0.11. Nikora and Goring (2000) also observed the reduction of κ-value from its traditional value (0.41) in flows on a mobile gravel bed.

7.1 Turbulent Bursting

Experimental evidences on the viscous sublayer by Kline et al. (1967), Corino and Brodkey (1969), and Grass (1971) revealed a viscous dominating flow characteristic that consists of large three-dimensional high- and low-speed fluid steaks. The near-bed flow has an extremely complex structure producing large turbulence (Nezu and Nakagawa 1993). The emission of low-speed fluid streaks entraining to high-speed fluid streaks initiates the process of turbulent burst. The sequence turbulence bursting is described by two significant features as ejections and sweeps, which play an important role on entrainment of sediments. During the ejections, the upward flow enlarges the shear layer and the associated small-scale flow structures to a wide region. The ejection process is prevalent as low-speed fluid streaks that oscillate in three-dimension lifts up from the bed and then collapse to entrain into the main body of flow. The ejected fluid streaks which remain as a result of retardation are brushed away by high-speed fluid approaching to the bed in a process called the sweeps. During sweeps, the downward flow generates a narrow, highly turbulent shear layer containing multiple small-scale eddies. The turbulent bursting process and the contributions from the conditional Reynolds shear stress towards the total shear stress can be described by a quadrant analysis.

7.2 Quadrant Analysis

To understand the characteristics of the bursting events, it is necessary to study the conditional statistics of the velocity fluctuations (u' and w') plotting them to quadrants on a $u'w'$-plane (Lu and Willmarth 1973). A *hole-size* parameter H is used discriminating the larger contributions to $-\overline{u'w'}$ from each quadrant leaving the smaller u' and w' corresponding to more quiescent periods (Nezu and Nakagawa 1993). The curve $|u'w'| = H(\overline{u'u'})^{0.5}(\overline{w'w'})^{0.5}$ determines the hyperbolic hole region. In this way, a clear distinction is achieved between the strong and the weak events for a small hole-size and only strong events for a large hole-size. The types of bursting events are characterized by four quadrants i (= 1, 2, 3 and 4). They are (1) *outward interactions or* $Q1$ *events* ($i = 1$; $u' > 0, w' > 0$), (2) *ejections or* $Q2$ *events* ($i = 2$; $u' < 0, w' > 0$), (3) *inward interactions or* $Q3$ *events* ($i = 3$; $u' < 0, w' < 0$), and (4) *sweeps or* $Q4$ *events* ($i = 4$; $u' > 0, w' < 0$). The hole-size $H = 0$ implies that all data of u' and w' are taken into account. The quadrant analysis provides an estimation of the fractional contributions $S_{i,H}(= \langle u'w' \rangle_{i,H} / -\overline{u'w'})$ to $-\overline{u'w'}$ from the bursting events for quadrant i outside the hole region of size H. The contribution $\langle u'w' \rangle_{i,H}$ to $-\overline{u'w'}$ from the quadrant i outside the hole of size H is estimated by:

$$\langle u'w' \rangle_{i,H} = \lim_{T \to \infty} \frac{1}{T} \int_0^T u'(t)w'(t)\lambda_{i,H}(z,t)\mathrm{d}t \tag{19}$$

where T = time of sampling; and $\lambda_{i,H}(t)$ is the detection function given by $\lambda_{i,H}(t) = 1$, if (u', w') is in quadrant i and if $|u'w'| \geq H(\overline{u'u'})^{0.5}(\overline{w'w'})^{0.5}$, and $\lambda_{i,H}(t) = 0$, otherwise. Here, $S_{i,H} > 0$ when $i = 2$ and 4 ($Q2$ and $Q4$ events), and $S_{i,H} < 0$ when $i = 1$ and 3 ($Q1$ and $Q3$ events). Hence, at a point, the algebraic summation of the contributions-from different bursting events to $-\overline{u'w'}$ for $H = 0$ is unity, that is, $\sum_{i=0}^{i=4} S_{i,0} = 1$.

7.3 Earlier Developments

The role of turbulent bursting corresponding to the sediment entrainment seems to have received increasing attention. Sutherland (1967) observed that the sediment threshold is associated with a near-bed eddy impact onto the bed particles to produce a streamwise drag force that is large enough enabling to roll the particles. The role of the turbulent structures on the sediment entrainment was investigated by Heathershaw and Thorne (1985) in tidal channels. They argued that the entrainment is not correlated with the instantaneous Reynolds shear stress but correlated with the near-wall instantaneous streamwise velocity. Field observations by Drake et al. (1988) on mobility of gravels in alluvial streams suggested that the majority of the gravel entrainment is associated with the sweep events which give rise to the motion of particles. These events occur during a small fraction of time at any particular location of the bed. Thus, the entrainment process is rather episodic with short periods of high entrainment together with long periods of relatively feeble or no entrainment. Thorne et al. (1989) observed that sweeps and outward interactions play an important role in sediment entrainment. It is the instantaneous increase in streamwise velocity fluctuations that generate excess boundary shear stresses, governing entrainment processes. Having studied the sediment entrainment by nonuniform flows over two-dimensional dunes, Nelson et al. (1995) reported that the near-bed turbulence can change considerably and hence the sediment entrainment; while the boundary shear stress remains almost unchanged. They observed that when the magnitude of the outward interactions increases relative to the other bursting events, the sediment flux increases albeit the boundary shear stress decreases.

7.4 Recent Developments

Sarkar (2010) studied the turbulence characteristics on immobile and entrainment threshold sediment beds having uniform sediment size of 4.1 mm. A summary of the results obtained by him is furnished below:

In Fig. 6, the distributions of nondimensional Reynolds shear stress $-\overline{u'w'}/u_*^2$, as a function of nondimensional height z/h, are shown. Here, h is flow depth; and the solid line in Fig. 6 represents gravity line that follows $-\overline{u'w'}/u_*^2 = 1 - z/h$ for free surface flows having a zero-pressure gradient. Near the bed, the experimental distributions of Reynolds shear stress for immobile and entrainment threshold beds

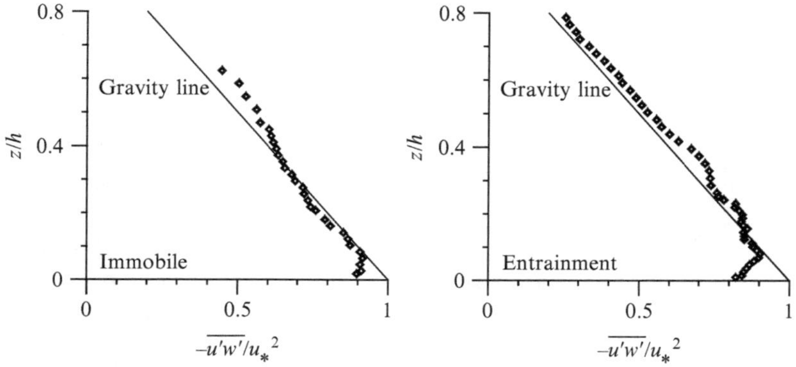

Fig. 6 Plots of $-\overline{u'w'}/u_*^2$ versus z/h for immobile and entrainment threshold beds

have a departure from the gravity line. On the other hand, in the upper flow zone, they are reasonably consistent with the gravity line, although they have a slight tendency to overestimate the gravity line. For $z/h < 0.1$ (near the bed), $-\overline{u'w'}/u_*^2$ for entrainment threshold beds diminishes more than that for immobile beds, although there is a general near-bed damping in the $-\overline{u'w'}$ due to roughness. The reduction in magnitude of $-\overline{u'w'}$ for entrainment threshold beds is attributed to the fact that a portion of the fluid turbulent stress is transferred to the bed particles to overcome the frictional resistance at the contacts of the entrained sediment particles. This is analogous to the concept of Grass (1970). The damping of the Reynolds shear stress can also be explained that the bed particles are associated with the provided momentum for the flow to maintain their motion.

Turbulent energy budget in two-dimensional flows is constituted by the turbulent production $t_P[= -\overline{u'w'}(\partial u/\partial z)]$ that is balanced by the summation of the turbulent dissipation ε, turbulent energy diffusion t_D $(= \partial f_{kw}/\partial z)$, pressure energy diffusion $p_D[= \partial(\overline{p'w'/\rho})/\partial z]$, and viscous diffusion v_D $[= -\upsilon(\partial^2 k/\partial z^2)]$; where $f_{kw} = 0.75(\overline{w'w'w'} + \overline{w'u'u'}$; p' = pressure fluctuations; and k = turbulent kinetic energy. In turbulent flows, the viscous diffusion v_D is insignificant. To evaluate ε, the relationship $\varepsilon = (15\upsilon/u^2)\overline{(\partial u'/\partial t)^2}$ was used. The pressure energy diffusion p_D was estimated as $p_D = t_P - \varepsilon - t_D$. Figure 7a, b illustrates the energy budget in flows over immobile and entrainment threshold beds having uniform sediment size of 4.1 mm. The nondimensional form of these parameters are T_P, E_D, T_D, $P_D = (t_P, \varepsilon, t_D, p_D) \times (h/u_*^3)$. In general, T_P increases near the bed with an increase in z/h up to $z/h > 0.05$ and then decreases rapidly becoming nearly constant for $z/h > 0.3$. The distributions of E_D have a distinct lag from those of T_P. The influence of a sediment entrainment is apparent in the near-bed distributions of T_P and E_D, where the lag is reversed, which means $E_D > T_P$. Essentially, the difference of T_P and E_D at any depth is balanced by the combination of T_D and P_D. In Fig. 7a, b, T_D decreases with an increase in z/h within the wall-shear layer and then it becomes almost invariant of z/h. On the other hand, P_D attends a positive peak at $z/h \approx 0.05$ and then gradually decreases with an increase in

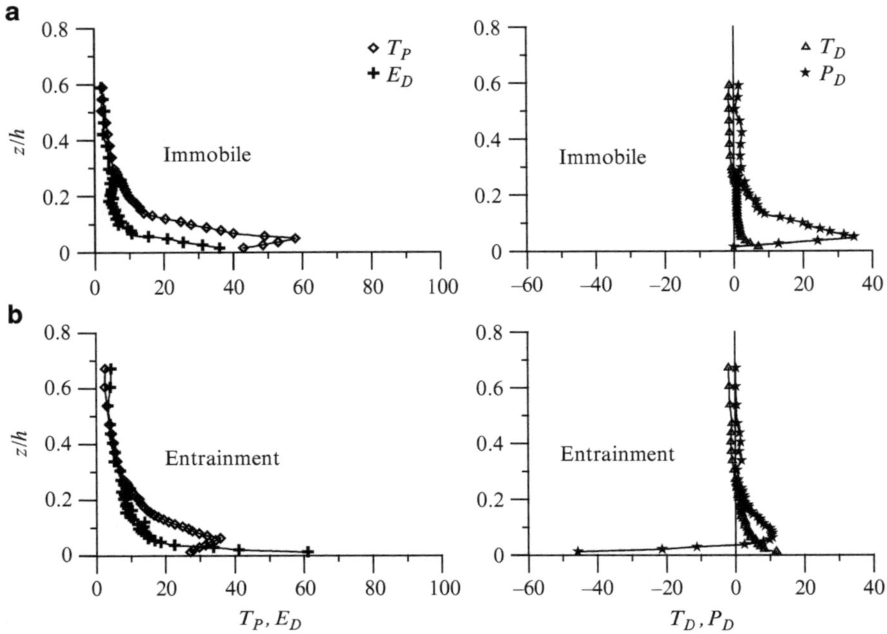

Fig. 7 Energy budget for (**a**) immobile and (**b**) entrainment threshold beds

z/h becoming a constant. The most interesting feature lies on the near-bed distributions of P_D in flows over entrainment threshold beds. It is apparent that the sediment entrainment is associated with a drastic changeover of P_D to a large negative value ($P_D \approx -45$). The negative value of P_D indicates a gain in turbulent production. It is therefore comprehensible that in near-bed flow zone over entrainment threshold beds, the turbulent dissipation exceeds the turbulent production and the pressure energy diffusion becomes considerably negative indicating a sediment entrainment.

The fractional contributions $S_{i,H}(z/h)$ towards the total Reynolds shear stress production from different bursting events, for the hole-sizes $H = 0$ and 2, are shown in Fig. 8a, b, respectively.

In Fig. 8a, for immobile beds, $Q2$ and $Q4$ events at the nearest point of the bed contribute about 75% ($S_{2,0} \approx S_{4,0} \approx 0.75$) to the total Reynolds shear stress production. On the other hand, $Q1$ events contribute moderately by 40% ($S_{1,0} \approx 0.4$), while $Q3$ events contribute minimal ($S_{3,0} \approx 0.1$). To be explicit, the arrival of low-speed fluid streaks from the near-bed region is revoked by the arrival of high-speed fluid streaks from the upper region. Thus, only a faster moving process is prevalent in the form of outward interactions $Q1$. In contrast, for entrainment threshold beds, $Q4$ events are the main mechanism to entrain sediments contributing about 90% ($S_{4,0} \approx 0.9$) towards the Reynolds shear stress production, while $Q2$ events contribute relatively less ($S_{2,0} \approx 0.6$). The tendency of $Q4$ events to dominate momentum transfer over a sediment bed is therefore strongly dependent upon the motion of surface particles.

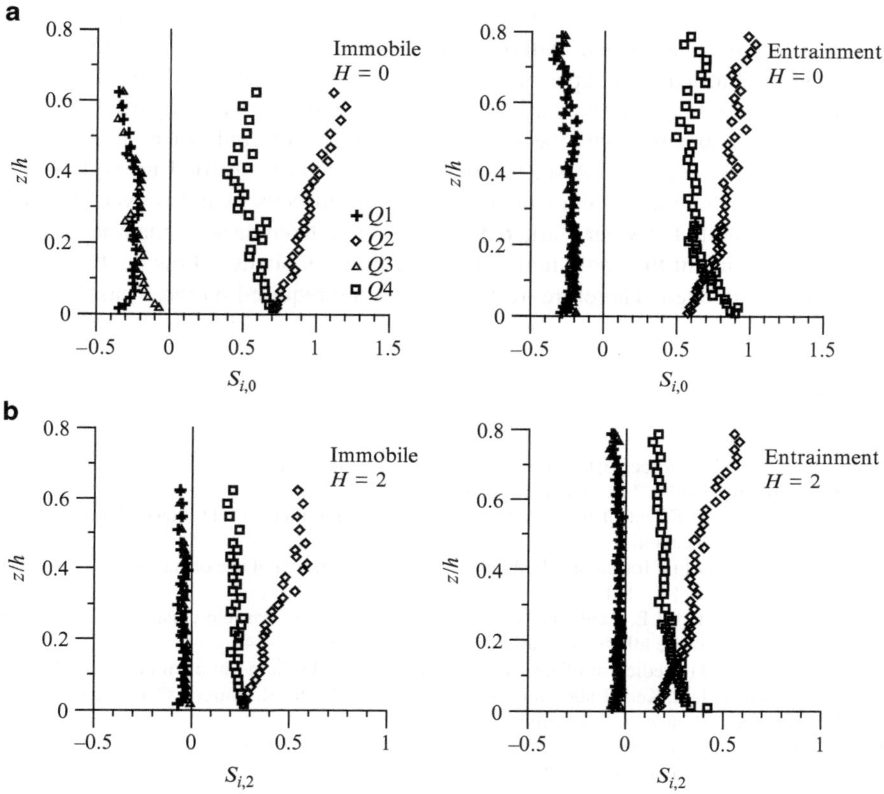

Fig. 8 Distributions of (**a**) $S_{i,0}(z/h)$ and (**b**) $S_{i,2}(z/h)$

It means the sediment motion is governed by the arrival of high-speed fluid streaks. But the contributions from $Q1$ and $Q3$ events are feeble ($S_{1,0} \approx 0.3$ and $S_{3,0} \approx 0.2$). Examining the Reynolds shear stress contributions from more extreme events occurring for hole-size $H = 2$ given in Fig. 8b, there remains a consensus that similar characteristics of most energetic $Q2$ and $Q4$ events prevail in the flow zone. Therefore, the quadrant analysis reveals that in the near-bed flow, ejections and sweeps in immobile beds rescind each other giving rise to the outward interactions, while sweeps are the dominant mechanism towards the sediment entrainment.

8 Closure

While plenty of researches have been carried out on the entrainment threshold of different sizes of sediments and all the investigations bring us a step nearer to an improved understanding of the sediment entrainment phenomenon in relation to the turbulence characteristics, there remains an inadequate attention to many cases.

For instance, the clear-cut role of turbulent coherent structure on sediment threshold is yet to be fully understood. Although there were some attempts, the flow fields were captured at least a couple of millimeters above the bed particles. Thus, the exact interaction between the particles and the fluid, in the level of particle micromechanics in association with the probabilistic feature of turbulence eddies, has not been completely revealed. Also, not many researchers have tried to explore the threshold of sediment entrainment for water worked beds. It is believed that the sediment threshold for water worked beds is different from those of manmade beds. Moreover, sediment threshold under the sheet flows or shallow flow depths seems to remain unattended. Therefore, further studies are required on these cases.

References

Aksoy S (1973) Fluid forces acting on a sphere near a solid boundary. In: Proceedings of 15th IAHR Congress, vol 1. Istanbul, Turkey, pp 217–224

Apperley LW (1968) Effect of turbulence on sediment entrainment. Ph.D. thesis, University of Auckland, New Zealand

Bagnold RA (1974) Fluid forces on a body in shear flow; experimental use of stationary flow. Proc R Soc Lond 340A:147–171

Brayshaw AC, Frostick LE, Reid I (1983) The hydrodynamics of particle clusters and sediment entrainment in course alluvial channels. Sedimentology 30:137–143

Brownlie WR (1981) Prediction of flow depth and sediment discharge in open channels. Report Number KH-R-43A, Keck Laboratory of Hydraulics and Water Resources, California Institute of Technology, Pasadena, California

Buffington JM (1999) The legend of A. F. Shields. J Hydraul Eng 125:376–387

Buffington JM, Montgomery DR (1997) A systematic analysis of eight decades of incipient motion studies, with special reference to gravel-bedded rivers. Water Resour Res 33:1993–2029

Cao Z (1997) Turbulent bursting-based sediment entrainment function. J Hydraul Eng 123:233–236

Cao Z, Pender G, Meng J (2006) Explicit formulation of the Shields diagram for incipient motion of sediment. J Hydraul Eng 132:1097–1099

Carstens MR (1966) An analytical and experimental study of bed ripples under water waves. Georgia Institute of Technology, School of Civil Engineering, Atlanta, Quarter reports 8 and 9

Chepil WS (1961) The use of spheres to measure lift and drag on wind-eroded soil grains. Proc Soil Sci Soc Am 25:343–345

Clifford NJ, McClatchey J, French JR (1991) Measurements of turbulence in the benthic boundary layer over a gravel bed and comparison between acoustic measurements and predictions of the bedload transport of marine gravels. Sedimentology 38:161–171

Coleman NL (1967) A theoretical and experimental study of drag and lift forces acting on a sphere resting on a hypothetical stream bed. Proceedings of 12th IAHR Congress, vol 3. Fort Collins, pp 185–192

Committee T (1966) Sediment transportation mechanics: initiation of motion. J Hydraul Div 92:291–314

Corino ER, Brodkey RS (1969) A visual investigation of the wall region in turbulent flow. J Fluid Mech 37:1–30

Dancey CL, Diplas P, Papanicolaou A, Bala M (2002) Probability of individual grain movement and threshold condition. J Hydraul Eng 128:1069–1075

Davies TRH, Samad MFA (1978) Fluid dynamic lift on a bed particle. J Hydraul Div 104:1171–1182

Dey S (1999) Sediment threshold. Appl Math Model 23:399–417

Dey S, Papanicolaou A (2008) Sediment threshold under stream flow: a state-of-the-art review. KSCE J Civ Eng 12:45–60
Dey S, Raikar RV (2007) Characteristics of loose rough boundary streams at near-threshold. J Hydraul Eng 133:288–304
Drake TG, Shreve RL, Dietrich WE, Whiting PJ, Leopold LB (1988) Bedload transport of fine gravel observed by motion picture photography. J Fluid Mech 192:193–217
Egiazaroff JV (1965) Calculation of non-uniform sediment concentrations. J Hydraul Div 91:225–247
Einstein HA (1950) The bed-load function for sediment transportation in open channel flows. US Department of Agriculture, Washington DC, Technical bulletin number 1026
Einstein HA, El-Samni EA (1949) Hydrodynamic forces on rough wall. Rev Mod Phys 21:520–524
Gessler J (1966) Geschiebetrieb bei mischungen untersucht an naturlichen, abpflasterungserscheinungen in kanalen. Nr. 69, Mitteilungen der Versuchsanstalt für Wasserbau und Erdbau, ETH Zurich, Germany
Gessler J (1970) Self-stabilizing tendencies of alluvial channels. J Waterway Harbors Div 96:235–249
Goncharov VN (1964) Dynamics of channel flow. Israel Programme for Scientific Translation, Moscow, Russia
Grass AJ (1970) Initial instability of fine bed sand. J Hydraul Div 96:619–632
Grass AJ (1971) Structural features of turbulent flow over smooth and rough boundaries. J Fluid Mech 50:233–255
Heathershaw AD, Thorne PD (1985) Sea-bed noises reveal role of turbulent bursting phenomenon in sediment transport by tidal currents. Nature 316:339–342
Ikeda S (1982) Incipient motion of sand particles on side slopes. J Hydraul Div 108:95–114
Iwagaki Y (1956) Fundamental study on critical tractive force. Trans JSCE 41:1–21
James C (1990) Prediction of entrainment conditions for nonuniform, noncohesive sediments. J Hydraul Res 28:25–41
Jeffreys H (1929) On the transport of sediments in stream. Proc Cambridge Philos Soc 25:272
Kennedy JF (1995) The Albert Shields story. J Hydraul Eng 121:766–772
Kline SJ, Reynolds WC, Straub FA, Runstadler PW (1967) The structure of turbulent boundary layers. J Fluid Mech 30:741–773
Kramer H (1935) Sand mixtures and sand movement in fluvial levels. Trans ASCE 100:798–838
Kurihara M (1948) On the critical tractive force, vol 4. Research Institute for Hydraulic Engineering, Report number 3
Lane EW, Kalinske AA (1939) The relation of suspended to bed materials in river. Trans Am Geophys Union 20:637
Leliavsky S (1966) An introduction to fluvial hydraulics. Dover, New York
Ling CH (1995) Criteria for incipient motion of spherical sediment particles. J Hydraul Eng 121:472–478
Lu SS, Willmarth WW (1973) Measurements of the structures of the Reynolds stress in a turbulent boundary layer. J Fluid Mech 60:481–571
Mantz PA (1977) Incipient transport of fine grains and flanks by fluids-extended Shields diagram. J Hydraul Div 103:601–615
McEwan I, Heald J (2001) Discrete particle modeling of entrainment from flat uniformly sized sediment beds. J Hydraul Eng 127:588–597
Miller MC, McCave IN, Komar PD (1977) Threshold of sediment motion under unidirectional currents. Sedimentology 24:507–527
Mingmin H, Qiwei H (1982) Stochastic model of incipient sediment motion. J Hydraul Div 108:211–224
Neill CR (1968) Note on initial movement of coarse uniform bed-material. J Hydraul Res 6:173–176
Nelson J, Shreve RL, McLean SR, Drake TG (1995) Role of near-bed turbulence structure in bed load transport and bed form mechanics. Water Resour Res 31:2071–2086

Nezu I, Nakagawa H (1993) Turbulence in open-channel flows. Balkema, Rotterdam, the Netherlands
Nikora V, Goring D (2000) Flow turbulence over fixed and weakly mobile gravel beds. J Hydraul Eng 126:679–690
Paintal A (1971) Concept of critical shear stress in loose boundary open channels. J Hydraul Res 9:91–113
Papanicolaou A, Diplas P, Dancey C, Balakrishnan M (2001) Surface roughness effects in near-bed turbulence: implications to sediment entrainment. J Eng Mech 127:211–218
Papanicolaou AN, Diplas P, Evaggelopoulos N, Fotopoulos S (2002) Stochastic incipient motion criterion for spheres under various bed packing conditions. J Hydraul Eng 128:369–380
Paphitis D (2001) Sediment movement under unidirectional flows: an assessment of empirical threshold curves. Coastal Eng 43:227–245
Reitz W (1936) Uber geschiebebewegung. Wasserwirtschaft und Technik, pp 28–30
Sarkar S (2010) Turbulence in loose boundary streams. Ph.D. thesis, Indian Institute of Technology, Kharagpur, India
Shields AF (1936) Application of similarity principles and turbulence research to bed-load movement, vol 26. Mitteilungen der Preussischen Versuchsanstalt für Wasserbau und Schiffbau, Berlin, Germany, pp 5–24
Soulsby RL, Whitehouse RJS (1997) Threshold of sediment motion in coastal Environments. Proceedings of combined Australasian coastal engineering and port conference, Christchurch, New Zealand, pp 149–154
Sutherland AJ (1967) Proposed mechanism for sediment entrainment by turbulent flows. J Geophys Res 72:6183–6194
Thorne PD, Williams JJ, Heathershaw AD (1989) In situ acoustic measurements of marine gravel threshold and transport. Sedimentology 36:61–74
USWES (1936) Flume tests made to develop a synthetic sand which will not form ripples when used in movable bed models. United States Waterways Experiment Station, Vieksburg, Technical memorandum 99-1
van Rijn LC (1984) Sediment transport, part I: bed-load transport. J Hydraul Eng 110: 1431–1456
Vanoni VA (1964) Measurements of critical shear stress. California Institute of Technology, Pasadena, Report number KH-R-7
Velikanov MA (1955) Dynamics of alluvial stream, vol 2. State Publishing House of Theoretical and Technical Literature, Russia (in Russian)
Watters GZ, Rao MVP (1971) Hydrodynamic effects of seepage on bed particles. J Hydraul Div 97:421–439
White CM (1940) The equilibrium of grains on the bed of a stream. Philos Trans R Soc 174A:322–338
Wiberg PL, Smith JD (1987) Calculations of the critical shear stress for motion of uniform and heterogeneous sediments. Water Resour Res 23:1471–1480
Wu FC, Chou YJ (2003) Rolling and lifting probabilities for sediment entrainment. J Hydraul Eng 129:110–119
Yalin MS (1963) An expression of bed-load transportation. J Hydraul Div 89:221–250
Yalin MS, Karahan E (1979) Inception of sediment transport. J Hydraul Div 105:1433–1443
Yang CT (1973) Incipient motion and sediment transport. J Hydraul Div 99:1679–1704
Zanke UCE (1977) Neuer Ansatz zur Berechnung des Transportbeginns von Sedimenten unter Stromungseinfluss. Mitteilungen Des Franzius-Institut, Technical University Hannover, Germany, Heft 46
Zanke UCE (2003) On the influence of turbulence on the initiation of sediment motion. Int J Sediment Res 18:17–31

Particle Imaging Velocimetry and Its Applications in Hydraulics: A State-of-the-Art Review

Cristiana Di Cristo

1 Introduction

In the last three decades, the advanced technologies related of optics, laser, electronics, video and computers have facilitated the development of the image-based techniques for flow visualization and quantitative measurements. Different image-based instruments, under the generic name of Particle Imaging Velocimetry (PIV), were developed and applied to a variety of fluid mechanics applications. PIV is widely applied in aerodynamics and hydraulic research because it is a nonintrusive technique able to provide quantitative measurement of fluid velocity over a relatively large surface with measurements documented at a large number of points simultaneously. PIV delivers whole-field velocity measurements, distinguishing the method from all the previous velocity measurement methods, such as the contemporary Laser Doppler Velocimetry and Hot Wire Anemometer that can only be used to gather single point measurements. With the PIV, the velocity is measured in an indirect way, by estimating the displacement of flow markers or patterns in small regions (interrogation areas) in a sequence of recorded images. Besides PIV, various image velocimetry methods were developed, such as the Particle Tracking Velocimetry (PTV), which suites better specific objectives of scientific investigations.

The chapter starts with a brief presentation of the historical development and of the basic principles of Particle Imaging Velocimetry (PIV) methods. The second part describes examples of PIV implementation in different hydraulic research applications. No attempt has been made herein to make the presentation of the methods exhaustive. The interested readers are directed to the comprehensive PIV literature where they can find extensive descriptions of the principles, components, and operational aspects (Adrian 1991; Raffael et al. 2007). A good source

C. Di Cristo
University of Cassino, via Di Biasio 43, Cassino (FR) 03043, Italy
e-mail: dicristo@unicas.it

P. Rowinski (ed.), *Experimental Methods in Hydraulic Research*,
Geoplanet: Earth and Planetary Sciences, DOI 10.1007/978-3-642-17475-9_3,

of keeping to date with the new developments in the area is the special issues on PIV, periodically published on *Experimental in Fluids Journal*. They are keeping the readers abreast with the most recent applications of PIV in different research fields.

PIV is widely used in experimental laboratory works, in particular for studying turbulent flows, but more recently it has been introduced also for field measurements. Another important application area is represented by the specific PIV methods developed in order to acquire measurements in two-phase flows, such as mixtures of water with air bubbles or sediments. The main goal of this chapter is to present a state-of-art of some representative applications of the PIV technique in hydraulic research.

2 Historical Development and Basic Principle of PIV Technique

The most rudimental form of PIV could probably be traced far back in history to the first time a person evaluated the velocity watching leaves floating and moving on the surface of a flowing stream. From this point of view, the PIV concept is old and very simple. The velocity is estimated from the movement of tracers present in the flow, assumed to closely follow the flow. In particular, the velocity vector is calculated using the velocity definition:

$$\vec{U} = s\frac{\vec{X}}{\Delta t} \tag{1}$$

where s is a scale factor, Δt is the time step, and $\vec{X}$ is the tracer displacement between two successive observations. In its modern form, a conventional PIV system has four major components: seeding, illumination, image recording, and image processing.

Historically, PIV first appeared in literature in the mid-1980s (Adrian 2005). Initially, researchers became interested in PIV because of its capability of studying turbulent flows. By its nature, turbulence is a phenomenon that occurs over a wide range of physical scales, extending from the largest scale of the flow to the Kolmogorov scale. So, a successful measurement technique must be able to measure over a wide range of physical scales in length and velocity. Turbulence occurs also at high velocity, so tracers must be small enough to follow the flow in presence of large local and randomly fluctuating accelerations. Moreover, since the tracer displacement is evaluated considering the particle position in two consecutive instants, a pulsed light is required. The duration of the illumination light pulse must be short enough so that the particles are "frozen" during the pulse, without streaks. The use of small particles, with small light scattering section, and the necessity of a short duration pulse imply the use of high intensity pulsed laser.

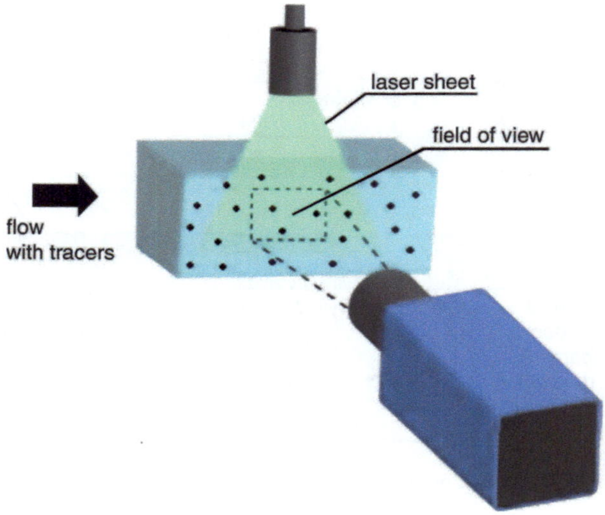

Fig. 1 Layout of a standard PIV system

The PIV systems, developed in order to achieve these characteristics, are now used for a wide range of fluid flow problems.

A standard PIV system now consists of a pulsed laser with optics, which creates an illuminated sheet in the flow with tracers, one digital camera, and a computer with a timer unit to control the system and store the data. A layout of a standard apparatus is shown in Fig. 1. 2D velocity fields are measured in the planar region, illuminated by the laser light sheet and captured by the field of view of one camera. So, in standard PIV, only the 2D components of real velocity vector in the laser sheet can be measured, but stereoscopic PIV (SPIV) systems, able to measure also the third component, are now available.

PIV measurements are realized in two main steps: recording and images elaborating phases. A wide variety of seeding particles, illumination coding sequences, recording techniques, and image analysis methodologies have produced many kinds of quite different PIV methods. In fact, the most important task in using PIV is to match all components in order to capture the desired flow features. Some trial-error testing is inevitably required to optimize the technique for a particular investigated situation. In the next sections, with reference to a standard 2D PIV system, a brief description of its components and of the main typologies of image analysis are reported.

2.1 PIV System Components and Recording Phase

A scheme of a 2D standard PIV system, with the main components, is shown in Fig. 2.

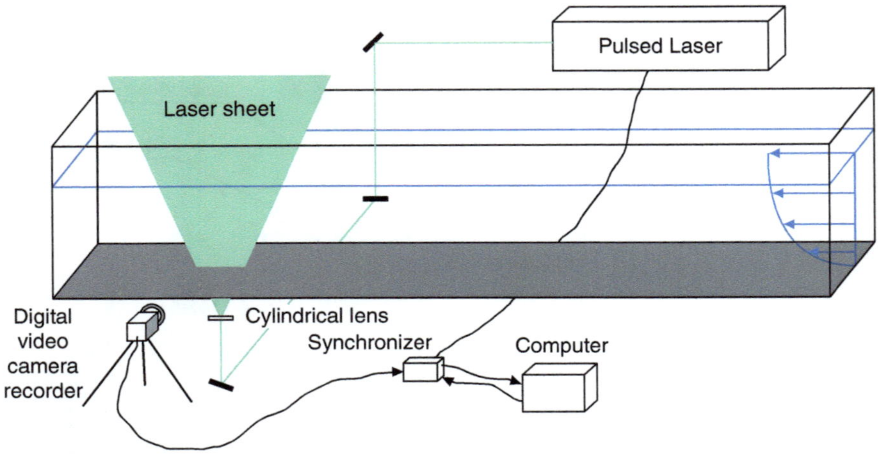

Fig. 2 Scheme of 2D standard PIV system

2.1.1 Tracer Particles

As mentioned before, PIV measures velocities indirectly, evaluating tracer movements. Ideal tracer particles follow perfectly the fluid, do not influence the flow, and do not interact with each other. For this reason, particle size, density, composition, and concentration are important factors when selecting seeds for PIV.

A fundamental requirement is that the flow has to be marked by particles selected in order to avoid significant discrepancy between fluid and particle motion. A convenient measure for the tendency of particles to follow the flow is the *relation time*, τ_p, defined as the time required for a particle at rest to be accelerated within about 63% of the fluid velocity (Hetsroni 1989), which can be estimated using the following expression:

$$\tau_p = \frac{\rho_p}{\rho} \frac{d_p^2}{18\nu} \tag{2}$$

where ρ_p and d_p are the density and the diameter of the particle, ρ and ν are the fluid density and kinematic viscosity, respectively. Even if the expression (2) is strictly correct only for Stokes' regime (i.e., not high flow velocities), also for larger Reynolds particle number it still furnishes a valid order of magnitude and a more precise estimation is not usually necessary. A smaller relation time is associated to a higher frequency response and to a greater capacity to follow rapid flow fluctuations. The τ_p value has to be compared with eddies' characteristic time of the flow under study. So, seeding particles should be small and light enough to follow the flow, but at the same time they have to generate a light signal, strong enough to be resolved by the imaging device. For this reason, light scattering properties of particles are very important too. In general, the light scattered by particles is a function of the ratio of the refractive index of particles to that of the surrounding medium, the particles' size and shape. In particular, the scattered light intensity increases with increasing particle diameter; moreover, particles with irregular shape surface are generally scattered well in all directions. So, a greater relative refractive index and a larger geometric size both help to improve the signal strength.

Sometimes the natural seeding already present in flow can act as markers. This solution, acceptable in some cases, such as field applications, can produce problems in the image analysis phase in many other situations. In fact, if particles of different sizes are present, as is usually the case, since large particles clearly dominate the PIV evaluation, it is difficult to give a sure estimate of the effective particle size and then of the lag velocity. Moreover, a large quantity of particles in a flow can increase the background noise on the recordings, reducing measurement quality. In almost all laboratory works, it is desirable to have clear water and add selected tracers in order to achieve sufficient image contrast and to control particle size. For liquid flow problems, like hydraulic applications, many solid particles with adequate characteristic can be found for satisfying all requirements. In particular, metallic and glass particles of few tens of microns are well suited for seeding water flows.

Other factors to consider are achieving homogeneous particle dispersion and maintaining particle life time in the flow. A good particle dispersion is related to how seeds are practically added to the flow. While for gas flows the supply procedure is very important (Melling 1997), for liquid flows this can easily be done by suspending solid particles into the fluid and mixing them in order to get a good distribution.

The tracer concentration in the flow is a very crucial aspect, which will be discussed in the section dedicated to image analysis, since it strongly influences the selection of the processing algorithm.

2.1.2 Light Source and Optics

In PIV applications, the light source is usually a laser (Fig. 2), *Light Amplification by Stimulated Emission of Radiation*, because of its ability to emit monochromatic light with high energy density, which can easily be bundled into thin light sheets. In some cases, white light source, for example, generated by xenon flash lamps, or natural light can also be used. Since white light cannot be collimated as monochromatic light, these solutions are adopted in particular applications only, like in large-scale and field measurements.

Every laser can be considered consisting of the following three components:

1. The laser material, which is an atomic or molecular gas, semiconductor or solid material
2. The pump source, which excites the laser material by the introduction of electromagnetic or chemical energy
3. The optical resonator, consisting of mirrors arrangements, which allows to amplify the light energy

A variety of lasers, using different laser materials, were proposed and used in the past, like helium-neon, copper-vapor, argon-ion, which are gas lasers, and ruby and Neodym-Yag, which are solid-state lasers.

In general, light pulses can be obtained with pulsed lasers or with continuous wave (CW) lasers, combined with a chopping system for producing light pulses and/or a shuttered recording camera. Actually, the choice of pulsed lasers, with their ability to produce high-power light with short pulse duration, is predominant. In particular, in standard PIV setups the commonly used light sources are actually pulsed solid-state Neodym-Yag (Nd:Yag) lasers. They are made by a Yag (Yttrium-aluminum-garnet) crystal and the beam is generated by Nd^{3+} ions. Nd: Yag lasers are widely diffused for their high amplification capacity and good mechanical and thermal proprieties. Now, Nd:Yag lasers are available in compact packages, with self-contained cooling supplies, easy to manage and positioning. They have a pulse energy in the range from 2 to 1,000 mJ; even for a standard type, the pulse energy is around 120 mJ. They emit primarily at 1,064 nm wavelength and its harmonics, but for safety reasons the laser emission is typically band pass filtered to isolate the 532 nm harmonics (green light), the only harmonic perceptible by naked eyes.

The illumination is characterized by two important parameters, which have been appropriately fixed: the duration of the illumination pulse, named *exposure time*, and the *delay time* between two consecutive pulses. The *exposure time* has to be short enough so that the motion of tracer particles is "frozen" during the pulse exposure to avoid "streaks" in the image, but not too short in order to guarantee a good illumination of seeds. The *delay time* has to be long enough to be able to determine the displacement of the same particle in the two consecutive images with a good resolution and without overlapping, but not too long, since the particle could go out from the light sheet. This is the most adjustable parameter during the recording phase, since it influences the maximum and the minimum velocity that can be measured. Moreover, the velocity vectors computed from (1) are "mean" values in the time interval Δt (*delay time*), so that they can be considered a good approximation of the "instantaneous" flow values only if the delay time is small enough. Finally, large Δt value increases also the out-of-plane errors, due to the fact that the particles present in the first image do not appear in the second one, since they went out of the illuminated sheet. In conclusion, these parameters have to be carefully fixed, considering the character of the flow under investigation.

The optics, consisting of a combination of mirrors, spherical and cylindrical lenses, generate a light sheet from the emitted laser beam. In particular, the mirrors are used to reflect the beam in the desired position, spherical lens expands the beam into a plane, while the cilindrical lens compresses the plane into a thin sheet. The modularity of the available optics permits to obtain a light sheet with fixed or adjustable thickness.

2.1.3 Recording Phase

One of the most important changes in PIV technique was the move from photographic to digital recording, named Digital-PIV (Willert and Gharib 1991),

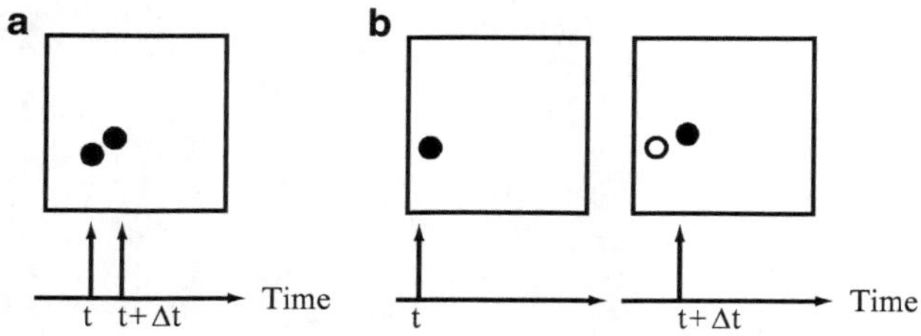

Fig. 3 Single frame–double exposure (**a**) and double frames–single exposure (**b**) methods

influencing also the recording modes. Even if different coding sequences can be adopted (double- or multi-exposure, see Raffael et al. 2007), in the following only double-exposure procedures, the actually usually adopted, are considered.

The recording modes can be separated in two main branches: single frame–double-exposure method (Fig. 3a), which captures the illuminated images into a single frame; double-frames–single-exposure method (Fig. 3b), which provides a single illuminated image for each illumination pulse. Historically, in conjunction with photographic cameras, a single-frame method was used, analyzing the images using an autocorrelation process. The principal distinction between the two types is that the single-frame method does not furnish information on the temporal order of the illumination pulses, with an ambiguity on the direction of the velocity vector. To overcome this ambiguity, additional costly efforts are required (i.e., rotating mirror, birefringent crystal, color coding, etc.). The double-frame–single-exposure method has the advantage that the direction of flow is automatically determined. Moreover, since it was observed that cross-correlation between two separately recorded images is superior to the autocorrelation of double exposures, the double frames–single exposure method is the one usually adopted in standard PIV systems.

High speed CCD cameras are now available. The camera and the laser are connected through a synchronizer, which is controlled by a computer and dictates the timing of the camera sequence in conjunction with the firing of the laser (Fig. 2).

2.2 *Image Analysis, Preprocessing and Postprocessing Procedures*

The algorithm for image analysis is related to the recording procedure and to the density of tracer particles images. Qualitatively three different situations can be distinguished: high, medium, and low density images (Fig. 4). In case of high image density, it is not possible to individuate single particles, because they overlap and form speckles. This situation is called Laser Speckle Velocimetry (LSV) and it is not considered here. When the image density is low, individual particles can be detected and images corresponding to the same particle can be recognized by visual

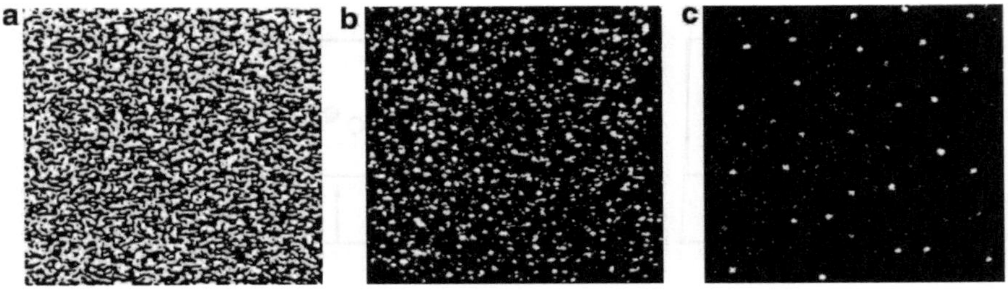

Fig. 4 Examples of high (**a**), medium (**b**), and low (**c**) density of tracer particles images

inspection. In case of medium density, the images of individual particles can still individuate, but it is not possible to identify image pairs. In this condition, the velocity is evaluated considering the displacement of small group of particles (patterns), assuming that they do not change their relative position in the group during the time interval between images. This medium density mode is the one properly named PIV.

In PIV mode, pattern displacements are evaluated by statistical means, computing two-dimensional correlations, autocorrelation or cross-correlation, on pairs of images. Autocorrelation was traditionally used with the double exposure mode, while cross-correlation is actually commonly used with double frames–single exposure method.

The general concept of cross-correlation technique is to individuate the corresponding particle patterns in the two images of the pair using a cross-correlation coefficient. In digital images, a *value* which represents the gray-level intensity is associated to each pixel. Considering the two images, I_1 and I_2 (Fig. 5), taken at time distance Δt, a grid of points at which the velocities are determined is defined. Then an *interrogation area*, A_1, of size $l_w \times l_z$, centered in $P(i, j)$, is considered. This area contains a group of particles, forming a pattern in the flow, characterized by a distribution of intensities. In the second images, a larger area, circular or rectangular, of dimension r, containing P and named *research area*, is selected. In this area, the interrogation area A_2 of size $l_w \times l_z$, corresponding, or in other words containing the same particle pattern, to the one of the first image is searched. In particular, different areas A_2, offset from the point P by the amount $\overline{PP_2}$, with $P_2(i + x, j + y)$, are considered and compared with A_1. The distance $\overline{PP_2}$, which furnishes the best correspondence, represents the particles movement in the time interval Δt and is used for computing the velocity vector.

The best correspondence can be selected by evaluating the cross-correlation coefficient:

$$C(x,y) = \frac{\sum\limits_{i=1}^{l_w}\sum\limits_{j=1}^{l_z}(I_1(i,j) - \mu_1)(I_2(i+x,j+y) - \mu_2)}{\sqrt{\sigma_1\sigma_2}} \tag{3}$$

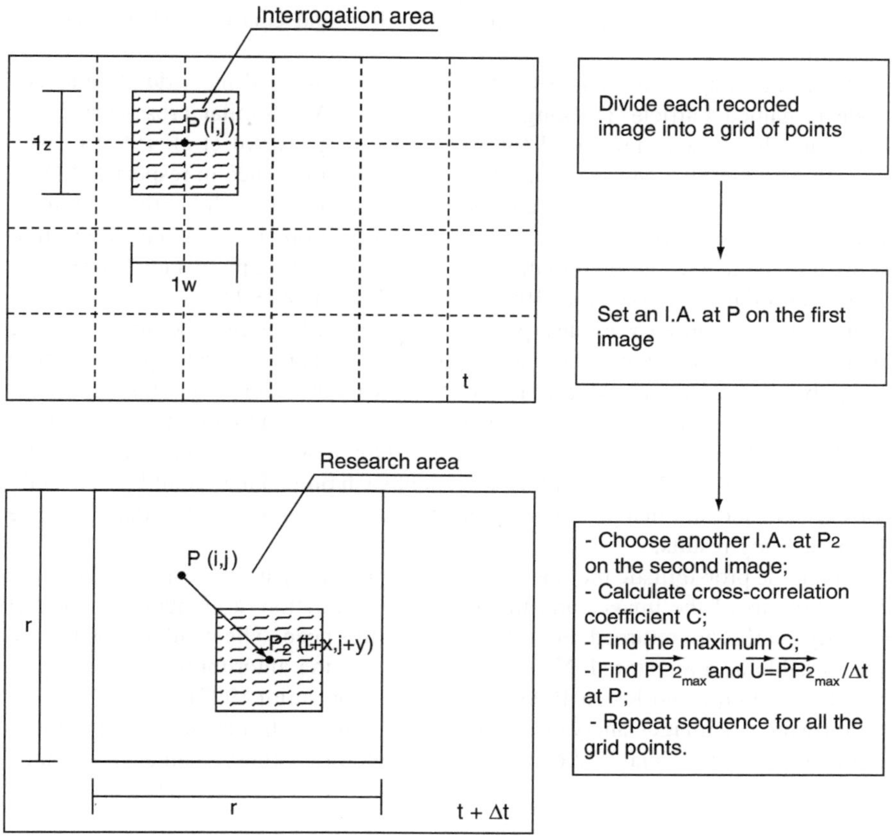

Fig. 5 General cross-correlation algorithm for PIV image processing

$I_1(i,j)$ and $I_2(i,j)$ being the pixel values in images 1 and 2; μ_1, μ_2 and σ_1, σ_2 the mean and variance values of the intensity in the interrogation areas A_1 and A_2, respectively. In particular, the algorithm finds the coordinates (x, y), such that $C(x, y)$ is maximized. The sequence of steps in image processing is also reported in Fig. 5. The correlation coefficient can be also calculated via the Fourier transform (Gui and Merzkirch 1998). Many different PIV algorithms, similar to the one presented, have been proposed (see for example, Cenedese et al. 1994; Stevens and Coates 1994).

For the quality of the results, an adequate choice of the delay time Δt, of the sizes of the interrogation and research areas and of the grid spacing, is very important. In fact, the ratios $r/\Delta t$ and $1/\Delta t$, multiplied for the pixel width, give the maximum and minimum velocity that the technique may resolve. Decreasing the interrogation area, the computational requirements decrease, but the statistical reliability of the data decreases too. To improve the resolution and the accuracy of the analysis, techniques for determining with sub-pixel accuracy the location of the maximum cross-correlation coefficients are available (Willert and Gharib 1991; Cowen and Monismith 1997).

When the image density is low, the particles are tracked individually and this mode is named Particle Tracking Velocimetry (PTV). Similar cross-correlation techniques have been developed for PTV mode (e.g., Lloyd et al. 1995; Cowen and Monismith 1997; Muste et al. 1998). The main difference in respect to PIV is that the velocity vectors are evaluated in the center of each particle, with an unstructured distribution across the image plane. To obtain a regular description of the flow field, several interpolation schemes have been implemented for regularizing vector data (e.g.,Wu et al. 1991; Agui and Jimenez 1997).

In both PIV and PTV modes, preprocessing procedures may be applied to the images in order to remove nonideal aspects and enhance image quality before to apply the processing algorithm. Well-known procedures are histograming-equalization, smoothing, hedge detection, background subtraction, thresholding, and binarization (Guezennec and Kiritsis 1990; Cowen and Monismith 1997; Crouser et al. 1997). The use and the effects of such procedures should be carefully considered, since an inappropriate application could degrade, rather than improve the processing results.

Given the probabilistic method used to determine the particle displacements, the imperfections of the images and the automatic evaluation, a number of incorrect velocity vectors are returned. For this reason, results need to be validated, so that no questionable data are stored. Numerous postprocessing algorithms have been proposed (e.g., Fuijta and Kaizu 1995; Cowen and Monismith 1997).

Software for images analysis, preprocessing, and postprocessing are produced by the major commercial PIV companies, but freeware versions can also be found.

3 Examples of Applications in Hydraulics

Nowadays, PIV and PTV are consolidated techniques to measure velocity fields for studying hydraulic phenomena. In particular, starting from the measurements of instantaneous velocities, turbulence characteristics, like mean velocities, turbulence intensities, Reynolds stresses, production term in transport equation for turbulence kinetic energy and more, can be obtained.

The initial application of PIV and PTV in hydraulics was primarily in experimental laboratory researches on turbulent flows (e.g., Landreth et al. 1988; Liu et al. 1991). PIV methods are still widely used in laboratory for studying hydrodynamics aspects in particular flow conditions, like in presence of rough (e.g., Campbell et al. 2005; Manes et al. 2007; Ferreira et al. 2010) or vegetated (e.g., Okamoto and Nezu 2009) beds, near hydraulic structures (e.g., Tsikata et al. 2009) and in complex phenomena, also combined with other measurement techniques (e.g., Zweifel et al. 2006). Considering that it is almost impossible to furnish a complete review of all applications of PIV methods in hydraulics, in the following sections, the attention is focused on the adaptation of such techniques to two particular situations of increasing interest in hydraulic research: for measurements in two-phase flows, with the possibility of discriminating between the phases; for measurements of free surface velocities in large laboratory and field scale flows.

3.1 Two-Phase Flows

The difficulty in investigating two-phase flows derives from the presence of the two different phases together and from the fact that usually they influence each other. In hydraulic applications, the mixtures of interest are essentially bubbly and sediment laden flows. The challenge in using PIV and PTV techniques is the possibility of discriminating between the two fractions. In particular, the main objective is to separate the particles seeding the liquid phase from the dispersed particles, bubbles or sediments, in order to measure water velocity. An additional feature is the measure of the velocities of both fractions simultaneously.

In many hydraulic phenomena, like for example, hydraulic jump, the presence of air bubbles represents a crucial aspect. Different methods have been proposed for discriminating between tracers and bubbles in PIV or PTV applications (Lindken and Merzkirch 2002). In a first kind of widely used methods, the discrimination is made at the recording stage, using optical separation criteria. In particular, fluorescent tracers are used in conjunction with an optical filter to ensure that only seeding particles are imaged and then PIV and PTV algorithms are applied (e.g., Cowen et al. 2003; Admiral et al. 2006; Lennon and Hill 2006). In some cases, a normal camera, without a filter, can be added in order to separately record bubble images. This is made when it is not possible to find a camera aperture setting adequate for both bubbles and tracers or when bubble reflections are too large and mask seeding particles scattering.

In a second type of methods, a single image with both phases is recorded and then preprocessing techniques are used in order to individuate bubbles and tracers signatures before applying PIV or PTV algorithms. Different criteria have been adopted for discriminating between the images of the two phases, among them the more often used ones are based on size (Hassan et al. 1992; Gui and Merzkirch 1996) or on gray-scale intensity differences (Sakakibara et al. 1996). Finally, the bubble signatures can be removed during the PIV or PTV processing stage. In fact, bubbles and water have different velocities and also different correlation peaks (Deen et al. 2002).

Among these techniques, the optical phase separation is the more diffused, but it requires expensive fluorescent tracer particles. Other methods, using standard PIV and/or PTV algorithms, have been proposed, as for example, by Seol et al. (2007).

In turbulent sediment laden flows, some important aspects can be investigated using the simultaneous measurements of sand and water velocities. In particular, it is possible to investigate on the presence of a slip velocity between the two phases, on the difference between turbulence characteristics of fluid and sediments, on the effect of sediments on the carrier fluid turbulence. For these applications, the PTV

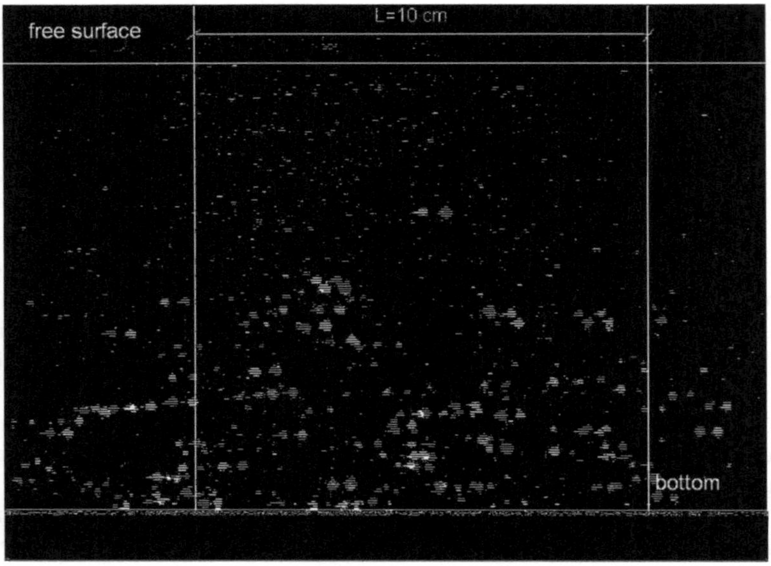

Fig. 6 Double exposure image with sediment and seeding particles (seeding diameter = 40 micron; sediment diamenter = 0.0680 mm)

technique is more suitable, discriminating in the processing stage between the two phases. In fact, since the size difference between sediment and seeding particles is usually considerable (Fig. 6), a size-based criterion can be adopted (Muste et al. 1998; Di Cristo and Muste 2002).

3.2 Large-Scale PIV

The information provided by imaging techniques is limited to their field of view, which in case of standard PIV system is up to several square centimeters. Starting from the nineties, image velocimetry concepts have been applied for documenting the free surface velocity field in laboratory flows, using imaged areas up to several square meters (Lloyd et al. 1995; Ettema et al. 1997; Muller et al. 2002). Moreover, Fuijta et al. (1998) adapted conventional image analysis technique for measurements in large body of water, with field of view hundreds square meters large. This kind of imaging technique application was named Large-Scale PIV (LSPIV).

LSPIV is now a suitable nonintrusive technique to measure instantaneous free surface velocity field in an entire physical model or in rivers. Then, from the measured velocity instantaneous values, temporal means can be computed. The general layout of a LSPIV system is reported in Fig. 7. While the LSPIV image

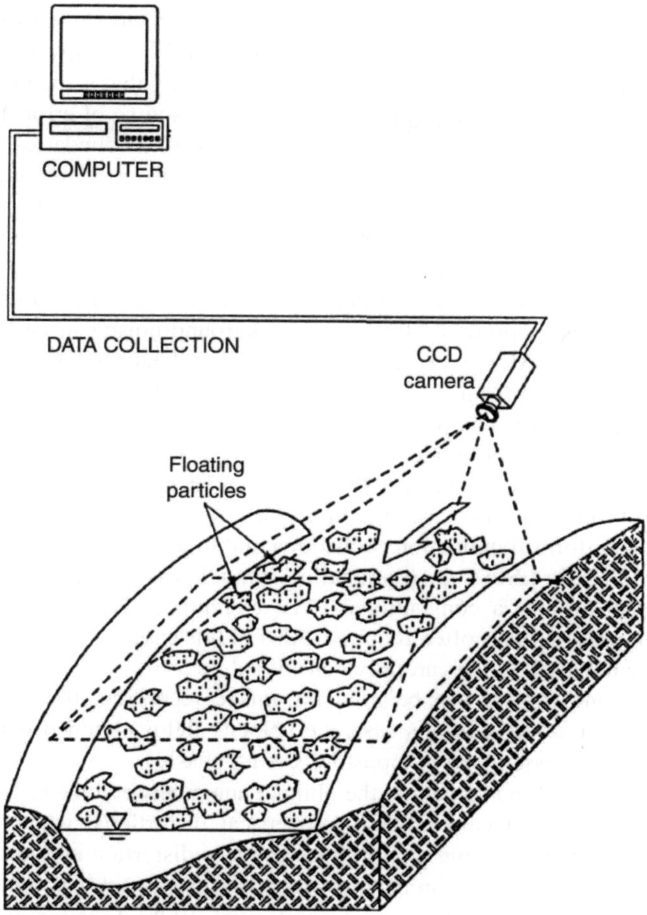

Fig. 7 Scheme of LSPIV system

concept is similar to conventional PIV, seeding and illumination are considerably different.

Sometime particles naturally present on free surface can be used as a tracer, but often they are not well distributed, so additional seeding particles need to be added to increase the efficiency of resolving velocity on the entire image. Given the large scale of flow and the necessity of measuring surface velocities, the tracer characteristics are quite different in respect to standard PIV and its choice is a very important aspect. First of all, seeding particles have to be lighter than water density; they have to be big enough, comparing with the pixel size, that in large-scale investigation is much larger than in standard PIV, considering the distance from the camera and the image plane; finally, they have to travel on the free surface with the flow velocity. Foams and straw (or other plant residues) are the two kinds of seedings usually adopted. A particular problem can be due to the fact that surface tension of water causes the tracer particles to coalesce, with the formation of groups. In fact, if two particles are attracted, each of them has a residual velocity that is inconsistent with the true flow velocity, which can produce a significant source of error. Because of the surface tension, tracer particles can also adhere to solid boundaries and roughness elements that protrude from the flow. The simplest way for reducing this phenomenon is the use of smaller tracer concentrations, but other methods are also available (Admiral et al. 2004). Moreover a good contrast between the tracers and the background color is desirable to easily analyze particle images. Whenever possible, like in laboratory investigations, the model or the water can be colored. In this way, with an appropriate threshold, background noise can be filtered in a preprocessing phase. Moreover, tracers need to be cheap and environmentally acceptable. Investigations on the effect of tracer material, seeding procedure, and concentration are reported by Muste et al. (2004).

Field measurements are made with natural light, while for indoor recordings strong illumination is required, usually realized with halogen, sodium-vapor, or other kinds of lamps. In this case, the lamps are to be strategically positioned in order to have uniform illumination of the area and to avoid light reflection on the free surface. Since in LSPIV the illumination is much less compared with the high intensity laser sheet of a conventional PIV, a preprocessing phase with image enhancement procedures is often required. Also in this case the use and the effects of such procedures should be carefully considered.

More than one camera can be used for image recording, if the size of the investigation area is large. In this case, the velocity fields obtained with different cameras have to be assembled in a postprocessing phase.

The large distance and sometime the oblique angle between the camera and the image plane can introduce lens and geometrical distortions, which should be removed before the processing phase. Since the lens distortion effects are usually negligible, simple correction procedures, based on a geometrical transformation applied to recorded images, can be used (Fuijta et al. 1998). Like for standard PIV, cross-correlation algorithms can be used to obtain instantaneous velocities and a post-processing phase for discarding erroneous vectors can be performed.

Given the large size of LSPIV imaged area and consequently of the interrogation areas, the spatial resolution of velocity measurements is limited. The flow scales smaller than the interrogation area cannot be solved, so LSPIV cannot measure small-scale turbulence, but it is intended to capture large-scale flow structures.

LSPIV was used to measure the surface velocity distribution in entire rivers (Fuijta et al. 1998), also during extreme events, using a mobile LSPIV unit (Kim et al. 2008), able to rapidly develop the measurement equipment in a desired site, or using video images from helicopters (Fuijta and Hino 2007b). In fact, the knowledge of surface velocities is useful for many purposes in fluvial hydraulics. They furnish important information in shallow water flows, for example, they were used for investigating on large coherent structures (Weibrecht et al. 2002) and mixing processes between a river and dead-water zones created by groin fields (Weibrecht et al. 2008). It is even possible to use surface velocities to estimate indirect quantities, for example, surface velocities are converted to depth integrated velocities by applying a 1/7 power law or other vertical distributions for computing the flow discharge (Fuijta et al. 1998; Hauet et al. 2008). Muste et al. (2004) suggest also the possibility of evaluating flow depth layer variation.

LSPTV technique is also used so that a small number of seeding particles is required. In particular, Admiral et al. (2004) compare the performances of LSPIV and LSPTV, applying both techniques to a large-scale model, demonstrating that, although both methods predict the overall distribution quite well, LSPTV works better near boundaries and in places with high velocity gradients. A particular configuration, named space-time image velocimetry (STIV), was recently developed for acquiring measurement with poor or no seeding (Fuijta et al. 2007a). It takes advantage of the specular reflection of the ambient light on the small waves present on the free surface.

3.3 Different PIV Setups

Classical 2D-PIV method is only capable of recording the velocity vectors in the light sheet plane and the out-of-plane velocity component is lost. For high three-dimensional flows, stereoscopic PIV (SPIV) has been developed (Nishino et al. 1989; Prasad and Adrian 1993; Gaydon et al. 1997). Using principles of the stereoscopic vision, SPIV reconstructs through equations the three-component velocity fields captured from two cameras. The equations have to be calibrated to account for the geometrical and optical characteristics of the stereoscopic system (Fouras et al. 2007; Grizzi et al. 2010).

Recent extensions of SPIV include dual-time SPIV (Perret et al. 2006) and multilayer PIV. In this latter method, more light sheets are generated by splitting the laser beam using beam splitter (Sanjou and Nezu 2009). Holographic PIV is another method for recovering the third velocity component (Coupland and Halliwell 1992).

4 Conclusions

The PIV techniques are actually widely used in hydraulics. Their main advantage is the possibility of nonintrusive velocity field measurements, with an elevate precision. Standard 2D-PIV methods are not able to measure the third velocity component, but now stereoscopic PIV is a well established extension of the traditional PIV in case of three-dimensional flow fields.

There are still some concerns related to the high prices and the operators' safety. Now more advanced systems are still prohibitively expensive, but standard PIV systems can be actually found to affordable price. For this reason, this technique is becoming very diffuse in hydraulic laboratories. The operators need to be well prepared to reach good quality results, but training courses and didactic material can be easily found.

The possibility of adapting the PIV technique for studying more complex flows is very exciting. Moreover, the applications to field measurements open the way to a more diffuse use, not only in the research field.

References

Admiral DM, Stansbury JS, Haberrman CJ (2004) Case study: particle velocimetry in a model of Lake Ogallala. J Hydraul Eng 130(7):599–607

Admiral DM, Musalem-Jara R, Garcia M, Nino Y (2006) Vortex trajectory hysteresis above self-formed vortex ripples. J Hydraul Res 44(4):437–450

Adrian RJ (1991) Particle-imaging techniques for experimental fluid mechanics. Annu Rev Fluid Mech 23:261–304

Adrian RJ (2005) Twenty years of particle image velocimetry. Exp Fluids 39:159–169

Agui JC, Jimenez J (1997) On the performance of particle tracking. J Fluid Mech 185:447–468

Campbell L, McEwan I, Nikora V, Pokrajac D, Gallagher M, Manes C (2005) Bed-load effects on hydrodynamic of rough-bed open-channel flows. J Hydraul Eng 131(7):576–585

Cenedese A, Doglia G, Romano GP, De Michele G, Tanzini G (1994) LDA and PIV velocimetry measurements in free jets. Exp Therm Fluid Sci 9:125–134

Coupland JM, Halliwell NA (1992) Particle image velocimetry: three dimensional fluid velocity measurements using holographic recording and optical correlation. Appl Opt 31:1005–1007

Cowen NL, Monismith SG (1997) A hybrid digital particle tracking velocimetry technique. Exp Fluids 22:199–211

Cowen EA, Sou IM, Liu PL, Raubenheimer B (2003) Particle image velocimetry measurements within a laboratory-generated swash zone. J Eng Mech 129(10):1119–1129

Crouser PD, Bethea MD, Merat FL (1997) Unattenuated tracer particle extraction through time-averaged, background image subtraction with outlier rejection. Exp Fluids 22:220–228

Deen N, Westerweel J, Delnoij E (2002) Two-phase PIV in bubbly flows: status and tends. Chem Eng Technol 25(1):97–101

Di Cristo C, Muste M (2002) Experimental study on liquid and suspended sediment turbulence characteristics in open-channel flows. In: 5th international symposium on engineering turbulence modelling and measurements. Elsevier, Mallorca, pp 979–988

Ettema R, Fujita I, Muste M, Kruger A (1997) Particle-image velocimetry for whole-field measurement of ice velocities. Cold Region Sci Technol 26(2):97–112

Ferreira RM, Amatruda M, Ricardo AM, Franca MJ, Di Cristo C (2010) Production and dissipation of turbulent kinetic energy in the roughness layer. Proceedings of the I European IAHR conference, Edinburgh

Fouras A, Dusting J, Hourigan K (2007) A simple calibration technique for stereoscopic particle image velocimetry. Exp Fluids 42:799–810

Fuijta I, Hino T (2007) PIV measurements of large-scale river surface flow during flood by using a high resolution video camera from a helicopter. Proceedings of Hydraulic Measurements and Experimental Methods Conference, ASCE-IAHR, Lake Placid, pp 344–349

Fuijta I, Kaizu T (1995) Correction method of erroneous vectors in PIV. J Flow Vis Image Process 2:173–185

Fuijta I, Muste M, Kruger A (1998) Large-scale particle image velocimetry for flow analysis in hydraulic engineering application. J Hydraul Res 36(3):397–414

Fuijta I, Watanabe H, Tsubaki R (2007) Development of a non-intrusive and efficient flow monitoring technique: the Space Time Image Velocimetry (STIV). Int J River Basin Manage 5(2):105–114

Gaydon M, Raffel M, Willert C, Rosengarten M, Kompenhans J (1997) Hybrid stereoscopic particle image velocimetry. Exp Fluids 23:331–334

Grizzi S, Pereira F, De Felice F (2010) A simplified, flow based calibration method for stereoscopic PIV. Exp Fluids 48:473–486

Guezennec YG, Kiritsis N (1990) Statistical investigation of errors in particle image velocimetry. Exp Fluids 10:138–146

Gui L, Merzkirch W (1996) A method for tracking ensembles of particle images. Exp Fluids 21(6):465–468

Gui L, Merzkirch W (1998) Generating arbitrary sized interrogation windows for correlation-based analysis of particle image velocimetry recordings. Exp Fluids 24:66–69

Hassan YA, Blanchat TK, Seeley CH, Canaan RE (1992) Simultaneous velocity measurements of both components of a two-phase flow using particle image velocimetry. Int J Multiph Flow 18(3):371–395

Hauet A, Kruger A, Krajewski WF, Bradley A, Muste M, Creutin J, Wilson M (2008) Experimental system for real-time discharge estimation using an image-based method. J Hydrol Eng 13(2):105–110

Hetsroni G (1989) Particles–turbulence interaction. Int J Multiph Flow 15(5,):735–746

Kim Y, Muste M, Hauet A, Krajewski W, Kruger A, Bradley A (2008) Stream discharge using mobile large-scale particle image velocimetry: a proof of concept. Water Resour Res 44: W09502. doi:0.1029/2006WR005441, 6 p

Landreth CC, Adrian RJ, Yao CS (1988) Double pulsed particle image velocimeter with directional resolution for complex flows. Exp Fluids 6:119–128

Lennon JM, Hill DF (2006) Particle image measurements of undular and hydraulic jumps. J Hydraul Eng 132(2):1283–1294

Lindken R, Merzkirch W (2002) A novel PIV technique for measurements in multiphase flows and its application to two-phase bubbly flows. Exp Fluids 33:814–825

Liu Z, Landreth CC, Adrian RJ, Hanratty TJ (1991) High resolution measurement of turbulent structure in a channel with particle image velocimetry. Exp Fluids 10:301–312

Lloyd PM, Stansby PK, Ball DJ (1995) Unsteady surface-velocity field measurement using particle tracking velocimetry. J Hydraul Res 33(4):519–533

Manes C, Pokrajac D, McEwan I (2007) Double averaged open-channel flows with small relative submergence. J Hydraul Eng 133(8):896–904

Melling A (1997) Tracer particles and seeding for particle image velocimetry. Meas Sci Technol 8:1406–1416

Muller G, Bruce T, Kauppert K (2002) Particle image velocimetry: a simple technique for complex surface flows. Proceedings of River Flow conference, Louvein, pp 1227–1234

Muste M, Fuijta I, Kruger A (1998) Experimental comparison of two laser-based velocimeters for flows with alluvial sand. Exp Fluids 24:273–284

Muste M, Xiong Z, Schone J, Li Z (2004) Validation and extension of image velocimetry capabilities for flow diagnostics in hydraulic modeling. J Hydraul Eng 130(3):175–185

Nishino K, Kasagi N, Hirata M (1989) Three-dimensional particle tracking velocimetry based on automated digital image processing. J Fluids Eng 111:384–391

Okamoto TA, Nezu I (2009) Turbulence structure and "Monami" phenomena in flexible vegetated open channel flows. J Hydraul Res 47(6):275–280

Perret L, Braud P, Fourment C, David L, Delville J (2006) 3-components acceleration field measurement by dual-time stereoscopic particle image velocimetry. Exp Fluids 40(5): 813–824

Prasad AK, Adrian RJ (1993) Stereoscopic particle image velocimetry applied to liquid flows. Exp Fluids 15:49–60

Raffael M, Willert C, Kompenhans J (2007) Particle image velocimetry – practical guide. Springer, Berlin

Sakakibara J, Wicker R, Eaton J (1996) Measurements of particle-fluid velocity correlation and the extra dissipation in a round jet. Int J Multiph Flow 22(5):863–881

Sanjou M, Nezu I (2009) Turbulence structure and coherent motion in meandering compound open-channel flows. J Hydraul Res 47(5):598–610
Seol DG, Bhaumik T, Bergmann C, Socolofsky SA (2007) Particle image velocimetry measurements of the mean flow characteristics in bubble plume. J Eng Mech 133(6):665–676
Stevens C, Coates M (1994) Application of a maximised cross-correlation technique for resolving velocity fields in laboratory experiments. J Hydraul Res 32(2):105–211
Tsikata JM, Tachie MF, Katopodis C (2009) Particle image velocimetry study of flow near trashrack models. J Hydraul Eng 135(8):671–684
Weibrecht V, Kuhm G, Jirka GH (2002) Large scale PIV-measurements at the surface of shallow water flows. Flow Meas Instrum 13:237–245
Weibrecht V, Socolofsky SA, Jirka GH (2008) Experiments on mass exchange between groin fields and main stream in rivers. J Hydraul Eng 134(2):173–183
Willert CE, Gharib M (1991) Digital particle image velocimetry. Exp Fluids 10:181–193
Wu Z, Zhu J, Chen S, Yang L, Xu H, Yang Y (1991) An image processing system for quantitatively analyzing the 2D fluid velocity-field image. Comput Fluids 20:359–371
Zweifel A, Hager WH, Minor HE (2006) Plane impulse waves in reservoirs. J Waterw Port Coast Ocean Eng 132(5):358–368

Turbulent Flow Hydrodynamics and Sediment Transport: Laboratory Research with LDA and PIV

Rui M.L. Ferreira

1 Introduction

1.1 Laser Doppler Anemometry and Particle Image Velocimetry

Laser Doppler Anemometry (LDA) and Particle Image Velocimetry (PIV) are indirect techniques to measure flow velocities. Both require the existence of small particles in the moving fluid, ideally in concentrations that do not change the properties of the fluid, and both measure specific statistics of the velocity of such particles, herein called tracing particles or seeding. Both techniques employ lasers, directly in the case of the former, making use of the Doppler effect as the laser light is scattered back to the receiver, and indirectly, in the case of the latter, as the laser is used to illuminate seeding particles in a plane section of the flow. Both techniques are among the least intrusive techniques to measure flow velocities. In fact, their intrusiveness is solely the result of seeding the flow with particles that may affect flow or fluid properties; in general, no material parts of the instrumentation are needed near the flow region under measurement.

LDA techniques are based on complex chemical and electronic units and very sensitive optical systems with little software complexity; on the contrary, PIV techniques, albeit requiring complex laser producing units, rely heavily on software for the analysis of raw data. In what concerns the study of water flows, both techniques are ideal for controlled research environments, as laboratories where the flows are confined in flumes, generally glass-walled, with optimal visualization conditions. Relatively to more intrusive techniques such as Acoustic Doppler Velocimetry, both techniques are applied with greater difficulty in field conditions, as the equipment is less robust and, in the case of the LDA, relying on complex cooling systems and demanding power sources.

R.M.L. Ferreira
CEHIDRO – Instituto Superior Técnico, UTL, Av. Rovisco Pais, Lisbon 1049-001, Portugal
e-mail: ruif@civil.ist.utl.pt

P. Rowinski (ed.), *Experimental Methods in Hydraulic Research*,
Geoplanet: Earth and Planetary Sciences, DOI 10.1007/978-3-642-17475-9_4,

LDA techniques have been operational since the 1960s (Yeh and Cummins 1964; reviews in Tropea et al. 2007), having been used in the scope of river hydraulics since the 1980s (Nezu and Nakagawa 1993; Gyr and Schmid 1997).

PIV techniques have been applied in hydraulics since the early 1980s (Lauterborn and Vogel 1984). Since it requires large storage space and large RAM memory, digital PIV techniques have benefited from the advance in computer architecture, especially since late 1990s. In the context of river hydraulics, PIV has been growingly used in the last 15 years (e.g., Tait et al. 1996).

1.2 Objectives and Structure of the Work

The main objective of the present chapter is to show applications of LDA and PIV techniques in the scope of fluvial hydraulics, within research work conducted at Instituto Superior Técnico, Technical University of Lisbon, Portugal.

The work is divided in three parts. The first is dedicated to the presentation of basic concepts of LDA and PIV measurements. The second part addresses the characterization of organized turbulence in open-channel flows over flat rough mobile beds and over flat rough fixed (armored) beds. Two data sets are compared to verify (1) if mobile bed flows have the same turbulent structures as flows over fixed beds and (2) if there are important feedback mechanisms between the organization of turbulence and the quality and quantity of the sediments moving in the bed. The third part is devoted to study and characterize the scour mechanism associated to a vertical cylinder mounted on a mobile bed. The instantaneous flow velocity in the scour hole in front of the cylinder, for a given instant in the developing phase of the scour hole, is measured and used to calculate mean and instantaneous out-of-plane vorticity. The analysis of the downflow and of the vortex system inside the scour hole allows for the identification of the most relevant scour mechanisms.

2 Basic Principles of LDA and of PIV Techniques

2.1 Principles of LDA Instrumentation

Laser Doppler Anemometry is a point-wise measuring technique of the velocity of tracers in viscous fluids. An LDA system features a continuous laser source, transmitting optics, receiving optics with photo-detector, analog filters, and signal amplifiers, a signal processing unit and memory unit (Durst et al. 1976), as seen in Fig. 1. Commonly used lasers comprise He-Ne (Helium–Neon), typically at 632.8 nm (red), Ar-Ion (ions of Argon), normally, in water, operated in the blue or green

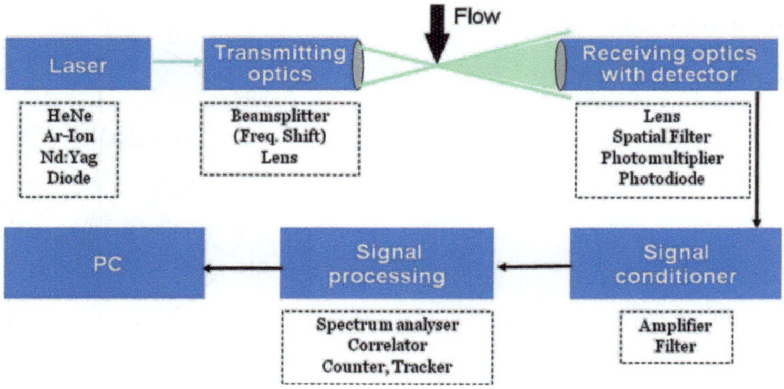

Fig. 1 General layout of an LDA system (Adapted from Dantec Dynamics LDA literature http://www.dantecdynamics.com/Default.aspx?ID=1046)

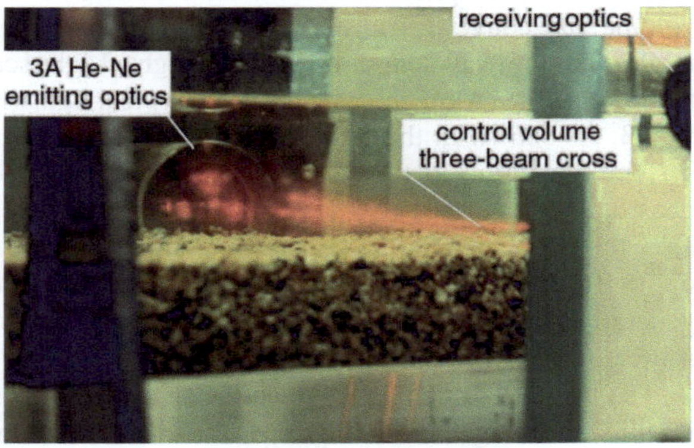

Fig. 2 A 3A He-Ne LDA, in foward scattering mode, measuring over a gravel bed (From Ferreira 2005)

spectra or Neodymiumdoped Yttrium Aluminum garnet (Nd:Yag). The former two are gas-lasers, i.e., coherent light produced by the excitation of a gas by a strong electric current. The latter is a solid-state laser, in which a crystalline medium is modified (doped) with small concentrations of impurities to change its electrical properties; the crystalline medium is optically excited (optically pumped, generally by diode lasers) to produce population inversion and, hence, emission of photons (Csele 2004).

The laser beam is split before reaching the transmitting optics. A Bragg cell is normally used for beam splitting and also to ensure that directional ambiguity is eliminated (see Fig. 5 below). The beams are focused in the transmitting optics in order to intercept in a measurement volume. One pair of beams of a given wavelength is necessary to measure one velocity component. Figure 2 shows a two-component He-Ne LDA system where two pairs of 632.8 nm laser beams intercept above a gravel bed. The system in Fig. 2 is operated in forward scattering mode,

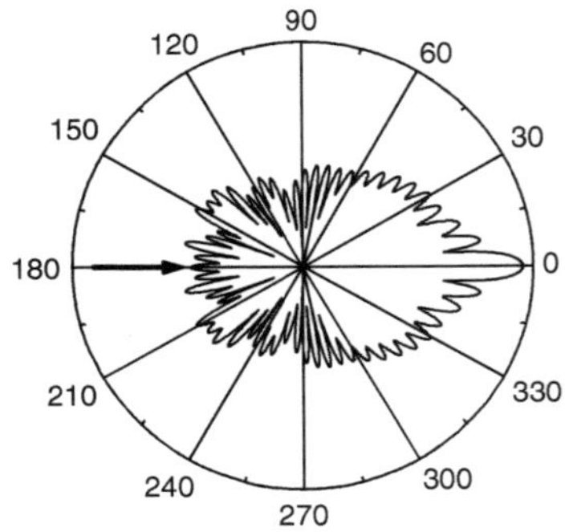

Fig. 3 Mie scattering for particles with diameter $d = 10\ \lambda$, where λ is the light wavelenght

i.e., the receiving optics is placed at an angle larger than 90° from the transmitting optics.

The receiving optics collects the light scattered by the particles in the measuring volume and conveys it to the photodiode. The diameter of the tracer particles is larger than the wavelength of the laser light. Light scattering is thus said to be in the Mie regime (Albrecht et al. 2003). A typical light scattering diagram is shown in Fig. 3. As shown in the figure, the forward direction is the most effective. LDA systems operating in the forward scatter mode are, however, not dominant, for constructive and operative reasons. Systems operating in backscattering systems have the receiving and the transmitting optics in one single probe, with the former optics focused by the manufacturer. In the forward scatter mode, the user is generally required to place and to focus the receiving optics, introducing extra source of measurement errors.

The Doppler principle by which the LDA operates is better explained with what is called the fringe model. Figure 4 illustrates the main principles. A fringe pattern is originated when two monochromatic (same wavelength) coherent (same phase for all waves) light beams are superimposed. Where the interfering light is in phase a peak (bright fringe) is generated; where it is out of phase, light is canceled out and a trough (dark fringe) occurs. In the LDA case, this fringe pattern occurs in the measurement volume. The fringe pattern is parallel to the axis of the transmitting optics.

The spacing between fringes depends on the wavelength of the crossing beams, λ, and on the angle that they form, θ. Simple geometrical considerations (Fig. 4) lead to the formula:

$$\delta = \frac{\lambda}{2\sin(\theta/2)} \tag{1}$$

A particle crossing the control volume with velocity u normal to the axis of the probe will generate a burst which, in the time domain, is a wave packet with frequency f

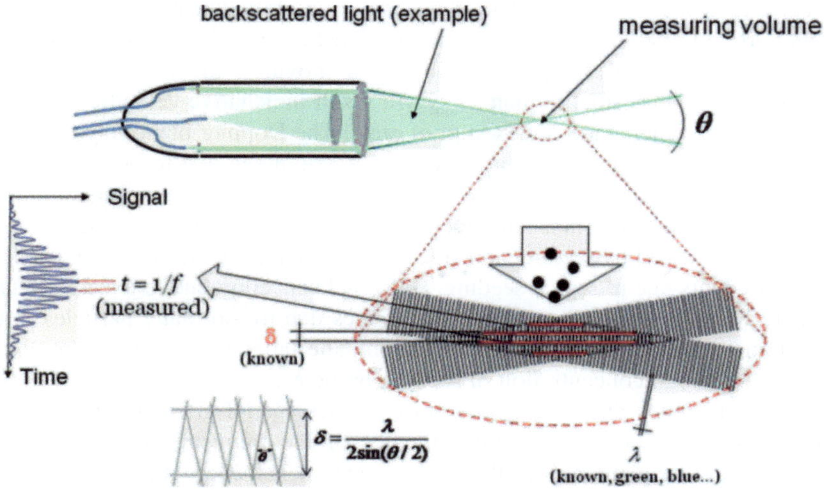

Fig. 4 Fringe model (Adapted from Dantec Dynamics LDA literature)

(Fig. 4). The interval between peaks is $t = 1/f$. Hence, the velocity of the particle that generated the burst is:

$$u = \frac{\delta}{t} = \frac{\lambda}{2\sin(\theta/2)} f \tag{2}$$

The key issue for the LDA signal conditioner and signal processing units is thus to identify bursts from background noise and to detect correctly the frequency f. These units are often condensed in the Real-time Signal Analyzer (RSA) unit. Omitting important details, the main operations undertaken in this unit are (1) subtraction of the Gaussian pedestal inherent to a coherent laser (the radial distribution of the beam intensity is approximately Gaussian), (2) amplification, (3) filtering of noise components (general low-pass filtering), (4) burst detection, for instance by thresholding in the time-amplitude domain, and (5) frequency estimation, normally by inverse-Fourier methods. The signal-to-noise ratio and the transit length are key parameters to verify the quality and to validate a Doppler burst. Naturally, the larger the number of fringes, the easier to detect the burst. The number of fringes, for an ellipsoid measuring volume, is $N_f = 8F \tan(\theta/2)/(\pi D_L)$. Hence, the larger the angle between beams and the larger focal distance, the larger the number of fringes.

The raw LDA signal is thus composed of irregular bursts, corresponding to particles that crossed the measuring volume, that are translated into as many instantaneous velocities. It is current to postprocess the velocity signal into evenly spaced velocity readings. This implies reconstruction for which there are several algorithms using polynomial interpolation (Tropea et al. 2007).

To distinguish between positive and negative velocities, a frequency shift in one of the beams is introduced in the Bragg cell. The latter is composed of a glass crystal subjected to acoustic vibration. The frequency of the shifted beam is changed by an amount proportional to the frequency of the acoustic wave, f_{bragg}. As a result, when the beams cross, the fringe pattern will appear moving with the shift frequency. Detected frequencies smaller than f_{bragg} will result in negative velocities (Fig. 5).

LDA measurements are not exempt of error. The Doppler burst may be hidden by noise associated to the instrumentation (electric noise due to poor insulation, noise introduced by the photodetector, noise introduced by the laser generation). In general, these sources of noise are uncorrelated and produce white noise.

Errors introduced by seeding and signal amplification include (1) signal spiking or absent counts due to scarce seeding, (2) signal canceling due to excess seeding, and (3) saturation and bias to small frequencies due to too many particles in the measurement volume and high amplification. The latter two are not common as they require a high concentration of seeding particles.

Poor beam focus due to lens aberration causes spatial gradients in the fringe pattern (Fig. 6), resulting in poor estimates of the frequency f. Errors due to positioning are frequent, especially in 3D arrangements where the user is required to match the measuring volumes of two different probes. Working in coincidence mode, for which a passing particle must generate a burst recognizable by the detectors associated to each of the three wavelengths, may drastically reduce the frequency count if the measurement volumes are not perfectly aligned.

Since LDA velocity measurements are actually particle velocities, errors are introduced, especially in turbulent flows, because particles do not follow the fluid exactly. The equation that describes particle motion in a viscous fluid was derived and solved by Crowe et al. (1998) and Hjemfelt and Mockros (1996). The leading

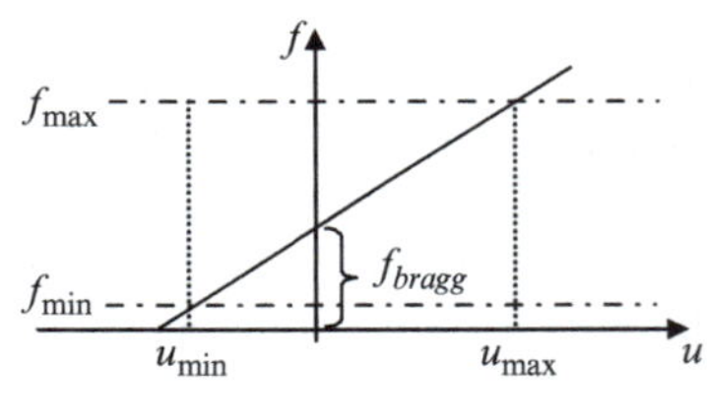

Fig. 5 Eliminating signal ambiguity through frequency shifting (Bragg cell principle)

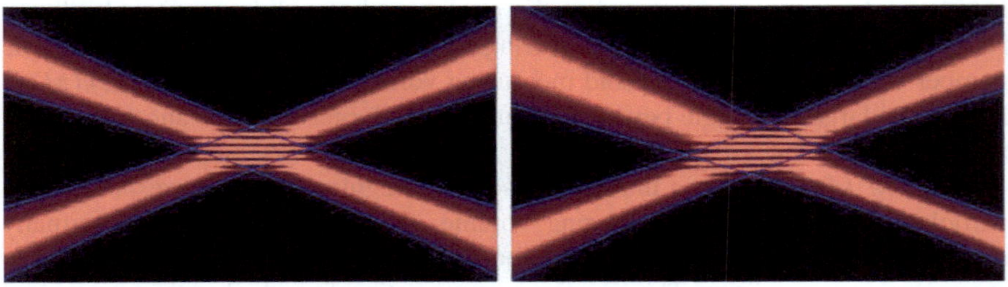

Fig. 6 Crossing of laser beams. *Left*: correct crossing with parallel fringe pattern. *Right*: incorrect crossing due to lens aberration (Fringe model from http://www.nat.vu.nl/environmentalphysics/REAL%20Experiments/LDA%20exp/LDA.html)

term is Stokes drag. Retaining only this term, the smaller turbulent scale measurable for a given velocity increment is:

$$\Delta r = \frac{1}{18} \frac{d(s-1)}{\nu^{(w)}} \Delta u \quad (3)$$

where S is the specific gravity of the seeding grains, $\nu^{(w)}$ is the kinematic viscosity of the fluid, and Δu is the velocity increment. To measure water flows, the diameter of seeding particles ranges normally from 5 to 50 μm and are denser than the fluid. Assuming $s - 1 = 1.65$, the smallest turbulent scale associated to a velocity increment of 0.1 susceptible to be measured with seeding particles of 50 μm would be 2.3×10^{-5} m. For current laboratory flumes this value would be well within the dissipative range and would not be a cause of significant errors.

The size of the measuring volume is important when discussing the quality of turbulent flow measurements. The dimensions of the measuring volume in the directions normal and parallel to the probe axis are:

$$\delta_x = \frac{4F\lambda}{\pi D_{\mathrm{L}} \cos(\theta/2)} \quad \text{and} \quad \delta_y = \frac{4F\lambda}{\pi D_{\mathrm{L}} \sin(\theta/2)} \quad (4)$$

respectively, where F is the focal distance and D_{L} is the diameter of the beam just before the lens. Control volumes too large may smooth out small turbulent scales, since particles with symmetric velocities belonging to the same eddy simultaneously present in the measurement volume may disturb burst detection. One other adverse effect of a large control volume is the generation of noise. If δ_y encompasses strong spatial gradients, particles with different velocities will cross the control volume and will be interpreted as turbulence even if the flow is laminar. Naturally, if the flow is turbulent, this type of noise will also be present but will be very difficult to detect as it may not be uncorrelated. A typical system with $F = 0.40$ m, $\theta = 30°$, $\lambda = 632.8$ nm, and $D_{\mathrm{L}} = 0.001$ m will exhibit $\delta_x = 0.0003$ m and $\delta_y = 0.0012$ m. The lateral size of the control volume, of the order of magnitude of 1 mm, would be a cause of some noise introduced in the turbulent signal.

The major favorable features of the LDA technique can be summarized as follows:

1. LDA is relatively low intrusive technique; given that the focal distances are large, no material pieces of instrumentation are required near the flow under measurement.
2. When measuring river and channel flows, sampling frequencies are generally high (it is easy to obtain 300 Hz), given that the flow is normally seeded with natural sediment tracers; the time discrertization of the velocity series is among the highest achievable in channel flows.
3. The spatial discretization of the velocity signal is good, relatively to acoustic Doppler techniques and other point-wise or profiling acoustic techniques; in fact, the measuring volumes of the latter are often one order of magnitude larger than that of the LDA.

2.2 Principles of PIV Instrumentation

Particle Image Velocimetry is an indirect technique for measuring instantaneous flow velocities in planar regions of the flow. It is thus neither a point-wise measurement technique nor a profiler although it can be used as so. Its main novelty, relatively to older techniques, is thus the possibility of calculating velocity gradients and thus out-of-plane vorticity.

PIV actually measures small displacements of tracer particles, illuminated by a laser sheet, recorded by a video camera. It can measure the two in-plane velocity components (2DPIV) or the three components (3DPIV), for which a second camera is required. Figure 7 shows a typical 2DPIV layout, featuring a laser, generally pulsed, with its optics, a video camera, both synchronized and controlled by software, a frame grabber and processing software.

Although more recent than LDA, there is a considerable body of literature that describes PIV hardware, methods and algorithms, more comprehensive than that available to Doppler techniques. This is probably due to the fact that the quality of PIV measurements depends mostly on data processing and less on hardware issues. Comprehensive reviews can be found in Raffel et al. (1998) and Tropea et al. (2007). Only a brief overview is offered in this section.

In the recent commercial systems, the laser source is generally a solid-state laser, for instance, a Nd:YAG (Neodymium-doped Yttrium Aluminum garnet). It is thus subjected to optical pumping, frequently with diode laser or white light. The Nd: YAG laser depends on the temperature. At standard operating temperatures, it emits at wavelength 1064 nm, in the infrared spectrum.

A pulsed laser is obtained by a quality switch (Q-switch) inside the cavity, operated in a software-controlled triggered mode. Q-switches are generally

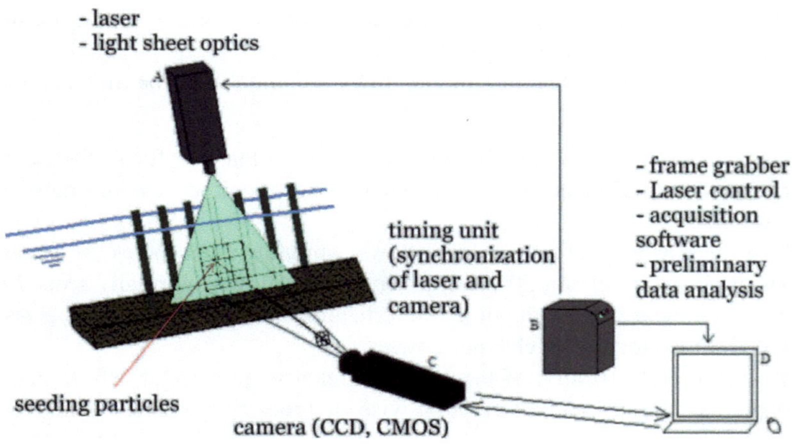

Fig. 7 Typical PIV layout

associated to double oscillators, which enables the user to adjust the separation time between the two illuminations of the tracer particles independently of the pulse strength (Fig. 7).

The infrared laser is subjected to phase matching, to convert infrared 1,064 nm to, generally, green light with 532 nm. The cooling system is an essential feature, as this process requires stable temperatures. The laser beams from each cavity are then conveyed to the output optics through a system of mirrors, reflectors, and shutters (Fig. 8). At the output optics, the laser beams are passed through prismatic lenses to generate laser sheets.

Most cameras are either CCD (charged-couple device) or CMOS (complementary metal oxide semi-conductor). The former operate with photoelectric effect at each pixel. Every pixel's charge is then transferred through the same output nodes, which is reliable (not prone to noise) but slow. In the latter (CMOS), each pixel has its own charge-to-voltage conversion node. It is thus highly sensitive and fast, allowing for time resolving PIV (high temporal resolution). However, CMOS cameras tend to be prone to noise.

The flow must be uniformly seeded with particles which, ideally, should have densities close to that of the water and diameters appropriate to the zoom level associated to the required level of detail. Conifer pollen particles have the density of the water but will eventually cause fungi infestations in the flume. Synthetic particles as the polyamide seen in Fig. 9 have been successfully used in water flows even if their density is larger than that of the water.

The motion of the seeding particles in a viscous fluid is affected primarily by their shape, size, and relative density, given a set of body and surface forces (e.g., gravity, drag, and lift). In a turbulent flow, body forces are considered negligible, given the small volume of the particles and the high values of inertial forces and drag (Melling 1997).

To evaluate the capacity of the particles of seeding to follow the flow and, hence, determine the smallest turbulent eddies, the solution of Hjemfelt and Mockros (1996), for the limit case $s \gg 1$ (Melling 1997), is employed:

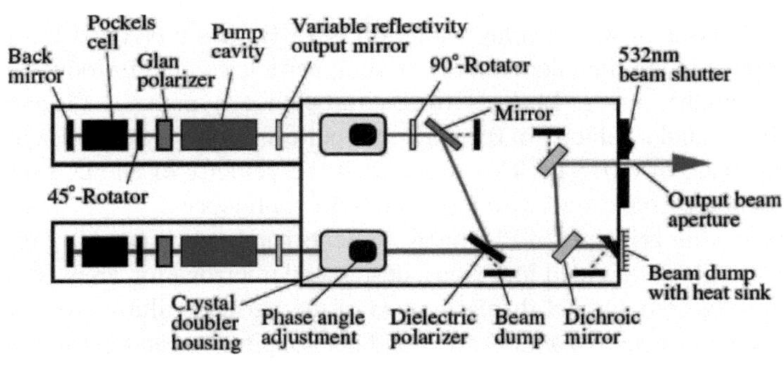

Fig. 8 Double cavity laser source and optics (From Raffel et al. 1998)

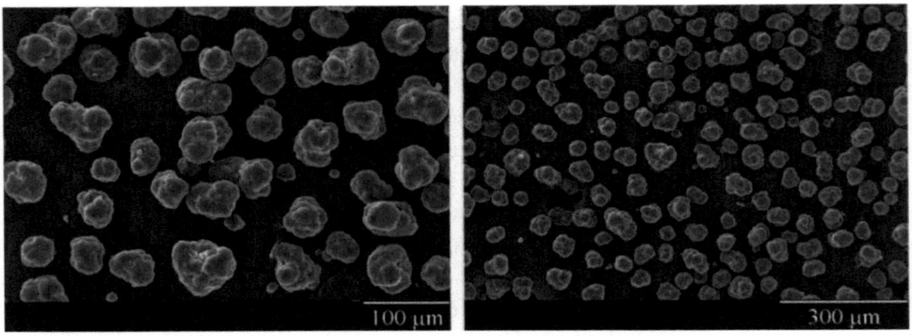

Fig. 9 Polyamide tracer particles used for seeding (Adapted from Ricardo 2008)

$$\frac{\bar{u}_{\mathrm{p}}^2}{\bar{u}_{\mathrm{f}}^2} \equiv r_{\mathrm{p}} = \left(1 + \frac{2\pi f_{\mathrm{c}}}{C}\right)^{-1} \tag{5}$$

where $\bar{u}_{\mathrm{p}}$ is the velocity of the seeding particles and $\bar{u}_{\mathrm{f}}$ is the velocity of the fluid,

$$C = \frac{18\nu^{(w)}}{sd^2} \tag{6}$$

and f_{c} is the turbulent frequency. Equation (5) expresses the ability of the seeding particle to follow the fluid turbulent micro-structure whose scale is f_{c} (expressed in terms of frequencies) as a function of its density and diameter, included in parameter C. The closer the ratio r_{p} is to unity, the better the tracking capability. In most cases, the difference of densities is not large. The application of (5) is justifiable for its conceptual simplicity as it retains the most important effects of drag and inertia (Melling 1997). It renders thus a conservative estimate of the tracking ability of the tracers. The graphical expression of (5) is seen in Fig. 10 for polyamide particles.

Melling (1997) proposed a threshold of acceptability of $r_{\mathrm{p}} = 0.95$ to quantify the tracking ability. Using this criterion, from Fig. 10 one concludes that particles of 50 μm are unsuitable to express turbulent scales higher than 13 Hz. Note that a common PIV system whose time resolution is 15 Hz has a Nyquist frequency of 7.5 Hz. Hence, in the time domain, the seeding particles would introduce errors. In the space domain, using Taylor's frozen turbulence hypothesis (Tennekes and Lumley 1972), and a velocity of 0.2 ms^{-1}, the cut-off frequency of 13 Hz represents a turbulent scale of 0.015 m. This means that the velocity of eddies smaller than 0.015 m would be measured with less than 95% confidence.

The measuring principle of PIV measurements can be illustrated in Fig. 11. The planar flow region is divided into areas, designated interrogation areas. If the mean vertical and the longitudinal displacements of the particles illuminated in a given interrogation area centered at (x,z) are found, the longitudinal and vertical velocities

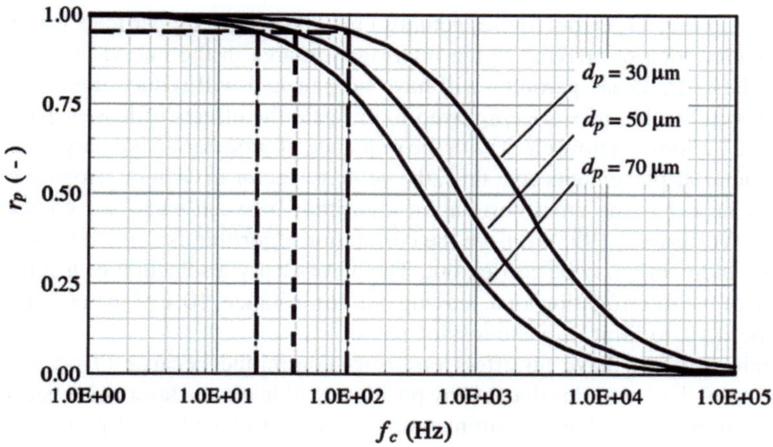

Fig. 10 Response of polyamide seeding particles in a turbulent flow

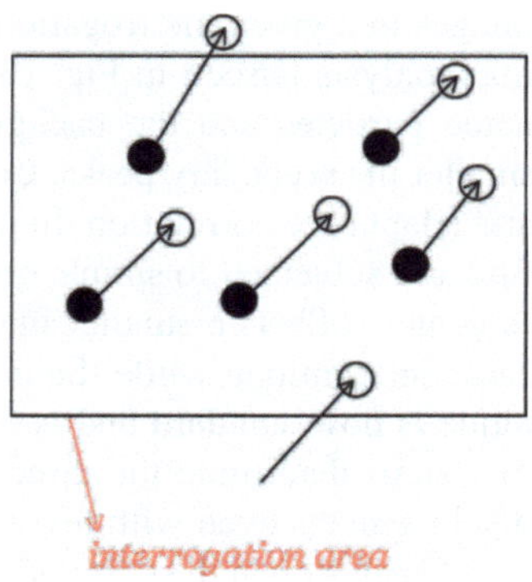

Fig. 11 PIV measurements require finding the initial and final positions of particles illuminated by two consecutive laser pulses within a given flow subregion, called interrogation area. This is mostly achieved by cross-correlation based algorithms

in that given interrogation area, relatively to orthogonal axes parallel to the sides, are given by:

$$
\begin{aligned}
u(x,z) &\simeq \frac{(x+\Delta d_x)-x}{\Delta t} \\
w(x,z) &\simeq \frac{(z+\Delta d_z)-z}{\Delta t}
\end{aligned}
\tag{7}
$$

The key issue in the PIV data processing is determining the mean planar displacements Δd_x and Δd_z in each interrogation area, with the smallest possible error. The standard way to compute these mean displacements is based on cross-correlation algorithms between the images resulting from two consecutive laser pulses.

As shown in Fig. 11, errors (noise, actually) may be caused by particles leaving the interrogation area or arriving to the interrogation area in the interval between the laser pulses, Δt. These particles will have no pair and will reduce the correlation peak. Several techniques are employed to mitigate the effects of loss-of-pairs.

Figure 12 shows a flow chart with the sequence of processes necessary to find the mean planar displacements and, thus, the velocity vector in each interrogation area. The key aspect is the cross-correlation algorithm between each two consecutive images (Westerweel 1993). Before the correlation algorithm is applied the optional step of windowing may be carried out. This is basically a selection of a subarea in the core of the interrogation area, leaving out, for instance, 10% of the area adjacent to the borders, to limit in-plane loss-of-pairs.

Windowing will have no effect on errors committed because of out-of-plane loss-of-pairs, i.e., particles that, at any position, will leave the laser sheet because of 3D motion. If out-of-plane is dominant relatively to in-plane loss-of-pairs, windowing is not recommended. Obviously, both types of loss-of-pairs can be minimized by a selection of a shorter time between pulses Δt.

Correlation-based algorithms actually compare gray levels of consecutive images in a given interrogation area and surrounding area. The result of a correlation analysis is seen in Fig. 13. Evidently, the better contrast between the illuminated particles and the background, the less noisy becomes the signal, i.e., the smaller the secondary peaks. Common algorithms comprise simple cross-correlation and adaptative correlation. In the latter case, a large interrogation area is employed first and subjected to simple cross-correlation. The correlation peak is then used to re-center (offset) a smaller interrogation, again subjected to correlation. The process can continue while there is enough illuminated seeding particles. This technique is now standard and is especially useful when flow gradients are large. The first steps determine the direction of the flow, and the last step finds the correct displacements even with few illuminated particles in the interrogation area.

At each step in the correlation process, filters may be applied to reduce noise and to enhance the peak width. If peak width is too low, subpixel interpolation may render integer values (the associated error becomes the size of the pixel). It is convenient that illuminated particles occupy an area of more than 4 square pixels.

The values of the displacements are lastly divided by the time between pulses to be converted into velocities. The process is repeated to all the interrogation areas in the image.

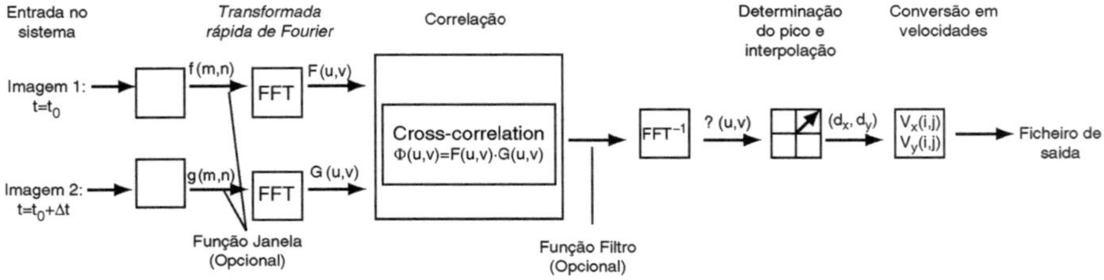

Fig. 12 Sequence of processes involved in the PIV data processing

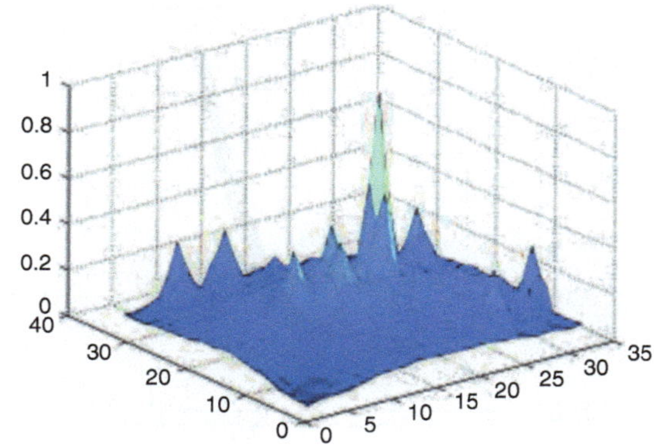

Fig. 13 Example of a 2D correlation plot. The coordinates of the peak correspond to the mean displacements of the tracer particles in the interrogation area. The relative height of second highest peak defines the signal-to-noise ratio and is a measure of the quality of the correlation

A last step, called validation, is necessary to ensure good results. Often, noisy interrogation areas will produce spurious velocity vectors. These vectors may be compared with their neighbors. If they fail a verisimilitude test, they are substituted, generally by the mode or the median of the surrounding eight vectors.

The number of instantaneous vectors is equal to the final number of the interrogation areas, i.e., each interrogation area produces one vector. However, interrogation areas may be overlapped, producing a denser vector map. Particles that were not considered in the correlation process because of in-plane loss-of-pairs may be retrieved with mild overlapping. However, this adds little new information, overlapping is generally oversampling for better visualization.

In the process of producing a PIV vector map, the key parameters to be controlled by the user are thus:

- *The (relative) size of interrogation area.* Too small areas will have few seeding particles and will promote in-plane loss-of-pairs. In the case of adaptative correlation, a rule-of-thumb is to have initial interrogation areas with more than 12 seeding particles and final interrogation areas such that the motion of the particles is 25% of the dimension of the area.
- *The type of correlation (cross-correlation, adaptative).* Adaptative is time consuming but is strongly recommended. Figure 14 shows the results of both types of correlation in a sensitivity analysis of the size of the interrogation area. In the case of the cross-correlation, large interrogation areas would produce considerable errors. Adaptative correlation is nearly insensitive under the same conditions.
- *The time between pulses.* Short intervals will minimize in-plane and out-of-plane loss-of-pairs. Loss-of-pairs cause noise and bias-to-zero errors (Sveen and Cowen 2004) and should be avoided. Figures 15 and 16, showing a synthetic image with background uncorrelated noise, illustrate this issue. In Fig. 15, one of the particles travels out of plane. The correlation is however strong enough so that the centered noise peak is dominated. In Fig. 16, there are in-plane as well as

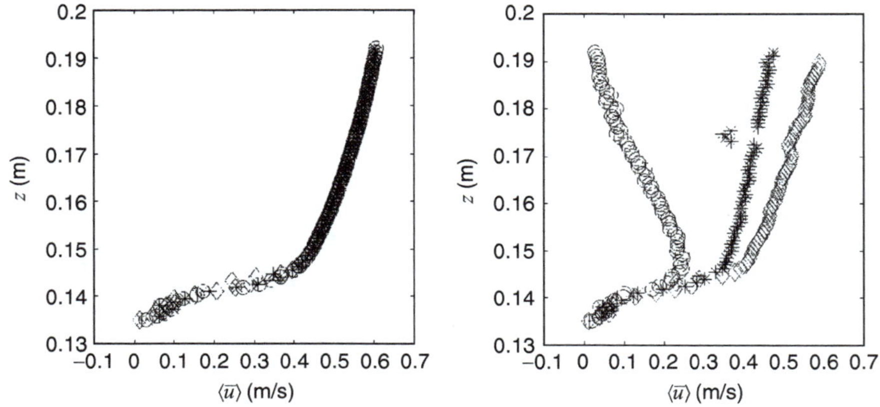

Fig. 14 Logarithmic profile in an open-channel flow with a gravel bed (discharge $Q = 19.2$ l/s). *Left*: results of adaptative correlation. *Right*: results of cross-correlation. Sensitivity to the size of the interrogation area, *circles*: 16 × 16 pixels; *asterisks*: 32 × 32 pixels; *diamonds*: 64 × 64 pixels

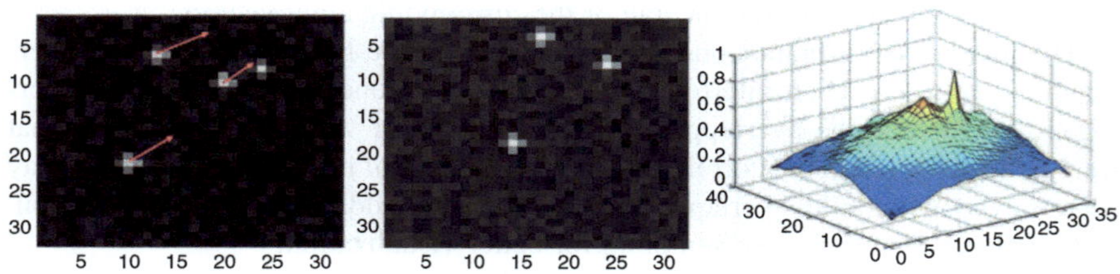

Fig. 15 *Left*: first PIV image and in-plane movement. *Center*: second image. *Right*: corresponding correlation plot. Noise peak is dominated by the true peak. Interrogation area is 32 × 32 pixels

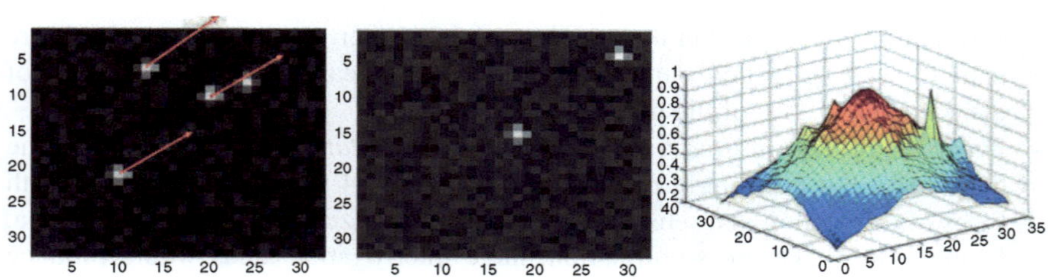

Fig. 16 *Left*: first PIV image and in-plane movement. *Center*: second image. *Right*: corresponding correlation plot. Noise peak dominates the true peak because there is too much loss of information between first and second image. Interrogation area is 32 × 32 pixels

out-of-plane losses. The information loss is too large, and the noise peak dominates the true peak. The corresponding vector would be approximately zero, exemplifying bias-to-zero. However, too short intervals may reduce precision if particles move less than its diameter. It is advisable that the time between pulses and the size of the interrogation area are such that the movement of the articles is about 25% of the latter.

- *The type of seeding particles*. They must have a high refraction index and it is advisable that its density is closer to that of the water. The diameter should be small but, to avoid errors in the determining the peak coordinates, the projected particle should occupy an area of 4–9 square pixels.
- *The number of seeding particles per interrogation area*. Flows should be abundantly and uniformly seeded (see thumb rule above). However, seeding concentrations should not be too high as to change with the properties of the fluid or of the flow. Especially delicate are flows with low Reynolds numbers of fluids with small viscosities (water, for instance) seeded with large concentrations of dense particles. In this case, an undesired density current of seeding particles may be formed.
- *The use (or not) of overlapping*. Useful for a better visualization of the flow field.
- *The use of windows/filters*. Useful to correct mild in-plane loss-of pairs and to enhance peak width.
- *The type of subpixel interpolation*. Current algorithms have been discussed by Westerweel (1993), Lourenco and Krothapalli (1995), or Kumar and Banerjee (1998), among others.
- *The type of validation algorithm*. In general, algorithms are based on comparison with neighboring vectors. They may be based on differentiability of the flow field and/or threshold criteria (reviews in Nogueira et al. 1997).

PIV is not a totally nonintrusive measuring technique. Interrogation areas may smooth out small turbulent scales (one should be aware of anisotropy suppression) and is prone to noise (from poor image contrast, loss-of-pairs, digital camera, and electronics) and to spikes (wrong correlation). Data treatment is time consuming and produces large amounts of data storage. However, PIV is also a very versatile technique, it can be used as a profiler and a point-wise measuring tool and simultaneously, as a visualization tool. The interest in this technique is demonstrated by the fact that 3D PIV and time-resolved PIV (high time frequencies of image collection) are among the fastest growing areas of research in measuring techniques. Field applications are also a growing topic of research.

3 Characterization of Turbulent Flows over Mobile and Immobile Rough Beds

3.1 Overview

A long established hypothesis is that the entrainment and the near-bed transport of sediment particles are directly related to organized turbulence. This hypothesis became fruitful after the pioneer work of Sutherland (1967) and benefited from the clarification of concepts such as bursting cycle or coherent structure (cf. e.g., Nelson et al. 1995). A thorough account of the significant achievements on this field can be found in the work of Nezu and Nakagawa (1993), Chaps. 7, 8, 9, 10, 11, and 12.

Nevertheless, and despite the conceptual progresses on the structure of turbulence, the discourse about near-bed sediment transport remained primarily dominated by considerations on mean bed shear stress and bed shear stress distribution.

Answers to fundamental interrogations about the nature of turbulent flows over mobile beds are yet to be obtained. Do these flows have the same turbulent structures and follow the same scaling laws as the flows over fixed beds? Are there important feedback mechanisms between the organization of turbulence and the quality and quantity of the sediments moving in the bed? If so, can we observe the echoes of such interactions on macroscopic parameters such as the mean bed shear stress?

In this section, a comparison between these sets of data with mobile and immobile (armored) beds is offered, in an attempt to support answers to the previous questions. In accordance to this strategy, the differences and similarities identified on the statistics of the bursting cycle are pointed out and discussed. It should be highlighted that, in order to illustrate the point that the same shear velocity, u_*, might be associated with different organized turbulence, each set of mobile bed data is compared with a set of fixed bed data with the same u_*.

3.2 Instrumentation and Experimental Procedure

The experiments were performed on a 12 m long and 40 cm wide recirculating and tilting flume. The flume floor was covered with a 6 cm deep sediment layer over its entire extension except the initial 120 cm and the final 60 cm, on which tiles with a rough upper face and the same depth were placed. The side-walls are made of glass, enabling visualization and laser measurements. In order to achieve and maintain uniform flow conditions, a manually operated vane was placed at the outlet. A smooth inlet transition constitutes the upstream end. The water is recirculated from a reservoir by means of a centrifugal pump. The sediment is collected at the downstream end, in a 250 μm wire-mesh cone, placed inside the main water reservoir, and recirculated by an independent centrifugal pump. The sediment circuit can be set to trap and dispose the sediment while still recirculating water.

The study required five kinds of measurements: (1) instantaneous velocity; (2) water depth and bed elevation; (3) bed texture; (4) water and sediment discharge; and (5) grain-size distribution of bed-load and of bed-surface composition.

Two orthogonal components of the instantaneous flow velocity were measured with a DANTEC 55X Modular Laser Doppler Anemometer composed of a forward scatter transmitting and receiver optics. The transmitting optics features a 20 mW, monochromatic He-Ne laser, capable of detecting positive and negative velocities. The signal is processed in a DANTEC 55 N20 Doppler Frequency Tracker and converted into a voltage output ready to be sampled on a personal computer. The dimensions of the control volume were $\delta_x = 0.0005$ m and $\delta_y = 0.0015$ m. The sampling software enabled the collection of 12,000 samples before writing to file. The LDA system was placed at distances from the inlet ranging from 80 to 125 hydraulic radii, depending on the water depth. The sampling frequency varied with the hydraulic parameters between 200 and 300 Hz (cf. Table 1).

The water depth and the bed elevation were measured with a 1.0 mm precision point-gage. In order to evaluate the bed texture, a laser bed-profiler was employed to measure the bed elevation along 60 cm lines with a definition of 4 samples/cm. Flow discharge was measured on a calibrated triangular weir placed in the hydraulic circuit. Sediment discharge was computed from samples collected downstream and dry-weighted. Bed-surface was sampled by scrapping off the initial 9 mm of core samples. The grain-size distribution of the bed-surface and bed-load were evaluated by means of dry sieving. For both fixed and mobile beds, the bed was permeable and composed of cohesionless sediment. The sediment grains were transported exclusively as bed-load, and no appreciable bed forms were registered. The flow was subcritical, steady, and approximately uniform for all flows.

A typical mobile bed test begins with laying out the bed and imposing a selected bed slope. Uniform flow is imposed by adjusting the downstream vanes. Bed-load samples are collected and water and bed elevations are monitored for about 8 h to ensure equilibrium sediment transport. Usable bed-load data is collected from this

Table 1 Conditions of mobile bed experiments

	Test	E0	E1	E2	E3
Q	$(\mathrm{m^3s^{-1}})$	0.0138	0.0135	0.0135	0.0130
h	(m)	0.098	0.070	0.065	0.056
i_0	(–)	0.0010	0.0028	0.0033	0.0050
u_*	$(\mathrm{m\ s^{-1}})$	0.030	0.041	0.044	0.049
Re	(–)	33,173	32,452	32,452	31,250
B/h	(–)	4.1	5.7	6.2	7.1
Fr	(–)	0.36	0.58	0.65	0.78
τ_{*84}	(–)	0.0120	0.0220	0.0246	0.0294
τ_{*c84} [a]	(–)	0.0329	0.0329	0.0329	0.0329
d_{84b}	(mm)	4.74	4.81	4.79	5.02
d_{84BL}	(mm)	–	2.66	2.86	3.73
ϕ_{BL}	(–)	–	23.2	37.8	70.5
f_{LDA} [b]	(Hz)	200	240	250	300

[a]Computed from the initial bed mixture; incorporates a hiding-exposure coefficient determined as in Wu et al. (2000).

[b]Computed from Nezu and Nakagawa (1993), p. 30.

stage on. Profiles of instantaneous velocity measurements are performed at four separate occasions in order to enable ensemble averages of turbulent characteristics. Bed texture and bed surface samples of the final water-worked bed are collected at the end of the test.

The experimental conditions for the fixed-bed tests were obtained from the mobile bed ones by letting the water-worked beds undergo a process of armoring. As a consequence of the armoring process, the bed slope decreased and the water depth increased. To maintain approximately the same mean bed shear stress, the downstream vanes were operated and the slope slightly adjusted until the same u_* (computed, on a first approach, from the balance of friction and gravity forces) was obtained. A more accurate value of u_* was later obtained from the Reynolds shear stress profiles.

The remaining procedural steps were identical to the mobile bed ones: velocity measurements were performed after the attainment of a uniform flow in the LDA reach, the final bed was sampled and a textural measurement was performed.

3.3 Description of the Experimental Tests

A total of seven experimental tests were performed in accordance to the procedures above. Test E0 was performed under subthreshold conditions. Tests E1, E2, and E3 were mobile bed experiments. Their main characteristics are shown in Table 1. The symbols in Table 1 are: Q = water discharge; h = water depth; i_0 = bed slope; $u_* = (\tau_0/\rho)^{0.5}$ = shear velocity; τ_0 = mean bed shear stress; ρ = water density; Re = hV/ν = flow Reynolds number; V = depth averaged longitudinal velocity; ν = cinematic viscosity of water; B = channel width; Fr = $V/(gh)^{0.5}$ Froude number; $\tau_{*i} = \tau_0/((\gamma_s - \gamma)d_i)$ nondimensional bed shear stress for the size fraction represented by d_i; $\tau_{*ci} = \theta_c \xi_i$ = critical shear stress; θ_c = Shields critical nondimensional shear stress; ξ_i = hiding-exposure factor; f_{LDA} = sampling frequency; $\phi_{BL} = (Q_{s0}^*/B)/((s - 1)gd_{50}^3)^{0.5}$ nondimensional sediment transport parameter; Q_{s0}^* = volumetric equilibrium bed-load discharge; s = specific gravity of the sediment particles; and g = acceleration of gravity.

As stated before, tests E1 to E3 were intended to have a fixed bed counterpart with the same u_* in order to bring to light eventual changes induced by the near-bed sediment transport in turbulence organization, undetectable by the evaluation of macroscopic mean parameters. Fixed bed tests E1D to E3D were thus performed upon the armored beds obtained from their mobile bed counterparts. The main characteristics of these tests are shown in Table 2.

The values of τ_{*84} and τ_{*c84} in Table 1 are predictions based on the composition of the initial bed. Tests E1 to E3 were designed to show size-selective sediment transport conditions. This intention is expressed by the condition $\tau_{*84}/\tau_{*c84} < 1.0$, which holds true for all tests. The coarser size fractions should, thus, remain immobile.

Table 2 Conditions of fixed bed experiments

	Test	E1D	E2D	E3D
Q	$(m^3 s^{-1})$	0.0135	0.0135	0.0130
H	(m)	0.071	0.069	0.070
i_0	(−)	0.0026	0.0030	0.0036
u_*	$(m s^{-1})$	0.042	0.044	0.049
Re	(−)	32,452	32,452	31,250
B/h	(−)	5.6	5.8	5.7
Fr	(−)	0.57	0.59	0.56
τ_{*84}[a]	(−)	0.0211	0.0224	0.0277
τ_{*c84}	(−)	0.0318	0.0316	0.0310
d_{84b}	(mm)	5.19	5.32	5.41
f_{LDA}	(Hz)	240	250	300

[a]Computed from the armored bed mixture; incorporates an hiding-exposure coefficient determined as in Wu et al. (2000).

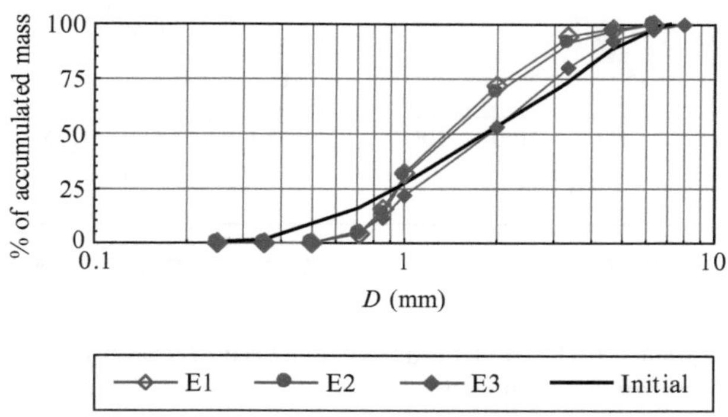

Fig. 17 Sedimentological aspects of the experimental tests. Grain-size distribution of bed-load and of initial bed

This prediction proved wrong. A surprisingly high bed-load rate for grains larger than the d_{84} of the initial bed was measured (see also MacAuley and Pender 1999; Ferreira 2005, pp. 139–151; Ferreira et al. 2007; among others). This size-fraction represented 1.7%, 2.1%, and 3.1% of the total bed-load of, respectively, Tests E1, E2, and E3. The presence of coarser grains in the bed-load is shown in Fig. 17, where the size distributions of the initial bed and of the bed-load for tests E1 to E3 are shown.

As a consequence of the armoring process, the nondimensional bed shear stress decreased to values below the critical value (cf. Table 2). One can conclude that the observation of τ_{*84} and τ_{*c84} works well as a verification but not as a prediction of bed-load composition and hiding functions. Later it will be discussed how the differences in the structure of turbulence can help to explain the entrainment of coarse grain-sizes even if its critical shear stress is larger than the applied shear stress.

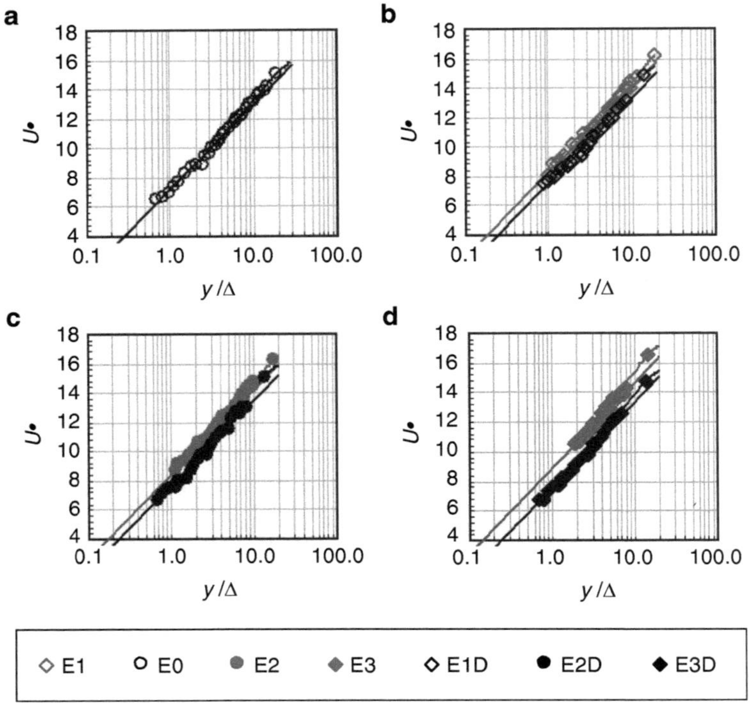

Fig. 18 Velocity profiles: (**a**) E0, (**b**) E1 and E1D, (**c**) E2 and E2D, and (**d**) E3 and E3D. Best fit lines: mobile bed ▬▬; fixed bed ▬▬

The hydrodynamical conditions are shown next. The time-averaged and ensemble averaged longitudinal velocity profiles are shown in Fig. 18. The ensemble average is a result of four independent measurements per point. The solid lines represent the log-wake law for rough beds:

$$u^+ = \frac{1}{\kappa}\ln\left(\frac{y}{k_s}\right) + B_r + \frac{2\Pi}{\kappa}\sin^2\left(\frac{\pi y}{2h}\right) \tag{8}$$

where $u^+ = u/u_*$; u = the longitudinal time-averaged velocity; y = height above some datum; k_s = scale of the roughness elements; Π = Coles' parameter; κ = von Kármán constant ($\kappa = 0.4$); and B_r = constant depending on roughness geometry.

The scale of the roughness elements was taken to be equal to the maximum amplitude in the bed texture series for the fixed bed tests. This method was not applied in the mobile bed tests since it is not possible to distinguish which grains are actually felt as roughness by the flow. In that case, k_s was related to the thickness of the bed-load layer and taken to be the d_{90} of the bed-load distribution.

3.4 Results

Bursting phenomena form a class of coherent structures, which are responsible for generating turbulent energy and shear stress (Grass 1971; Nakagawa and Nezu 1977). If u' and v' are the longitudinal and normal instantaneous velocity fluctuations, the complete bursting cycle is composed by an ejection event (QII, $u < 0$ and $v > 0$) and a sweep event (QIV, $u > 0$ and $v < 0$), mediated by outward (QI, $u > 0$ and $v > 0$) or inward interactions (QIV, $u < 0$ and $v < 0$).

A conditional sampling technique was employed to identify the events of the bursting cycle. The proposed detection criterion was based on the u' vs. v' threshold quadrant technique, chosen for its conceptual simplicity. In its standard version (Nakagawa and Nezu 1977; Franca and Lemmin 2006; Hurther et al. 2007), a sweep occurs when, in quadrant IV, $|u'\ v'| > \sigma_{\mathrm{h}}^{(IV)}$ and an ejection is identified when, in quadrant II, $|u'\ v'| > \sigma_{\mathrm{h}}^{(II)}$, where $\sigma_{\mathrm{h}}^{(IV)}$ and $\sigma_{\mathrm{h}}^{(II)}$ stand for the detection thresholds.

This procedure may lead to faulty characterization: a given turbulent event, seen by direct visualization, may not be detected as so but as a train of separated smaller events. This is illustrated in Fig. 19 where a clear sweep event, occurred between $t = 6.6$ s and $t = 6.7$ s, features a persistent positive value of u' and an *essentially* negative value of v'. Because of the fluctuation in the value of v', this event would be recognized as train of five events of smaller duration resulting in underestimation of the values of the mean of the duration and of the period.

A simple method to enhance the results of the quadrant threshold method is based on the observation that the persistence of u' is well correlated with the persistence of the event. The detection of the initiation of ejections or sweeps remains that of the standard quadrant threshold method but the event is considered to last while the absolute value of u' is above a given threshold, σ^+ for

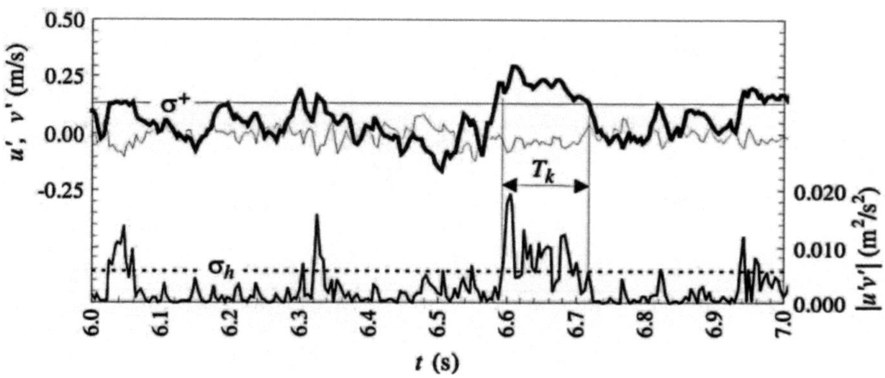

Fig. 19 *Top*: detail of u' (——) and v' (——) series. *Bottom*: detail of a $|u'v'|$ series. Data series belonging to the E1 data set

sweeps and σ^- for ejections. The domains of occurrence of each event, once detected, are:

1. Outward interactions, $Q_{\text{out}} = \left\{ u', v' \in \mathbb{R} : u' > 0 \wedge v' > \frac{\sigma_{\text{h}}^{(I)}}{u'} \wedge u' < \sigma^+ \right\}$
2. Ejections, $Q_{\text{ej}} = \left\{ u', v' \in \mathbb{R} : u' < 0 \wedge \left\{ v' > \frac{\sigma_{\text{h}}^{(II)}}{|u'|} \vee |u'| > \sigma^- \right\} \right\}$
3. Inward interactions, $Q_{\text{in}} = \left\{ u', v' \in \mathbb{R} : u' < 0 \wedge |v'| > \frac{\sigma_{\text{h}}^{(III)}}{|u'|} \wedge |u'| < \sigma^- \right\}$
4. Seeps, $Q_{\text{sw}} = \left\{ u', v' \in \mathbb{R} : u' > 0 \wedge \left\{ |v'| > \frac{\sigma_{\text{h}}^{(IV)}}{u'} \vee u' > \sigma^+ \right\} \right\}$

After a survey of the data, it was adopted $\sigma_{\text{h}}^{(I)} = \sigma_{\text{h}}^{(II)} = \sigma_{\text{h}}^{(III)} = \sigma_{\text{h}}^{(IV)} = \sigma_{\text{h}}$ with $\sigma_{\text{h}} = H u_{\text{rms}} v_{\text{rms}}$, where u_{rms} and v_{rms} are the turbulent intensities associated to the longitudinal and normal velocities, respectively, and H is the hole size. Figure 20 shows the variation of the fraction of transported moment with the hole size for two experiments. The increase of the hole size filters out small events and contributes to counter the fragmentation of coherent events. A too large value of H would, however, eliminate, small legitimate events (note that, at $H = 2.5$, events in the first and third quadrants almost disappear and the respective statistics are unreliable).

The compromise eventually employed, $H = 1.7$, was determined heuristically and was found to be close to the half-value threshold level (Nezu and Nakagawa 1993, p. 182). The values of σ^+ and σ^- depend on the value of σ_{h}. It was adopted $\sigma^+ = 2.5u_{\text{rms}}$ and $\sigma^- = -2.5u_{\text{rms}}$.

The domains of persistence of each of the events can be seen in Fig. 21 for two sets of data taken at $y/k_{\text{s}} = 0.8$, where y is the normal (upwards) coordinate.

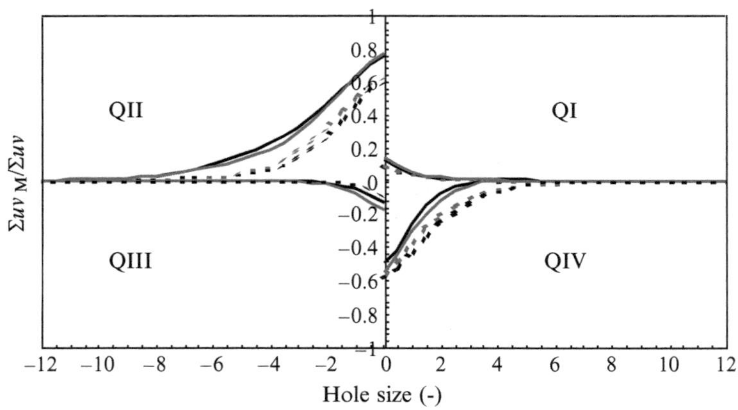

Fig. 20 Fraction of transported momentum in one of the realizations of tests E1 and E1D. Mobile bed, $y/h = 0.52$ ——; mobile bed, $y/h = 0.07$ ——; fixed bed, $y/h = 0.51$ ⋯⋯; fixed bed, $y/h = 0.07$ ⋯⋯

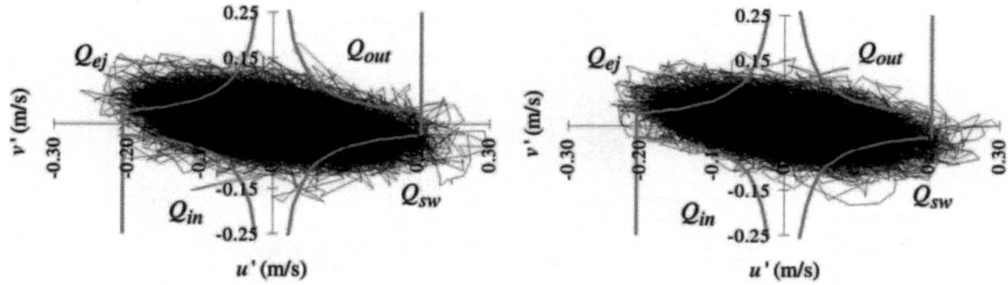

Fig. 21 Quadrant plot of the instantaneous velocity data with thresholds. *Left*: data set E1 (mobile bed). *Right*: data set E1D (immobile bed). Gray lines (———) stand for the $|u'v'|$ and $|u'|$ thresholds

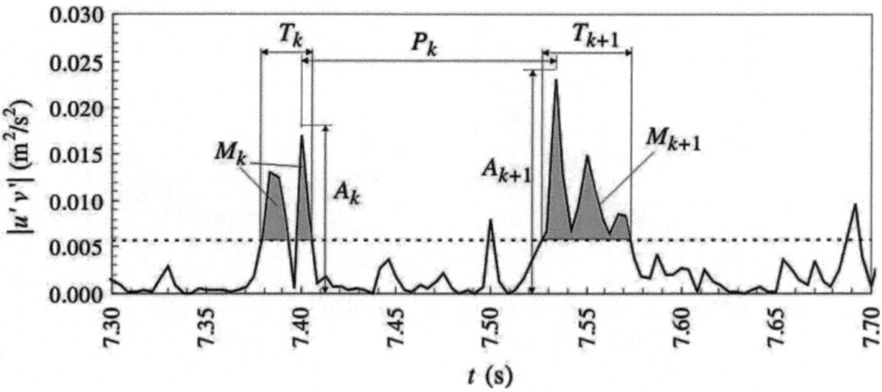

Fig. 22 Detail of a $|u'v'|$ series with definition of the parameters that characterize sweep and ejection events

The data plotted in Fig. 21 were treated to eliminate spurious points – values that differ more than 600% from the average of neighboring points. In data sets of 12,000 entries, bad data were no more than 22 points.

Four parameters will be used to characterize the sweep and the ejection events of the bursting cycle: the duration, T_{k}, the transported momentum, M_{k}, the maximum shear stress, A_{k} (absolute value), and the period, P_{k}, computed as seen in Fig. 22.

For each data series, a record of the detected events is created and analyzed. All events with a duration smaller than $2\mathrm{d}t$, where $\mathrm{d}t = 1/f_{\mathrm{LDA}}$ is the sampling interval, were eliminated. The N_j valid events in a given data series j are used to compute the average value $\bar{X}^{(j)}$.

Three data series were collected in each test of series E and ED. The ensemble averages of the parameters that characterize the events are $\langle \bar{X} \rangle = \frac{1}{3}\sum_{j=1}^{3} \bar{X}^{(j)}$, where $\langle \bar{X} \rangle$ stands for $\langle \bar{T} \rangle \equiv T$, the duration of the event, $\langle \bar{M} \rangle \equiv M = -\int_T |u'v'|\,\mathrm{d}t$, the transported momentum $\langle \bar{A} \rangle \equiv A$, the maximum instantaneous shear stress or $\langle \bar{P} \rangle = P$, the period between events. To simplify the notation, the brackets and the overbar will be dropped.

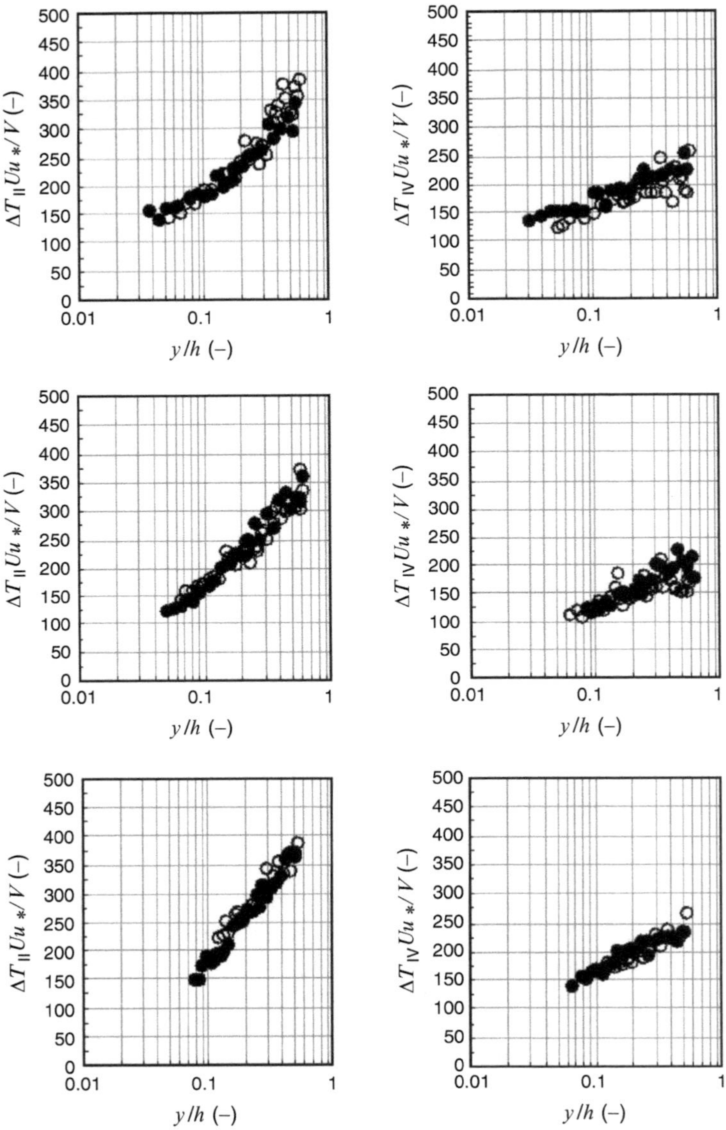

Fig. 23 Nondimensional duration of the events. QII on the *left*; QIV on the *right*. From *top* to *bottom*: E1/E1D; E2/E2D; E3/E3D. Mobile bed: *circle*; fixed bed *dark circle*

Since this study is primarily concerned with the events that produce shear stress, only sweeps and ejections will be studied. The first parameter to be discussed is the duration of the event. Profiles of TUu_*/v, where U is the time-averaged longitudinal velocity, are shown in Fig. 23. Apart from small and unbiased deviations, possibly

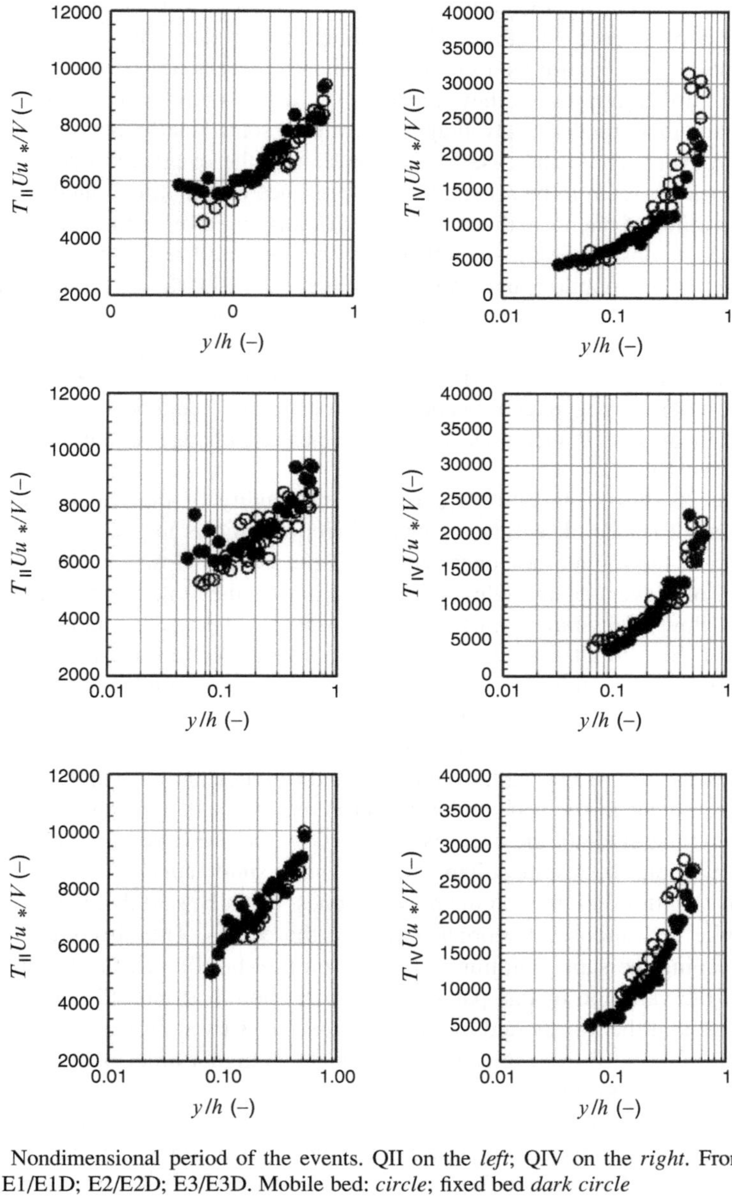

Fig. 24 Nondimensional period of the events. QII on the *left*; QIV on the *right*. From *top* to *bottom*: E1/E1D; E2/E2D; E3/E3D. Mobile bed: *circle*; fixed bed *dark circle*

explained from a misestimating of the water viscosity, the duration of the events appears independent from the type of boundary.

The spacing of the events, revealed by their period, is analyzed next (Fig. 24). The period is computed as the average of the series of time intervals that mediate the extreme value of each two consecutive events (Fig. 22). It is thus much influenced by event fragmentation. The first observation worth being made is that the period of sweeps (quadrant IV) is much larger than that of the ejections (quadrant II). This feature seems to indicate that the criterion for detecting ejections could be further improved: they appear to be a less organized kind of bursting motion, and there might be advantages to coalesce some small intensity events into one large event. As in the case of the duration of the event, the period of these events does not seem to be affected by the presence of a mobile boundary.

The profiles of the normalized momentum transported by each event are shown in Fig. 25. Its observation reveals that over the measured reach there seems to be little effect of the mobile boundary (cf. Nikora et al. 2001). However, small changes may pass unnoticed due to data scattering. For this reason, the ratio M_{IV}/M_{II} will be analyzed with more detail (Fig. 26). The ratio sweep-magnitude to ejection-magnitude depends on the roughness Reynolds number k_s^+ (Nezu and Nakagawa 1993, p. 184). Sweep events become stronger as the roughness increases. This effect attenuates as y/h increases, a behavior related to changes in the skewness of u and v distributions.

Figure 26 shows the ratio M_{IV}/M_{II} for the totality of the experimental tests performed. It stands clear that the sweep/ejection ratio is larger in the fixed bed tests than in the mobile bed ones with the same u_*. This result indicates that the existence of a layer of finer sediment traveling among the protuberant coarser grains trigger a flow beahviour propper, in some aspects, of smoother boundaries. The analysis of Fig. 26 also reveals that the ratio sweep/ejection might decrease with increasing Froude number. In fact, the lowest values correspond to the Tests E3 and E3D, with the largest Froude numbers.

Figure 27 shows the vertical distribution of the average maximum value of $|-uv|$, calculated as seen in Fig. 22. A fundamental difference between mobile and fixed beds is revealed: in the wall region the profiles clearly depart, revealing that the maximum magnitude of the event is increased by the presence of a mobile bed. In the outer region, the results collapse visibly into a single curve.

Analyzing this feature with the previous ones, it can be said that the presence of the mobile bed imposes a reorganization of the events. They become "sharper" albeit keeping the same duration at the chosen threshold level. Further investigations on the variation of this parameter with H should be carried out. Under the hypothesis that the transported momentum does not vary significantly, changes on the shape of the $u'v'$ series must produce effects on higher order moments, namely the skewness and kurtosis.

The last feature worth mentioning about uv_{max} is that sweep and ejection events are equally affected from the presence of the mobile bed. Since ejections are related to the uplifting of low-speed flow and the occurrence of streaky structures in plan view, this finding should encourage further study on the formation and characteristics of such streaky structures under generalized sediment transport.

It is concluded that the structure of the profiles is much the same for fixed and mobile bed. Yet, subtle differences can easily pass unnoticed, hidden in the data scattering.

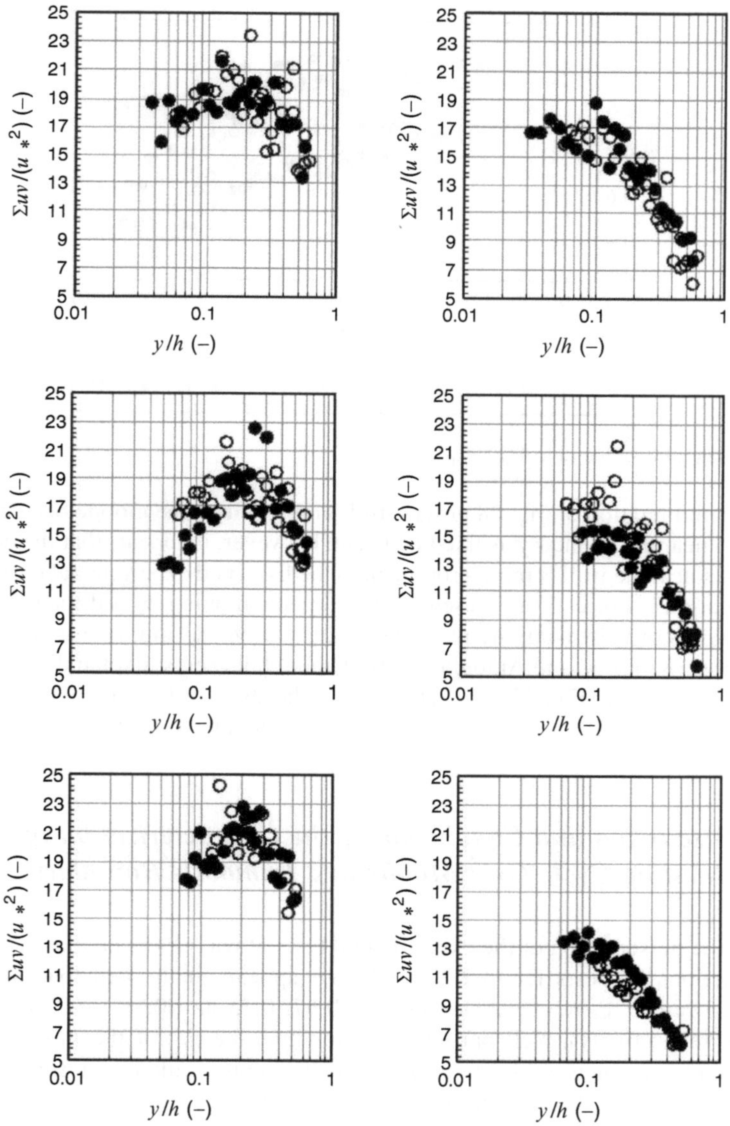

Fig. 25 Nondimensional transported momentum. QII on the *left*; QIV on the *right*. From *top* to *bottom*: E1/E1D; E2/E2D; E3/E3D. Mobile bed: *circle*; fixed bed *dark circle*

A different and more useful organization of data is shown in Fig. 28. It is the turbulent energy flux, obtained from the third moments ${v_{med}}^3$ and $u^2\, v_{med}$ assuming that $w^2 v_{med} \approx v^2 v_{med}$, following Nezu and Nakagawa (1993, p. 79) (w is the transverse instantaneous velocity fluctuation).

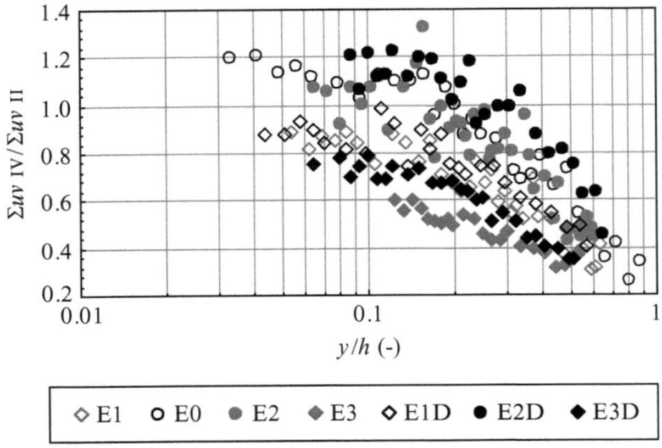

Fig. 26 Variation of the ratio $M_{\mathrm{IV}}/Mv_{\mathrm{II}}$

The interpretation of Fig. 28 proves itself ambiguous since mobile bed data does not go as near the boundary as fixed bed data. However, it appears that the decrease of the flux near the wall is more prominent in the fixed bed experimental tests. If this interpretation is correct, the presence of sediments moving as bed load decreases the skewness of v, by damping its negative fluctuations, as in the case of smoother boundaries (Krogstad and Antonia 1999). Although it seems that the differences in the $uv_{\max}$ plots (cf. Fig. 27) are reflected by the third order moments, the exact mechanism by which the energy flux is changed remains to be explained.

3.5 Discussion and Conclusion of the Characterization of the Bursting Cycle Over Mobile and Immobile Boundaries

During the presentation of the experimental tests, it was observed that the coarser size fractions in Tests E1 to E3 should not appear on the bed load since the predicted ratio τ_{*84}/τ_{*c84} was below 1.0 (0.67, 0.75 and 0.89 for Tests E1, E2 and E3, respectively). Yet, particles coarser than the d_{84} of the initial mixture were found in nonnegligible quantities. This fact alone could be interpreted as the result of the application of an inappropriate hiding-exposure factor and that the mobility of the larger grains was underestimated. If this interpretation was true, an armored bed would be difficult to attain given the high mobility of the largest particles.

It was observed that an armored bed was indeed attainable. The final values of the ratio τ_{*84}/τ_{*c84} were 0.66, 0.70, and 0.89, the same order of magnitude as for the mobile bed tests. It can be concluded that the grain mobility of the largest grains did not remain constant during the armoring process. A similar finding was presented by Macauley and Pender (1999) and observed by many other researchers. The introduction of a fine sediment mixture on an armored bed induces the

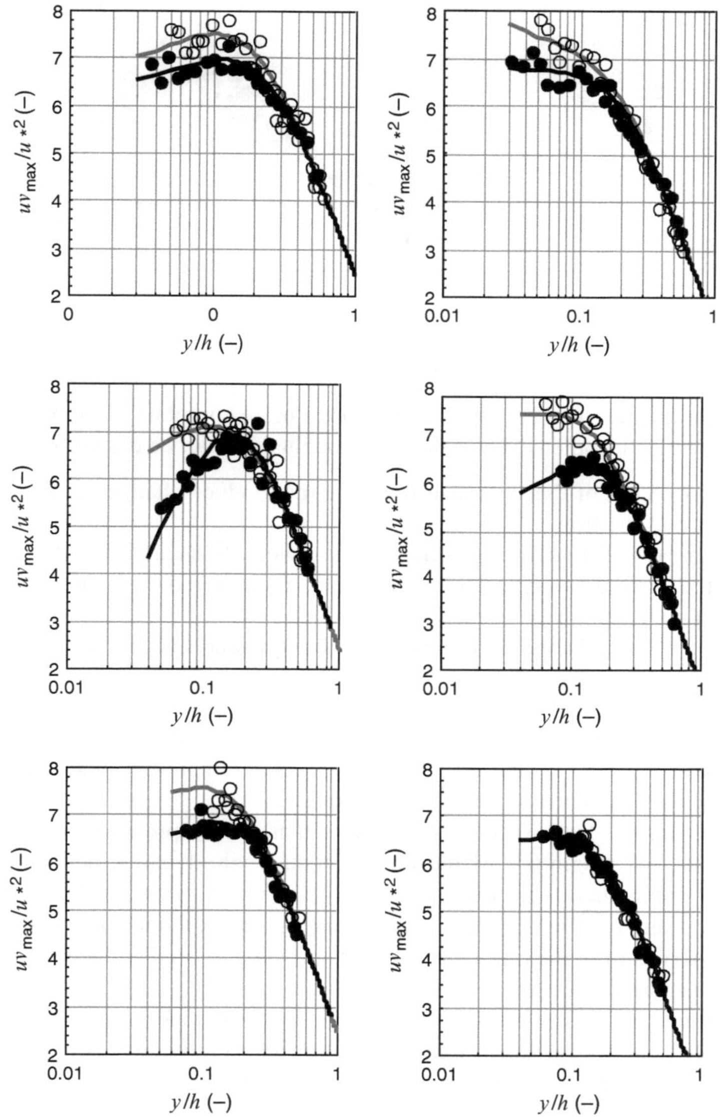

Fig. 27 Nondimensional $uv_{\max}$. Best fit lines are also presented. QII on the *left*; QIV on the *right*. From *top* to *bottom*: E1/E1D; E2/E2D; E3/E3D. Mobile bed: *circle*; fixed bed *dark circle*

movement of coarser size fractions. They attributed the observed behavior to increased velocity fluctuations as the fine sediment fills the pores of the rough bed. Yet, those authors had only measurements of longitudinal turbulence intensities to justify their belief.

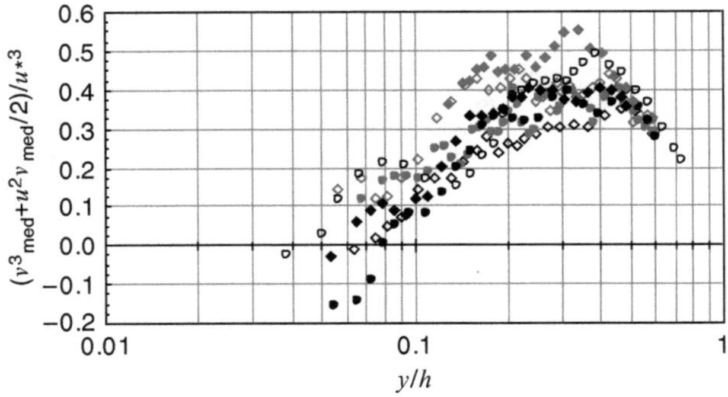

Fig. 28 Turbulent energy flux. Experiments identified as in Fig. 11

The results presented in this study provide a more complete set of parameters to interpret such phenomena. The following conclusions can be drawn from the text:

1. The movement of the finer size fractions may induce changes in the turbulent structure of the flow.
2. The fundamental structural differences between fixed and mobile beds were detected on the sweep/ejection ratio and, more important, on the averaged maximum instantaneous stress, uv_{max}.
3. Under mobile bed conditions, uv_{max}, for both quadrants II and IV, is larger in the wall region (cf. Fig. 27).
4. The observation of the sweep/ejection ratio reveals differences in the transported momentum, namely that the sweep event looses importance to ejections in the presence of a mobile bed (cf. Fig. 26); this feature represents a smoothening of the bed, consequence of the reduced porosity.
5. Combining the last two points, one sees that, in the presence of fine mobile fractions, the magnitude of the instantaneous hydrodynamic actions upon the largest grains increases, thus increasing its entraining probability; the apparent largest grain mobility for the coarsest grains is thus explained; as a consequence, the observation of the ratio τ_{*i}/τ_{*ci} is not sufficient to predict the composition of the transported sediment.
6. As pointed out before by Gyr and Schmid (1997), it appears that the interaction between near-bed sediment transport and turbulent structures is reflected by an increased degree of coherence of the latter; this observation should explain the differences found in the distribution of the third moments; the main link should be the decrease of the skweness of the distribution of v in the presence of fine mobile sediments; as a result, the mobile bed data show a less steeper decrease in the turbulent energy flux near the bed (cf. Fig. 28).

The last point highlights the fact that a consistent phenomenological frame, capable of becoming a predicting model, is required to understand the interactions between turbulent flow structures and the transport of sediment grains. It is prudent that it should be attempted only in the presence of further data concerning other granulometric mixtures.

4 PIV Study of the Flow in the Scour Hole of a Cylinder

4.1 Overview

Enforcing the structural integrity of bridges requires the capacity to predict pier local scour morphology. The complex nature of the interaction between turbulent flow hydrodynamics, local morphology, and sediment transport in the near field of the pier justifies the important volume of research performed on these topics.

The contribution of the present work is mostly in the characterization of the turbulent flow in the scour hole in front of a bridge pier, idealized as a wall-mounted cylinder of vertical axis, intercepting the full depth of an open-channel flow with a developing boundary layer. Such a cylinder changes the pressure field around it. The adverse pressure gradient upstream the cylinder modifies the pattern of the approaching flow, which becomes downward in the upstream face of the obstacle, while the remaining flow goes around the cylinder in the downstream direction (Morton 1987; Raudkivi 1998). If the pressure gradient becomes sufficiently strong, the interaction between the downward and the approaching flow results in flow separation near the base of the cylinder, where it develops, in a turbulent flow, a complex and unsteady vortex system known as the horseshoe vortex. This phenomenon has already been studied, but the contribution of the horseshoe vortex in the scour development is not totally clarified.

In this study, the local scour mechanism is again studied and characterized. The main goal of the present work is the visualization and characterization of the downward-flow and of the turbulent flow field in the separated region in front of the cylinder. In particular, the downward-flow and the vortex system inside a scour hole in its early stages of development are investigated in order to identify the most relevant scour mechanisms.

The work was essentially laboratorial and involved the use of Particle Image Velocimetry (PIV) for the visualization and quantification of the velocity field. Velocity measurements were performed in the vertical plane of symmetry in front of the cylinder, in vertical planes at 30° and 45° from the symmetry plane. Measurements in horizontal planes were also performed on selected experiments.

4.2 Theoretical Considerations

In front of the cylinder, the separation area is characterized by a complex structure, eventually dominated by horseshoe vortex systems. Morton (1987) demonstrated that the vorticity mechanism is governed by the pressure field. The component of the vorticity normal to a vertical plane in the flow direction depends on the pressure gradient in the flow direction:

$$\upsilon^{(w)} \partial_z \eta\big|_{z_0} = \frac{e_y}{\rho^{(w)}} \partial_x p \tag{9}$$

where $\upsilon^{(w)}$ is the fluid cinematic viscosity, η is the vorticity component in the transversal direction (y), and p is the total pressure. The vorticity field is solenoidal. Hence, the vortex intensity, defined as:

$$I_{\mathrm{v}} = \int_{\Omega} \omega \cdot \mathrm{d}A = \int_{G} \eta \mathrm{d}A \qquad (10)$$

is conserved in any section around the cylinder. The term ω represents the vector of the vorticity.

Shen et al. (1969) studied the mechanism of local scour near cylindrical piers and suggest that the horseshoe vortex generated upstream the pier is due to the concentration of vorticity, induced by the obstacle, already existent in the flow field. They believe that this horseshoe vortex is the primary mechanism of local scour.

Qadar (1981) explored the interaction between horseshoe vortex and the temporal increase of scour depth. He proposed that the horseshoe vortex is approximated, in the symmetry plane, by a circle whose radius increases with the depth of the scour hole, as represented in Fig. 29.

Muzzammil and Gangadhariah (2003) investigated the horseshoe vortex in the plane of symmetry through a visualization technique that employs suspension of clay. They observed that the vortex shape is elliptical, instead of circular as proposed by Qadar (1981). The dimensions of the horseshoe vortex were experimentally measured and found to depend on the scour depth, diameter of the cylinder and Reynolds number. They pointed out the existence of a cusp and two distinct slopes in the scour hole, in the plane of symmetry, as shown in Fig. 30.

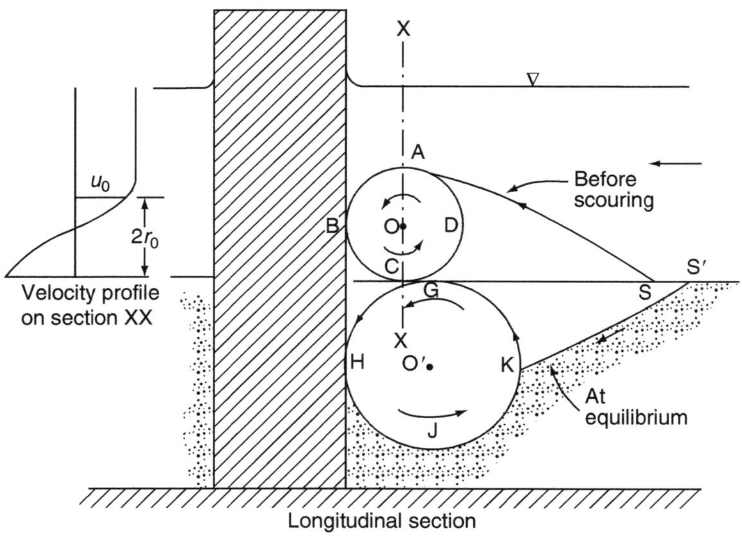

Fig. 29 Evolution of the horseshoe vortex with the increase of the scour hole (Adapted from Qadar 1981)

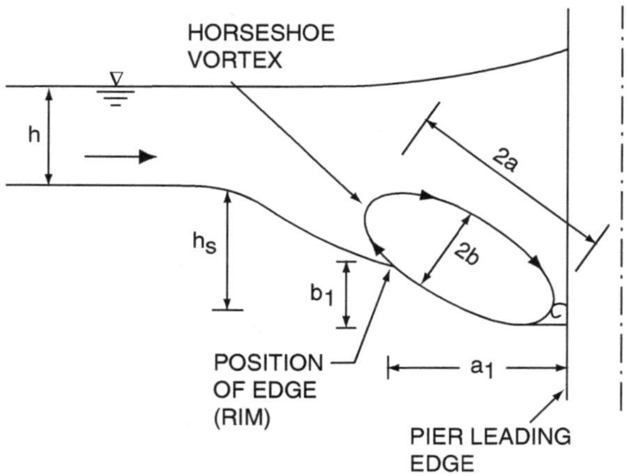

Fig. 30 Scheme of horseshoe vortex upstream the cylinder in the scour hole (Adapted from Muzzammil and Gangadhariah 2003)

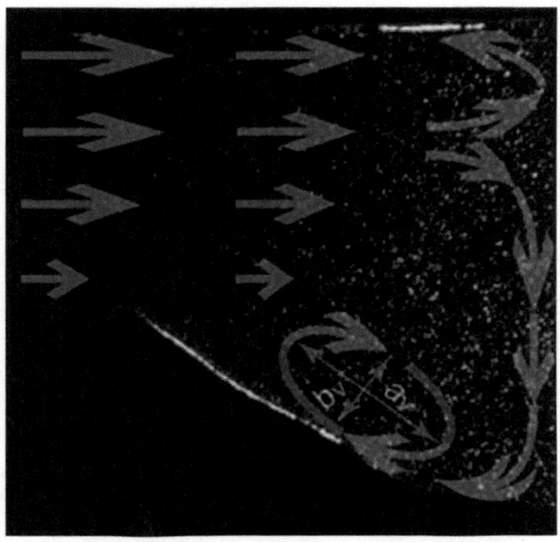

Fig. 31 Schematic flow pattern inside the scour hole in front of the pier (Adapted from Unger and Hager 2007)

Unger and Hager (2007) investigated the characteristics of the internal flow upstream a half-cylinder placed on the side wall of a channel. They performed a detailed temporal description of the horseshoe vortex with PIV. The placement of the cylinder at the lateral wall of the channel allows better visualization of the flow phenomena and a better quality of the data obtained. However, the turbulent boundary layer existent along the wall influences the flow behavior, so the results cannot be directly compared with those obtained in the plane of symmetry of the channel. These authors suggest that the equivalent vortex is located at the point of intersection of the two slopes in the scour hole, over the cusp (Fig. 31), rather than located in the greater slope, as defendend by Muzzammil and Gangadhariah (2003).

They concluded that both the horseshoe vortex inside the continuously increasing scour hole and the downflow in front of the pier are the governing scour agents.

4.3 Experimental Facilities and Instrumentation

The experiments were conducted in a 8.0 m long and 0.70 m wide rectangular flume. The flow discharge was controlled by a flow meter installed in the pumping circuit, and the flow depth was controlled by the adjustment of the downstream gate. A horizontal mobile bed reach 2.00 m long was placed 4.55 m downstream of the flume inlet (Fig. 32a). In the upstream reach of the mobile bed, uniform quartz sand was glued on the flow bed, in order to increase its roughness. The bed material, used both on mobile and fixed bed, is characterized by a median diameter d_{50} of 0.837 mm and geometric standard deviation σ_D of 1.48. The PIV laser source was attached on a mobile carriage, settled on the longitudinal rails of the channel (Fig. 32b). The cylinder (PVC), with 0.048 m diameter, was placed in the symmetry axis of the channel (Fig. 32c). To investigate the effect of the roughness in the development of the turbulent boundary layer, velocity measurements on the approach flow were accomplished before and after the placement of granular material with 25 mm of characteristic diameter, in the upstream region of the flow

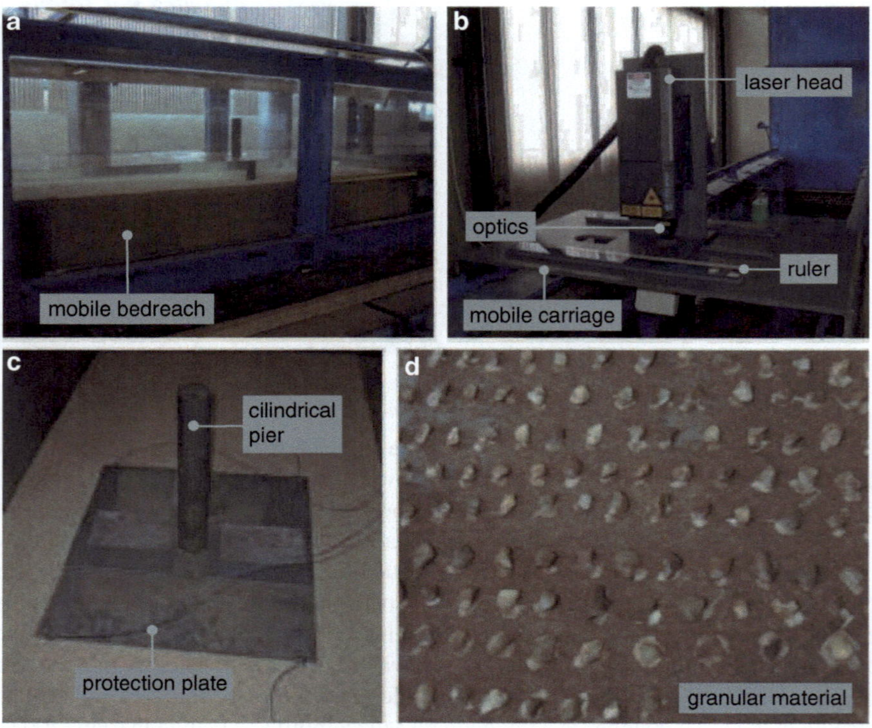

Fig. 32 Experimental details (**a**) mobile bed reach, (**b**) laser source, (**c**) cylinder, and (**d**) disposal of granular material at the upstream reach of the channel

bed (details *in* Nogueira et al. 2008). Figure 32d shows the arrangement of the rough elements.

The Particle Image Velocimetry (PIV) was used to measure the two-dimensional instantaneous velocities of the flow. The double-pulsed Nd:YAG laser illuminates a given section of the flow which is recorded by a digital camera. The particles in the fluid scatter the light incident, which is detected by the camera. The frames are then split in interrogation areas, being possible to calculate the displacement vector for each one of these areas, using spatial correlation techniques.

Calibration was performed by recording a triangular scale (Fig. 33a). The cylinder placed in the flow field causes oscillation of the flow surface upstream the obstacle so, to inhibit optical problems due to the impact of the laser light with surface waves, the water surface was covered with an acetate sheet (Fig. 33b).

The velocity measurements were performed under clear water conditions, being the flow discharge established based on the concept of critical flow velocity. Three different criteria (cf. Raudkivi 1998) were used to estimate the value of critical velocity, being adopted the value $U_{cr} = 0.286$ m/s correspondent to a flow discharge of 20 l/s. Table 3 summarizes the flow variables and channel parameters of the experiments.

The variables in Table 3 are Q, the flow discharge; h, the water depth; U, the depth average critical velocity; B, the channel width; and D is the cylinder diameter.

A scour hole was allowed to develop for 5 min around the cylinder placed in the mobile bed. The scour hole can be seen in Fig. 34a. Subsequently, the sediment bed was fixed with spray varnish to prevent the movement of the bed material during the measurements although maintaining the roughness characteristics of the bed.

The tests were performed in a superficial region of the scour hole, in the symmetry axis and 30° from it, and inside the scour hole, 45° from the symmetry axis. Measurements in a horizontal plane, 5 mm from the bed, were also performed. Further details on the experimental arrangements can be found in Fig. 34. Figure 35

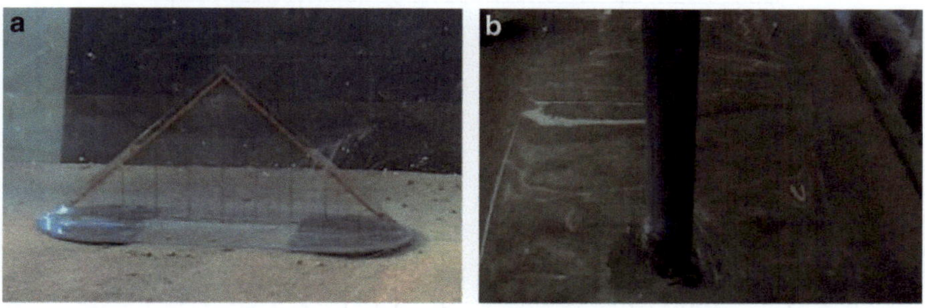

Fig. 33 Additional material used in experiments. (**a**) graduated object used for lens-focusing and calibration purposes, and (**b**) acetate sheet placed on the flow surface, downstream view

Table 3 Flow variables and channel parameters

D_{50} (mm)	Q (m^3/s)	h (m)	U (m/s)	B/D (−)	B/h (−)
0.837	0.20	0.10	0.286	14.58	7.00

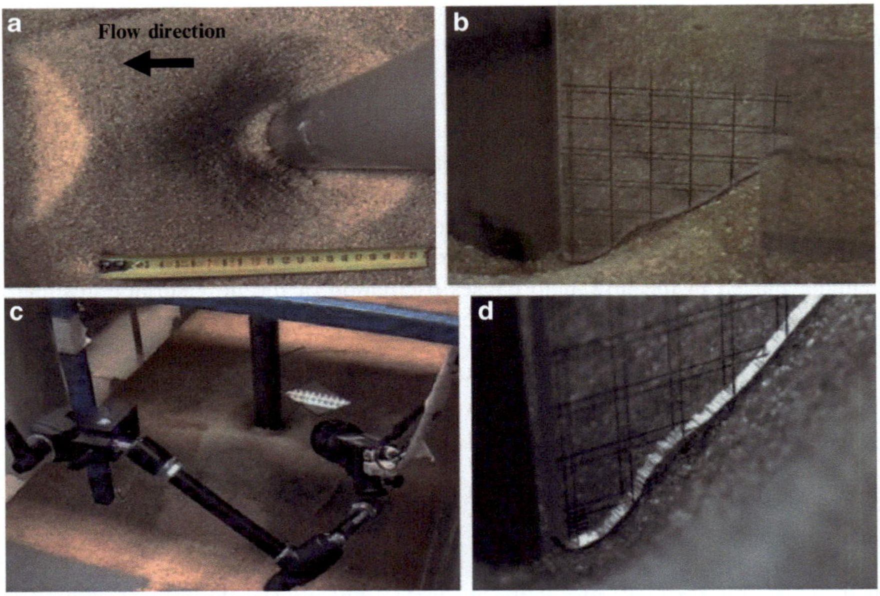

Fig. 34 Measurements inside the scour hole. (**a**) Scour hole – downstream view, (**b**) graduated object for image calibration, (**c**) camera positioning, and (**d**) image acquired for calibration

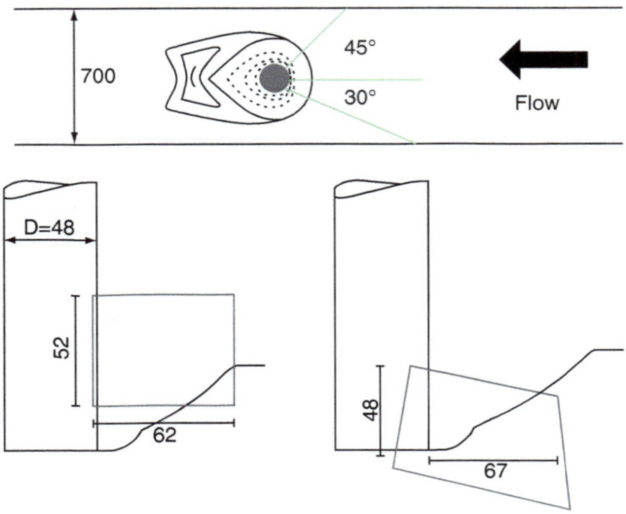

Fig. 35 Vertical measuring planes in the scour hole. *Top*: view from above. *Dashed line* stands for negative bed elevations. *Bottom*: side view, location of images and their dimensions. *Left*: vertical plane in the symmetry axis and at 30°. *Right*: inside the scour hole, vertical plane at 45° from the symmetry axis. Dimensions in milimeters

shows the location of the vertical measuring planes and the images dimensions relatively to the cylinder.

4.4 Results and Discussion

The dimensions of the horseshoe vortex were estimated based on the vorticity maps, as shown in Fig. 36, being the vortex equivalent diameter D_v determined by $D_V = \sqrt{ab}$. This procedure was applied to the greatest instantaneous vortex observed in each test and to the equivalent vortex, which is characterized by the separated flow region that shows a circulation pattern with nonzero vorticity and resulting from the time-averaged vortical instantaneous structures registred in that flow region.

The time-averaged vorticity map and streamlines obtained before the beginning of the scour hole are shown in Fig. 37. Figure 38 shows instantaneous vorticity maps in two vertical positions of the measuring plane.

The time-averaged vorticity map shows the presence of an equivalent horseshoe vortex. It is observed that the vortex location, relatively to the cylinder, depends on the boundary layer thickness (details in Nogueira et al. 2008). With an increasing thickness of the boundary layer, the longitudinal dispersion of the vortex location also increases (cf. Baker 1980). However, the instantaneous vortexes are apparently independent of the thickness of the boundary layer (Fig. 38).

The time-averaged velocity components u and w obtained at the upper part of the scour hole are presented in Fig. 39, in the symmetry axis, and in Fig. 40, in a plane 30° from the symmetry axis (see also Fig. 35), where u_p is the longitudinal direction

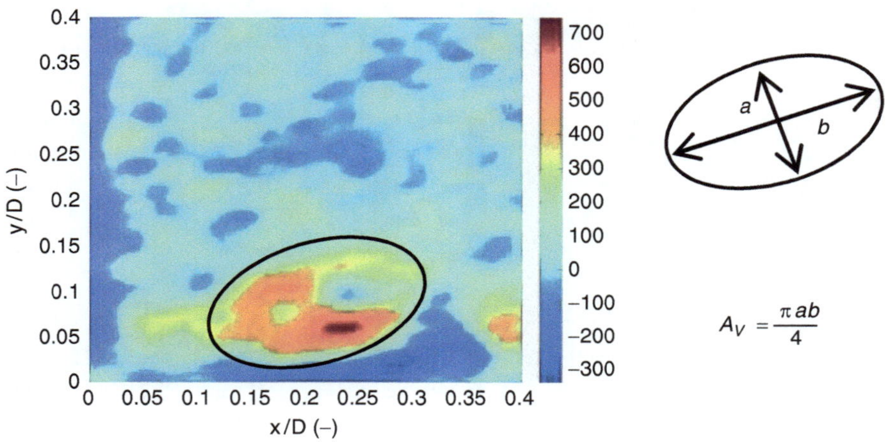

Fig. 36 Determination of the dimensions of the horseshoe vortex

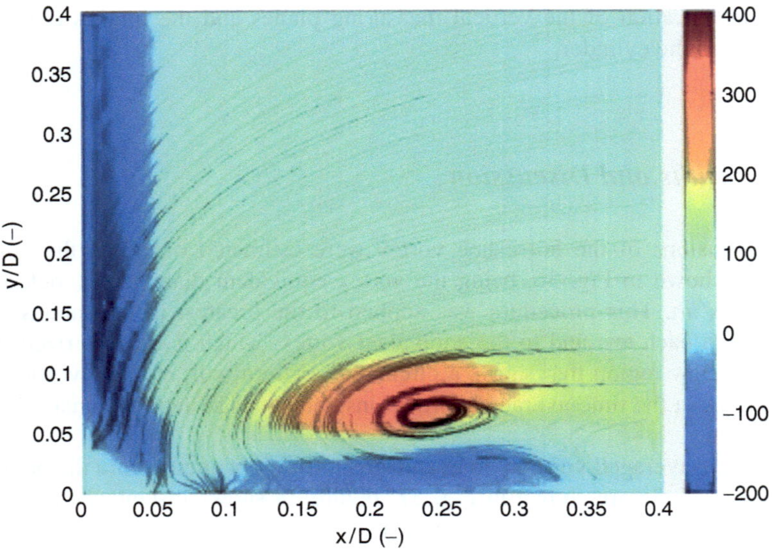

Fig. 37 Time-averaged vorticity maps and streamlines

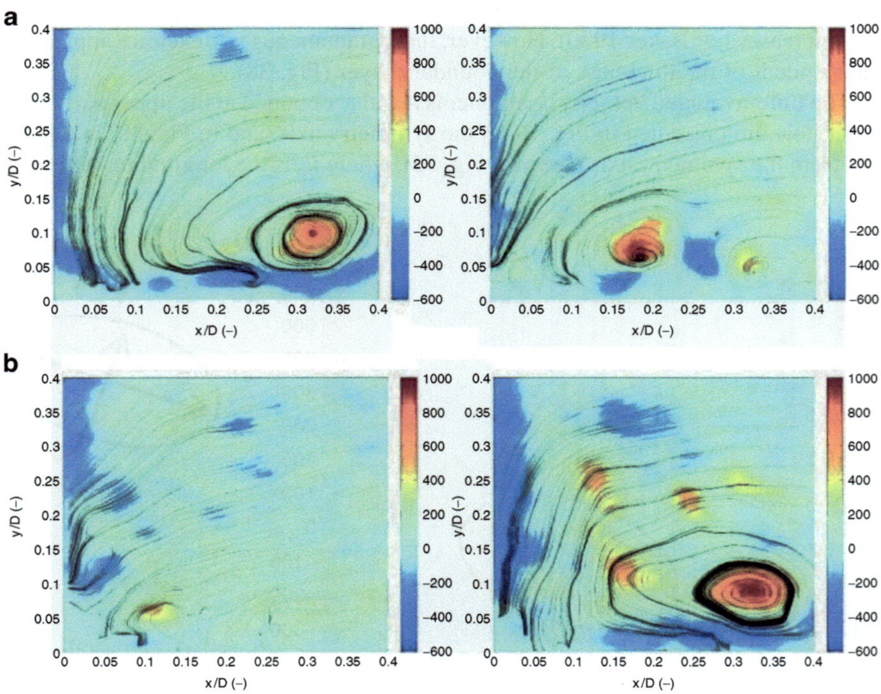

Fig. 38 Instantaneous vorticity maps. (**a**) vertical plane at the axis, and (**b**) vertical plane 30° from the symmetry axis

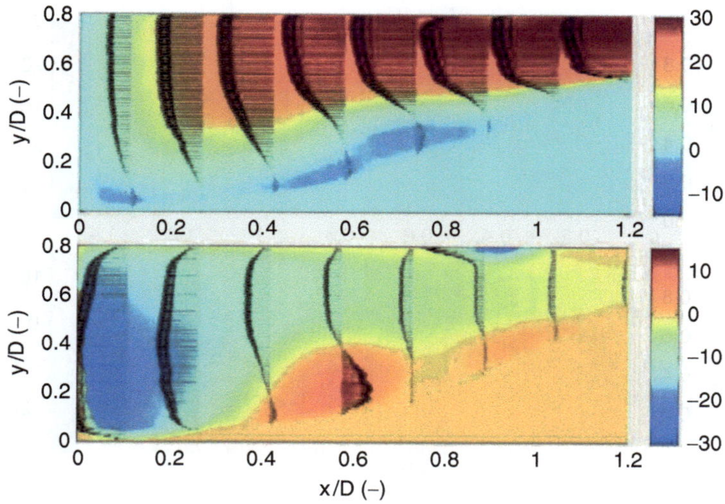

Fig. 39 Time-averaged velocity components in test CC, symmetry axis. On *top*: longitudinal component *u*; *bottom*: vertical component *w*. Velocities in cm/s

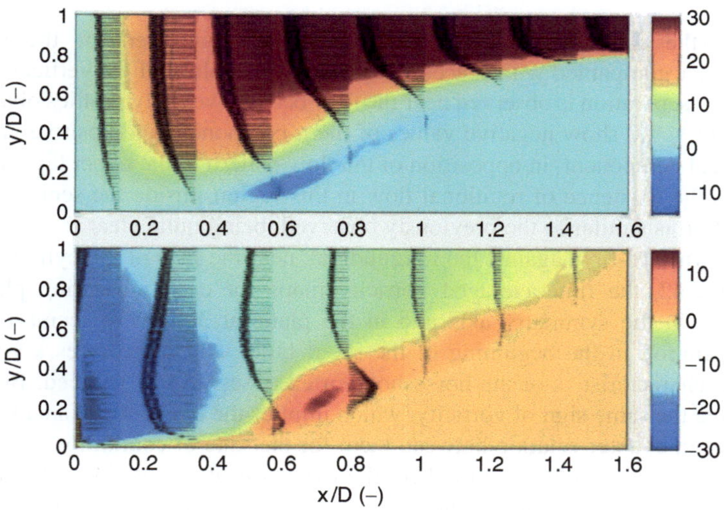

Fig. 40 Time-averaged velocity components in test CC, at 30° from the symmetry axis. On *top*: component u_p; *bottom*: vertical component *w*. Velocities in cm/s

in the image plane. Figure 41 shows the velocity field inside the scour hole, in a vertical plane at 45° from the symmetry axis.

In Fig. 39, in the upstream region, it is noted the dominance of the *u*-component, being the vertical component practically negligible. As the flow approaches the

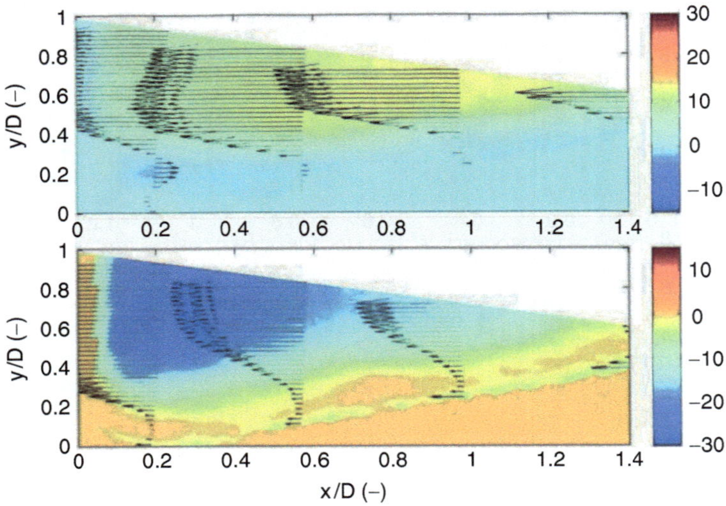

Fig. 41 Time-averaged velocity components in test CC, in the scour hole, at 45° from the symmetry axis. On *top*: longitudinal component u_p; *bottom*: vertical component w. Velocities in cm/s

cylinder, the situation reverses: the longitudinal component of the velocity decreases, accompanied with successively increasing values of the vertical component. Flow separation is observed near the flow bed: the velocity profiles obtained at $0.4 \leq x/D \leq 1.0$ show negative values of the u-component and positive values of the vertical component, in opposition of the flow pattern in the upper layers, which indicates the existence of rotational flow in this region. Inside the scour hole, the flow pattern is similar to the previously observed, being quite clear the dominance of the downward-flow against the longitudinal, near the face of the cylinder.

In Fig. 42, the time-averaged vorticity maps for each measuring plane are presented. In the symmetry axis and in the plane at 30°, there seems to occur flow separation in the beginning of the scour hole, which generates a vortex of different characteristics of the horseshoe vortex observed in fixed bed. Both vortexes have the same sign of vorticity, which implies the existence of small vortical structures in counter-rotation between them for the sake of continuity.

Inside the scour hole, near the bed, the existence of two regions of maximum vorticity is noticed. The instantaneous maps show the intermittent behavior of the horseshoe vortex (Fig. 43). Near the cylinder face, negative values are observed of vorticity and of higher intensity than what was detected near the bed. This leads to the conclusion that, inside the scour hole, the vorticity generated in the boundary layer which develops along the wall of the cylinder, due the presence of the downflow, is more permanent and relevant than the vorticity generated near the bed (Fig. 43).

As a summary, it can be said that the downflow upstream the cylinder plays a major role in the mechanism of local scour near the obstacle. The down-flow acts as

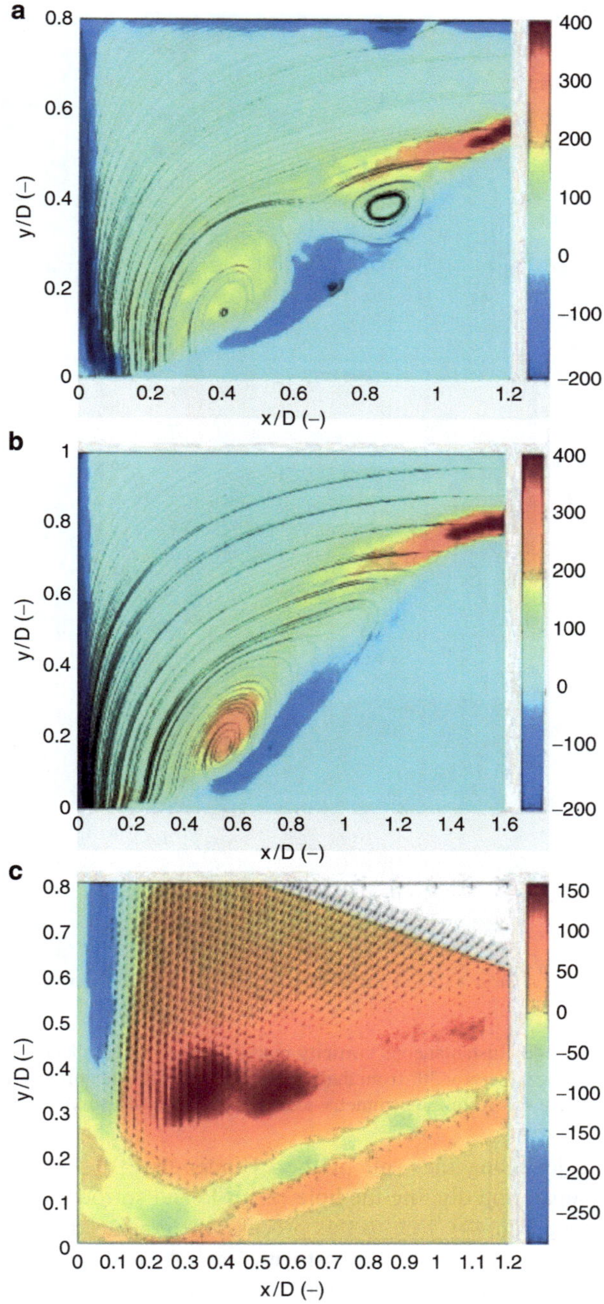

Fig. 42 Time-averaged vorticity maps. (**a**) the vertical plane of symmetry (with streamlines), (**b**) vertical plane at 30° from the symmetry axis (with streamlines), and (**c**) the scour hole, at a vertical plane 45° from the symmetry axis (with velocity vectors). Vorticity in s^{-1}

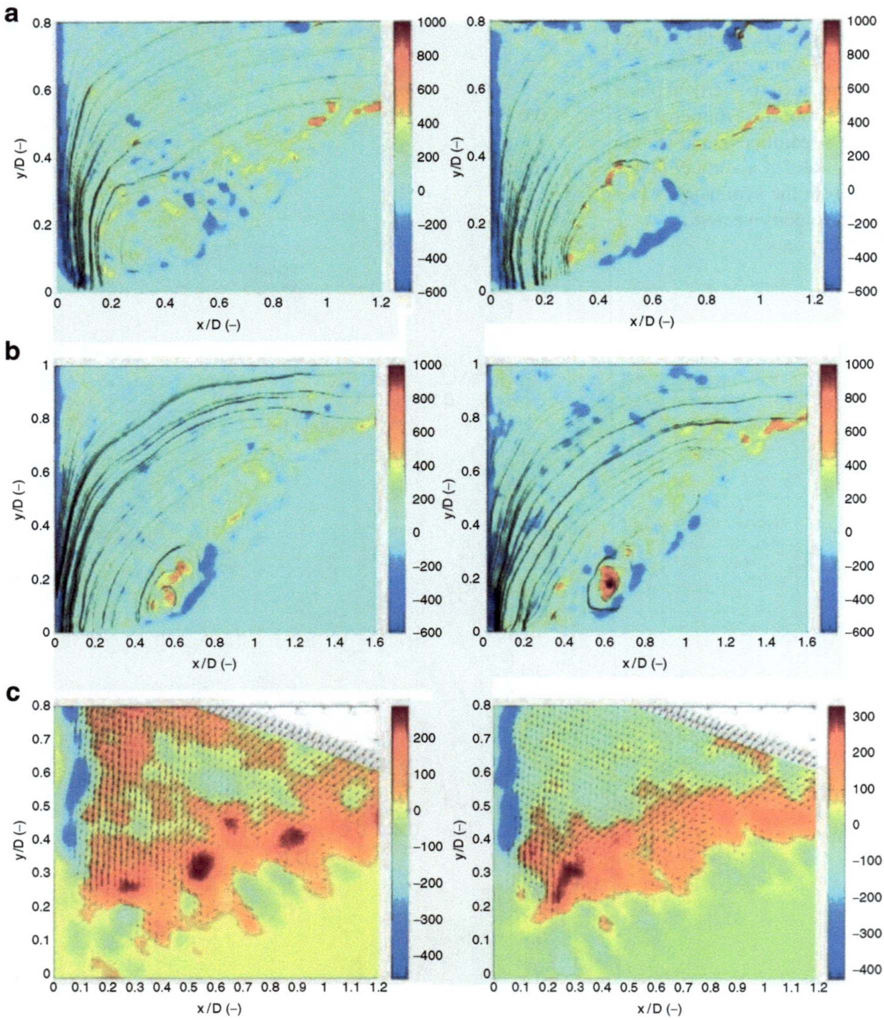

Fig. 43 Instantaneous vorticity maps. (**a**) the vertical plane of symmetry (with streamlines), (**b**) vertical plane at 30° from the symmetry axis (with streamlines), and (**c**) the scour hole, at a vertical plane 45° from the symmetry axis (with velocity vectors). Vorticity in s^{-1}

a jet along the wall of the cylinder that, by reaching the flow bed, undergoes separation digging the hole (zone 1) and surrounding the cylinder near the bottom (see Fig. 44). Due to this process, the displaced sediments are mostly transported downward, by the downstream flow. A small part is transported, inside the scour hole, to higher topography levels by the downward-flow which incorporates the horseshoe vortex. It was seen that the horseshoe vortex moves frequently towards the cylinder, interacting strongly with the downflow.

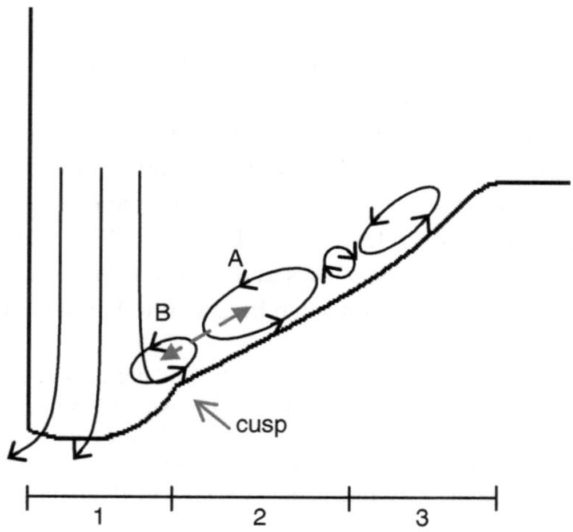

Fig. 44 Proposed flow structure in the scour hole in front of the cylindrical pier

The interaction between the downflow and the horseshoe vortex inside the scour hole leads to the formation of a cusp, separating the region in the scour hole mainly shaped by the downward-flow and the region shaped by the horseshoe vortex and the separation vortexes. Indeed, the bed topography between zones 1 and 2, i.e., the observed cusp, would thus be formed by the upward movement of the jet flow that incorporates the horseshoe vortex and the flow upward from the bottom of the vortex. In zone 3, the separate flow is the result of adverse gradient of pressure which results from the sudden deepening of the bed. Among the areas 2 and 3, there are structures in counter-rotation.

This model is compatible with the one proposed by Unger and Hager (2007), but not with those by Qadar (1981) or Muzzammil and Gangadhariah (2003). It follows that the scour mechanism depends fundamentally on the downward flow and not on the intensity of the horseshoe vortex as defended by Shen et al. (1969).

5 Conclusion

This text was prepared for the 30th International School of Hydraulics, in Wiejce, Poland, September 2010. It presents a sample of work performed at Instituto Superior Técnico, Technical University of Lisbon, Portugal using LDA and PIV techniques in the scope of fluvial hydraulics.

The work was divided in three parts, the first of which devoted to the presentation of basic concepts of LDA and PIV. The second part addressed the characterization of organized turbulence in open-channel flows over flat rough mobile beds and over flat rough fixed (armored) beds. In the third part, the scour mechanism associated to a vertical cylinder mounted on a mobile bed was characterized and discussed.

The first application, the characterization of turbulent open-channel flows over flat rough mobile beds and over flat rough fixed beds, required detailed turbulent measurement, with a high temporal resolution, hence the option for LDA. The second application, describing the flow within the scour hole of a wall-mounted vertical cylinder on a mobile bed, required fine spatial details of the vortical structures that develop inside the scour hole. The need to calculate vorticity and to observe a considerable large flow region conditioned the choice of PIV.

Details on raw data processing and data treatment were offered in order to highlight the relevance of LDA and PIV techniques to understand the interaction between turbulent flows and sediment transport. Specific phenomenological conclusions and discussions can be consulted at the end of Sects. 3 and 4 of the present text.

Acknowledgments This research was partially supported by the Portuguese Foundation for Science and Technology (FCT) through the project PTDC/ECM/65442/2006. Investigation of local scour around a cylinder was performed with Helena Nogueira and Mário J. Franca. The author acknowledges Paweł M. Rowiński, Institute of Geophysics, Poland, for the invitation to lecture at the 30th International School of Hydraulics, September 2010.

References

Albrecht H-E, Borys M, Damaschke N, Tropea C (2003) Laser Doppler and phase Doppler measurement techniques. Springer, Berlin

Baker CJ (1980) The turbulent horseshoe vortex. J Wind Eng Ind Aerodyn 6(1–2):9–23

Crowe C, Sommerfeld M, Tsuji Y (1998) Multiphase flows with droplets and particles. CRC, Boca Raton

Csele M (2004) Fundamentals of light sources and lasers. Wiley, Hoboken. ISBN 0-471-47660-9

Durst F, Melling A, Whitelaw JH (1976) Principles and practice of laser Doppler anemometry. Academic, London

Ferreira RML (2005) River morphodynamics and sediment transport. Conceptual model and solutions. Ph.D. thesis, Instituto Superior Técnico, Technical University of Lisbon, Lisbon

Ferreira RML, Franca MJ, Leal JGAB (2007) Laboratorial and theoretical study of the mobility of gravel and sand mixtures. In: Dohmen-Janssen CM, Hulscher SJMH (eds) Proceedings of the 5th RCEM, Cambridge University Press, Cambridge, UK, pp 531–539

Franca MJ, Lemmin U (2006) Detection and reconstruction of coherent structures based on wavelet multiresolution analysis. River Flow 2006 1:181–190

Grass AJ (1971) Structural features of turbulent flow over smooth and rough boundaries. J Fluid Mech 50:233–255

Gyr A, Schmid A (1997) Turbulent flows over smooth erodible sand beds in flumes. J Hydr Res 35(4):525–544

Hjemfelt AT, Mockros LF (1996) Motion of discrete particles in a turbulent fluid. Appl Sci Res 16(1):149–161

Hurther U, Lemmin U, Terray EA (2007) Turbulent transport in the outer region of rough-wall open-channel flows: the contribution of large coherent shear stress structures (LC3S). J Fluid Mech 574:465–493

Krogstad PA, Antonia RA (1999) Surface roughness effects in turbulent boundary layers. Exp Fluids 27:450–460

Kumar S, Banerjee S (1998) Development and application of a hierarchical system for digital particle image velocimetry to free-surface turbulence. Phys Fluids 10(1):160–77

Lauterborn W, Vogel A (1984) Modern optical techniques in fluid mechanics. Annu Rev Fluid Mech 16:223

Lourenco L, Krothapalli A (1995) On the accuracy of velocity and vorticity measurements with PIV. Exp Fluids 18(6):421–428

Macauley AN, Pender G (1999) The response of an armoured bed to the infiltration of fine sand in a compound channel. In: Proceedings 1st IAHR symposium on river, coastal and estuarine morphodynamics, Genova, 6–10 Sept 1999, pp 141–149

Melling A (1997) Tracer particles and seeding for particle image velocimetry. Meas Sci Technol 8:1406–1416

Morton BR (1987) Trailing vortices in the wakes of surface-mounted obstacles. J Fluid Mech 175:247–293

Muzzammil M, Gangadhariah T (2003) The mean characteristics of horseshoe vortex at a cylindrical pier. J Hydraul Res 41(3):285–297

Nakagawa H, Nezu I (1977) Prediction of the contribution to the Reynolds stress from bursting events in open-channel-flows. J Fluid Mech 80(1):99–128

Nelson JM, Shreve RL, McLean SR, Drake TG (1995) Role of near-bed turbulence in bedload transport and bedform mechanics. Water Resour Res 31:2071–2086

Nezu I, Nakagawa H (1993) Turbulence in open-channel flows. International Association for Hydraulic Research, Monograph Series. Balkema, Rotterdam

Nikora V, Goring D, McEwan I, Griffiths G (2001) Spatially averaged open-channelflow over rough. J Hydraul Eng 127(2):123–133

Nogueira J, Lecuona A, Rodriguez PA (1997) Data validation, false vectors correction and derived magnitudes calculation on PIV data. Meas Sci Technol 8(12):1493–1501

Nogueira H, Franca MJ, Aducce C, Ferreira RML (2008) Bridge piers in mobile beds: visualisation and characterization of the surrounding and approaching flows. River flow 2008. International conference on fluvial hydraulics,vol 3. Çesme, Turkey, 3–5 Sept 2008, pp 2397–2406, ISBN 978-605-60136-2-1

Qadar A (1981) The vortex scour mechanism at bridge piers. Proc ICE Res Theory 71(3):739–757

Raffel M, Willert CE, Werely ST, Kompenhans J (1998) Particle image velocimetry. A practical guide. Springer, Berlin

Raudkivi AJ (1998) Loose boundary hydraulics. A.A. Balkema, Roterdam

Ricardo AMC (2008) Caracterização do escoamento turbulento em canais com vegetação emersa rígida Aplicação ao estudo da resistência hidráulica. PhD thesis, Instituto Superior Técnico, TU Lisbon (in Portuguese)

Shen HW, Schneider VR, Karaki S (1969) Local scour around bridge piers. J Hydraul Div Proc Am Soc Civil Eng 95(HY6):1919–1940

Sutherland AJ (1967) Proposed mechanism for sediment entrainment by turbulent flows. J Geophys Res 72:191–178

Sveen JK, Cowen EA (2004) Quantitative imaging techniques and their applications to wavy flows, Chapter 1. WSPC, University of Oslo, Norway, available in http://www.math.uio.no

Tait SJ, Willetts BB, Gallagher MW (1996) The application of particle image velocimetry to the study of coherent flow structures over a stabilizing sediment bed: facts, mechanism and speculation. In: Ashworth PJ, Bennet SJ, Best JL, McLelland SJ (eds) Coherent flow structures in open channels. Wiley, Chichester, pp 185–201

Tennekes H, Lumley JL (1972) A first course in turbulence. MIT Press, Cambridge. ISBN ISBN 0262200198

Tropea C, Yarin A, Foss J (2007) Handbook of experimental fluid mechanics. Springer, Berlin

Unger J, Hager W (2007) Down-flow and horseshoe vortex characteristics of sediment embedded bridge piers. Exp Fluids 42:1–19

Westerweel, J. 1993. Digital particle image velocimetry – theory and application. Ph.D. thesis, Technical University of Delft, Delft, the Nertherlands

Wu W, Wang SSY, Jia Y (2000) Nonuniform sediment transport in alluvial rivers. J Hydraul Res 38(6):427–433

Yeh Y, Cummins H (1964) Localised fluid flows measurements with a He-Ne laser spectrometer. Appl Phys Lett 4:176–178

Studies for Nuclear and Pumped-Storage Power Plant Żarnowiec

Wojciech Majewski

1 Introduction

In the 1970s, there was considerable shortage of electric power in Poland. It was proposed to build a unit consisting of nuclear and pumped-storage power plant. Nuclear power plant would supply basic electric energy and energy for pumping without energy loss for transmission. The place for power plants was chosen in the vicinity of Lake Żarnowiec, which would be the reservoir of cooling water for nuclear power plant and simultaneously lower reservoir of pumped-storage power plant. There was a convenient place to construct artificial upper reservoir on moraine hills whose elevation exceeded 100 m water level in the lake and the distance from the lake was about 1 km. The capacity of nuclear power plant consisted of four reactors 440 MW each, which resulted in the discharge of cooling water of 90 m^3/s with temperature increase $\Delta t = 9.2°C$. Operation of pumped-storage power plant (four reversible Francis units, 700 MW) would cause daily water exchange between the lake and upper reservoir of about 14 hm^3, resulting in the discharge of about 600 m^3/s. Lake Żarnowiec is situated in Pomeranian province, about 5 km from the Baltic Sea.

To provide data for the design and operation of both power plants, the following studies were initiated. They included lake and the neighboring region in natural conditions:

- Hydrologic state of the lake and its catchment including water balance
- Hydrodynamic and thermal regime of the lake
- Ecological and fishery state of the lake
- Meteorological conditions in the region of the lake

W. Majewski
Institute of Meteorology and Water Management, 61 Podleśna str, Warsaw 01-673, Poland
and
Committee of Water Resources Management, Polish Academy of Sciences, Kościerska 7, Gdańsk 80-328, Poland
e-mail: wojciech.majewski@imgw.pl; wmaj@ibwpan.gda.pl

P. Rowinski (ed.), *Experimental Methods in Hydraulic Research*,
Geoplanet: Earth and Planetary Sciences, DOI 10.1007/978-3-642-17475-9_5,

Subsequently, the following studies were carried out:

- Thermal regime of the lake under the influence of heated water discharge
- Possible hazard to the nuclear power plant resulting from a breakdown of the upper reservoir of the pumped-storage power plant

In this chapter, special attention will be directed to thermal regime of the lake, which is very important for hydro-chemical and hydro-biological state of the lake in natural conditions but also with the operation of nuclear plant. An important problem was the water balance of the lake taking into account additional evaporation caused by heated water discharge from the nuclear plant.

2 General Information on the Lake Żarnowiec

Lake Żarnowiec is of glacial origin; it is situated in the Piaśnica River valley between two forest grown moraine hills elevated more than 100 m above the lake water level. The average water surface of the lake is 14.30 km^2, and the volume is 120 hm^3. The length of the lake is 7.6 km, and the maximum width is 2.6 km. The maximum water depth is 19.4 m, and the average depth is 8.4 m.

In natural conditions, average water level is 1.34 m above sea level (asl). The minimum observed water surface level was 0.92 m asl, and the maximum was 1.78 m asl. The residence time of the inflow to the lake is about 1.7 years. During vegetation season, water was abstracted from the lake to irrigate meadows situated north from the lake. Wind induced currents in the lake are of the order 0.1 m/s. Wind generated waves reach a height of 0.5–0.6 m. During winter, the lake is covered with ice, which may achieve a thickness of 0.5 m. Maximum number of days with ice on the lake was 133, and the average was about 80 (Majewski 1983).

There are two surface inflows to the lake and one outflow directly to the sea. The total surface inflow to the lake is estimated at 1.6 m^3/s. Because high moraine hills are situated on both sides of the lake, there is a significant inflow of ground water to the lake, which is estimated as 0.6 m^3/s. Outflow from the lake to the sea is in natural form and its value changes with water level in the lake.

3 Catchment of the Lake

The catchment of the lake is 249 km^2 and belongs to the catchment of Piaśnica River (310 km^2). Lake catchment consists of the catchment of Piaśnica Górna (88 km^2), catchment of the Bychowska Struga (122 km^2), and direct catchment of the lake (39 km^2). The length of Piaśnica Górna is 16.7 km with the average slope 3.3‰, and the length of Bychowska Struga is 20.1 km with average slope of 2.2‰. The catchment of the Lower Piaśnica has the surface of 61 km^2. The catchment of the lake is occupied by agricultural land, meadows, forests, and wasteland.

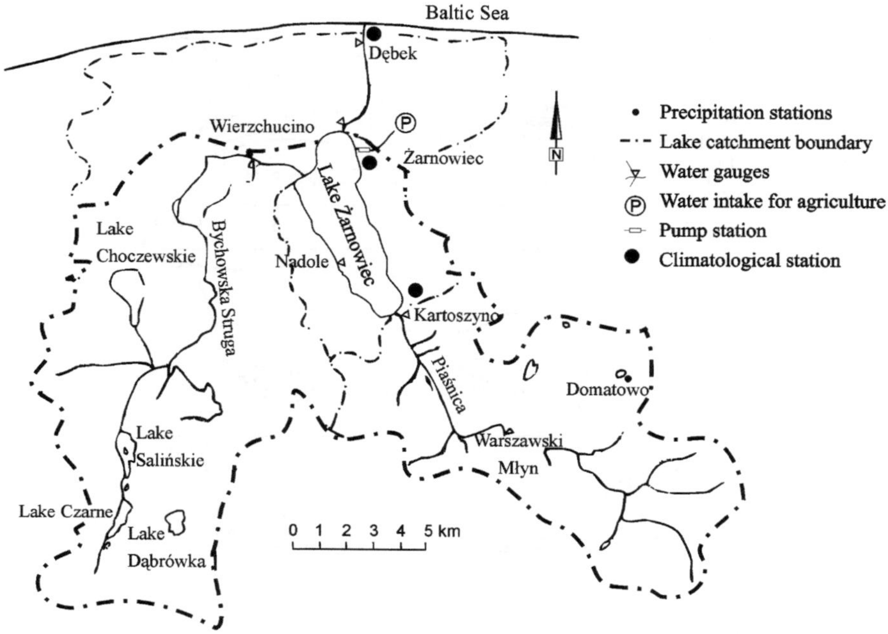

Fig. 1 Catchment of Lake Żarnowiec

The average multiyear precipitation over the catchment is estimated as 629 mm, which is very close to the average precipitation for the whole area of Poland. The lake catchment is shown in Fig. 1.

Design and operation of the nuclear power plant required several meteorological data. Three climatologic stations were established (Fig. 1). The first is close to the place of nuclear plant, the second at the end of the lake on its eastern shore, and the third is at the sea shore near the outflow of Piaśnica River to the sea. The following measurements were carried out at these stations:

- Wind speed and direction
- Air temperatures and humidity
- Solar radiation and albedo
- Evaporation from lake surface

4 Thermal Regime of the Lake in Natural Conditions

Thermal regime was one of the most important lake characteristics. Water temperatures were measured on the shore stations of the lake and also in several verticals during monthly expeditions. Thirty-seven verticals were established in longitudinal and perpendicular cross sections of the lake. In deep verticals,

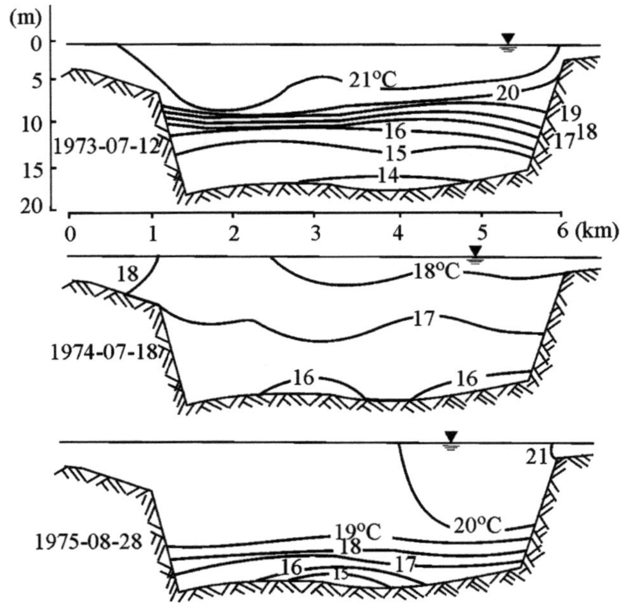

Fig. 2 Water temperatures along the longitudinal cross section of the lake

temperatures were measured in several points along the depth. In July and August, thermal stratification in the lake was observed as well as surface temperature variations. Temperature differences in deep verticals achieved in July and August were in the range of 5–7°C, while thermal stratification disappeared during other months. Variations of lake surface temperatures did not exceed 2–3°C with higher temperatures close to the shore where the depths are smaller. Analysis shows that average annual water temperature is 8.7°C, maximum 23.3°C and minimum is 0.0°C. In general, it can be stated that water temperatures in the lake follow air temperatures (Majewski et al. 1988a). Examples of water temperature distributions along the longitudinal lake cross section in July and August 1973, 1974, and 1975 are shown in Fig. 2.

5 Thermal Regime of the Lake Caused by the Discharge of Cooling Water

The layout of the designed system of cooling water discharge on the lake is presented in Fig. 3. Cooling water is conveyed by means of surface channel along the lake to its northern part. Here a dike is proposed across the lake, from which cooling water will be uniformly distributed over the lake surface. Bottom intake of cooling water is located in the southern part of the lake near the nuclear power plant.

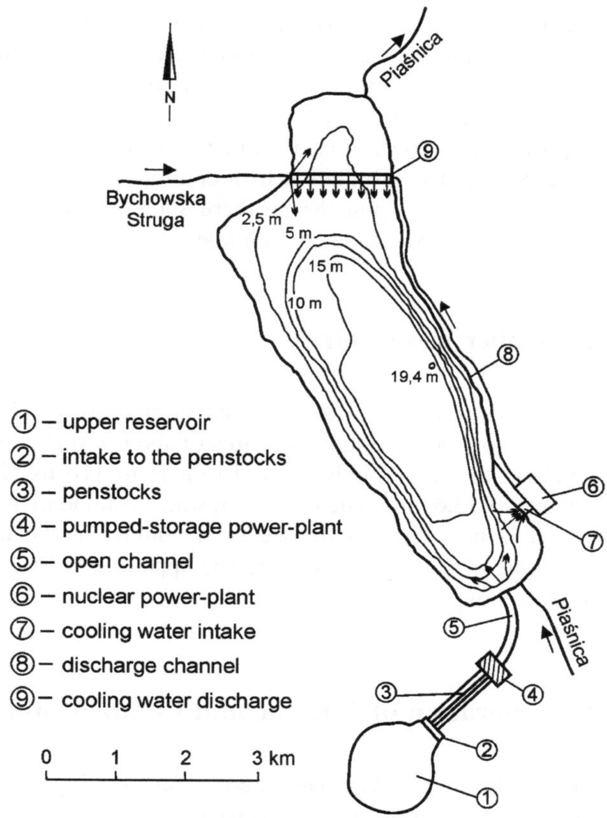

Fig. 3 Layout of the designed system of cooling water

The operation of this system was investigated on the hydrothermal model, where cooling water discharge and withdrawal was simulated. In the laboratory conditions, it was not possible to simulate heat losses from water surface and this was accomplished by means of calculations (Majewski and Widomski 1977). Heat exchange between water surface and atmosphere was determined for various water temperatures and meteorological conditions. These calculations proved that the designed cooling water system is working correctly from technical point of view and also that permissible lake surface temperatures were not exceeded.

6 Changes of the Lake to Secure the Operation Requirements of Both Power Plants

Future exploitation of the lake required introduction of several changes in the lake layout. Daily changes of lake water level due to the operation of pumped-storage power plant were estimated at 1.0 m. It was also expected that increase of maximum water level in the lake is indispensable in order to increase lake water volume because of the increased evaporation from heated water discharge. It was determined that the maximum water level in the lake will be 2.70 m asl and cannot be exceeded. Minimum water level in the lake was assumed at 0.70 m asl, which was imposed by the operation of turbines of the pumped-storage power plant. These changes required installation of a control weir at the outflow from the lake and additional dykes along the northern shore of the lake. Lake water volume and surface area were estimated within levels 0.70 and 2.70 m asl.

7 Upper Reservoir

The upper reservoir of the pumped-storage power plant was constructed on the nearby moraine hill and has the area of about 1 km^2. Operational variation of water levels in the reservoir is about 16 m, from 110 to 126 m asl. This gives water volume of 13.6 hm^3. The maximum water volume in the upper reservoir is 16 hm^3, and it is necessary to foresee such volume in the lake in case of emergency emptying of the upper reservoir. The upper reservoir is water tight due to a special asphalt lining.

8 Operation of Nuclear and Pumped-Storage Power Plants

Pumped-storage power plant operates on a weekly cycle, and there are differences between consecutive days of the week. Maximum discharges in turbine cycle reach about 400 m^3/s, while in pump cycle maximum discharges reach up to 500 m^3/s. This results in the daily water level variation in the lake of about 1.0 m and currents near the outlet channel reaching up to 1.0 m/s, which results in considerable mixing in lake volume. In this situation, it is difficult to expect thermal stratification along lake depth or important temperature variation over the lake surface.

Nuclear power plant with four units will have a maximum installed capacity of 1760 MWe. With general efficiency of the nuclear power plant of 34%, the waste heat discharged to the lake amounts to 3420 MWt. This situation will result in discharge of cooling water of about 90 m^3/s with temperature rise in the condensers of $\Delta t = 9.2°C$.

9 Water Balance of the Lake Under the Influence of Both Power Plants

One of the main problems of the future operation of both power plants was the water balance of the lake taking into account small inflow to the lake and additional evaporation from lake surface due to heated water discharge. Simulation of this

Table 1 Data for water balance calculations in 1966

Month	Surface inflow (m^3/s)	Groundwater inflow (m^3/s)	Precipitation (mm/day)	Natural evaporation (mm/day)	Irrigation (m^3/s)	Additional evaporation (mm/day)
I	2.30	0.43	0.77	0.39		5.59
II	2.00	0.58	0.81	0.39		5.59
III	2.28	0.62	1.39	1.22		5.59
IV	1.81	0.79	1.13	1.45		6.02
V	1.49	0.61	1.52	2.28	2.58	6.64
VI	1.32	0.72	0.63	3.73	1.34	7.13
VII	1.39	0.58	2.32	3.80	1.46	7.31
VIII	1.59	0.64	1.94	3.84	0.75	7.17
IX	1.50	0.58	0.80	2.39	0.44	7.01
X	1.42	0.53	1.42	1.78		6.51
XI	1.65	0.53	0.73	0.67		5.84
XII	2.08	0.47	1.29	0.33		5.59

process was carried out by means of the mathematical model (Majewski et al. 1988b). The following requirements had to be fulfilled:

- Maximum water level in the lake cannot exceed the elevation of 2.70 m asl.
- Minimum water level in the lake cannot be lower than 0.70 m asl.
- Minimum biological outflow from the lake is 0.40 m^3/s.
- Water for irrigation during vegetation period must be secured.

Calculations were carried out for one calendar year (1966) with a time step of 1 h. Averaged monthly values of the following model parameters were determined:

- Surface and ground water inflow (m^3/s)
- Precipitation (mm/day)
- Natural evaporation and additional evaporation (mm/day)
- Irrigation requirements (m^3/s)

These data are presented in Table 1.

Results of calculations are presented in Fig. 4. There is no problem during months January–April. However, problems appear during months May–October due to higher evaporation and water required for irrigation. To improve this situation, the outflow from the lake was decreased to the minimum biological value of 0.4 m^3/s. However, it is shown (Fig. 4) that the minimum water level (0.70 m asl) was exceeded by about 0.10 m.

10 Possible Failure of the Upper Reservoir

The upper reservoir is constructed artificially on the nearby moraine hill. The total water volume of the reservoir is 16 hm^3. In case of breaching, the reservoir embankment water which will flow out can endanger the nuclear power plant and

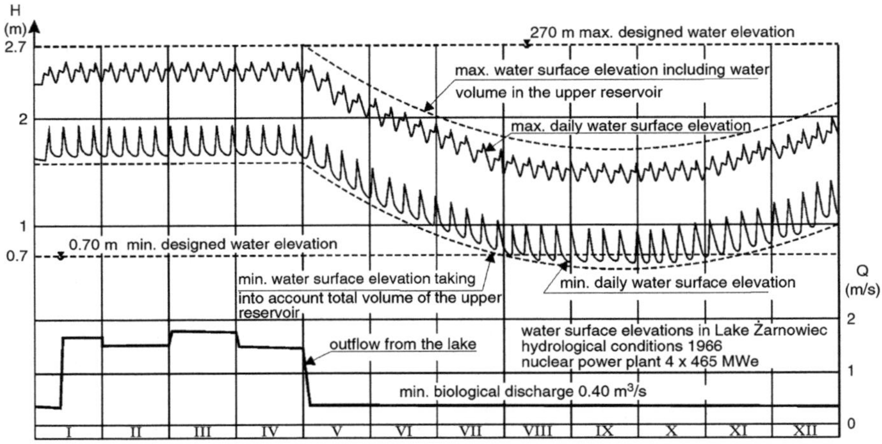

Fig. 4 Results of the calculations of lake water balance

its operation, which may bring considerable hazard. Special study was carried out on the hydraulic model. The breach of the embankment in several places was investigated. Measurements were carried out on the hydraulic model in the scale 1:250 without vertical scale distortion (Hydroproject Włocławek Section 1978). Various effects were investigated. The main result was the change of the previous layout of cooling water system. Water intake was moved in the north direction and designed as bottom intake. Discharge of cooling water was located near the area of nuclear power plant. This new layout is shown in Fig. 5

11 State of the Lake After 10 Years Operation of Pumped-Storage Power Plant

After 10 years of the operation of pumped-storage power plant, comprehensive studies of the lake were carried out (Majewski 1996, 2004).

12 Conclusions

The idea of constructing two power plants, which will utilize Lake Żarnowiec as the lower reservoir for pumped-storage power plant and as cooling water reservoir for nuclear power plant, required numerous data for the design and future operation of both power plants. The source of information included:

- Information from previous measurements and records.
- Stationary measurements carried out as standard data.

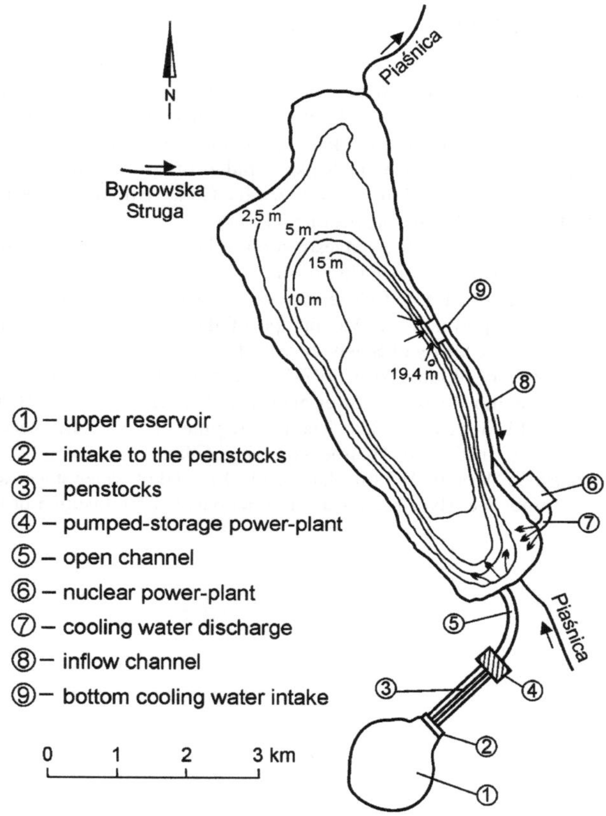

Fig. 5 New layout of cooling water system

- Data obtained from measurements during expeditions and on specially installed stations.
- Realization of these measurements required participation of specialized teams and good coordination of their activity was necessary.
- Several design decisions required calculations based on appropriate data and special mathematical models were developed as well as studies on hydraulic models were carried out.
- Hydrothermal model was used to study the efficiency of designed cooling system; hydraulic model was used to investigate the effect of the break-down of the upper reservoir of pumped-storage power plant.
- Mathematical model was developed to study heat exchange between water and atmosphere for the discharge of cooling water to the lake.
- Mathematical model was developed for the estimation of water balance of the lake and tested for a particular year.
- Field measurements and studies by means of hydraulic, hydrothermal, and mathematical models required several years and a lot of effort.

References

Hydroproject, Section Włocławek (1978) Hydraulic model investigations of the threat to nuclear power plant Żarnowiec by the catastrophic outflow of water from the upper reservoir of the pumped-storage power plant, Internal Report (in Polish)

Majewski W (2004) Comprehensive studies of lake Żarnowiec. In: Proceedings of XXIV International School of Hydraulics, Jastrzębia Góra, Hydraulic problems in environmental engineering, Institute of Hydro-engineering, Gdańsk

Majewski W (ed) (1983) Investigation of lake Żarnowiec for pumped-storage and nuclear power plants. PWN, Poznań (in Polish)

Majewski W (ed) (1996) The state of lake Żarnowiec after 10 years of the operation of pumped-storage power plant. Monograph of the committee of water resources management of the Polish Academy of Sciences, vol 11

Majewski W, Widomski A (1977) Lake thermal regime in connection with the design of nuclear and pumped-storage power plants. XVII IAHR Congress, Baden Baden

Majewski W et al (1988a) Thermal regime of lake Żarnowiec, Hydrotechnical Transactions, vol 50. Institute of Hydroengineering, Gdańsk

Majewski W et al (1988b) Mathematical model of water stages and outflow discharge from lake Żarnowiec, Hydrotechnical Transactions, vol 50. Institute of Hydroengineering, Gdańsk

Integrated Experimental and Computational Hydraulic Science in a Unique Natural Laboratory

J.R. Manson, B.O.L. Demars, and S.G. Wallis

1 Introduction

This chapter is concerned with the initial stage of investigations into stream metabolism, which can be studied by quantifying the fate and transport of dissolved oxygen. In this, for a given stream reach we would perform an oxygen mass balance to determine the rate of endogenous oxygen production (photosynthesis) and consumption (respiration). Before undertaking the oxygen mass balance, however, we need to determine the flow and transport characteristics of the stream, such as discharge, lateral inflow, velocity, surface gas exchange, mixing rates, relative volumes of flowing versus stagnant stream water, etc. Some of these quantities are required in order to perform the oxygen mass balance; others are determined because we hypothesise that they directly influence stream metabolism and therefore will be useful correlates. Most of these characterising parameters may be found through the analysis of tracer experiments. In such experiments, we release a discrete (known) mass of tracer and then measure tracer concentration-time profiles at locations downstream of the tracer release site. In this study, we obtain these parameters by fitting a computational stream transport model to the tracer data. This requires an efficient and accurate numerical solution of the advection–dispersion–transient storage equation and a reliable optimisation approach. The aim of this chapter is to describe our integrated experimental and computational approach and to present the findings for the study site in question. Because we performed these experiments twice (in early spring and in late summer), we are able to discuss

J.R. Manson (✉)
Richard Stockton College of New Jersey, Pomona, USA
e-mail: Russell.Manson@stockton.edu

B.O.L. Demars
Macaulay Land Use Research Institute, Aberdeen, UK

S.G. Wallis
Heriot-Watt University, Edinburgh, UK

P. Rowinski (ed.), *Experimental Methods in Hydraulic Research*,
Geoplanet: Earth and Planetary Sciences, DOI 10.1007/978-3-642-17475-9_6,

seasonal changes and seasonal constants in the flow and transport parameters. The emphasis of this chapter is on the transient storage parameters.

2 A Unique Natural Hydraulics Laboratory

The study watershed is situated in south-west Iceland (64°05′ N, 21°30′ W) on the mid-Atlantic ridge between the North American and Eurasian tectonic plates and is characterised by intense volcanic and geothermal activity (Arnason et al. 1969; Franzson et al. 2005). Heating of the stream water is by steam from boiling geothermal water reservoirs, which heats up the upper cold groundwater that feeds the streams (Arnason et al. 1969). The water chemistry is very similar between streams despite large temperature differences. Precipitation, which exceeds 3,000 mm per year, infiltrates the porous volcanic bedrock (Einarsson 1984) and numerous small permanent streams, mostly groundwater fed, emerge from the valley side and discharge into the River Hengladalsá (Friberg et al. 2009).

Before human settlement in Iceland (900 AD), birch woodland and scrub covered the area with unbroken heathland vegetation continuing up to 500–600 m elevation.

The catchment comprises a higher plateau region connected to a lower floodplain by a narrow river gorge. The lower floodplain is surrounded on all sides by steep valley sides. Apart from the moss cover and sparse grassland of the plateau and plain, there are extensive areas stripped of vegetation and soil where rocks of volcanic origin protrude. Allochthonous organic matter input to the streams is therefore considered minimal beyond the dissolved organic carbon coming from the groundwater. Long-term landscape change is the only known anthropogenic pressure on the streams investigated (Simpson et al. 2001).

Sixteen streams in the same watershed were available for experiments with flows ranging from 1 to 16 L/s, typical stream widths varied from 0.25 to 1.25 m and depths were of the order of 0.1–0.4 m.

3 Tracer Experiments

For each experiment, YSI600xlm multi-parameter sondes (YSI, Yellow Spring, USA) were placed at two longitudinal stations (typically 60 m apart) in the study stream and set to record conductivity at fixed 10 s intervals. Pre-weighed NaCl was fully dissolved in 2 L of stream water and then immediately released into the stream at some distance upstream of the upper station. The mixing zone was generally sufficiently long (~11 m) for complete cross-sectional mixing to take place before the upper station. In the shortest streams, additional deflectors and pools were created upstream of the upper station to increase mixing. Any naturally occurring background conductivity signal was subtracted from the observations prior to the modelling. A typical conductivity versus time graph is shown in Fig. 1. Although there were 16 streams in

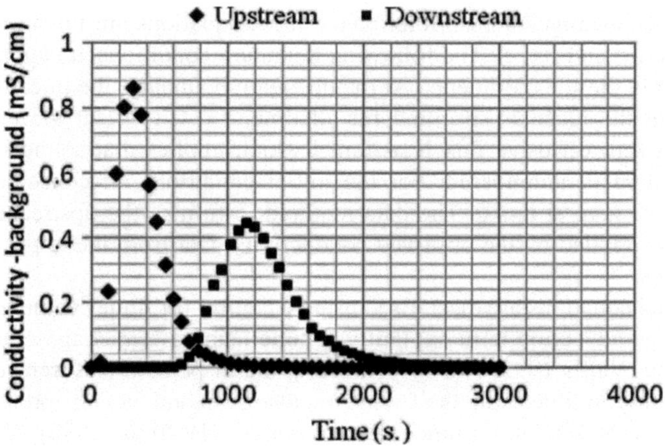

Fig. 1 Typical tracer result showing the spreading of the tracer cloud due to dispersion and transient storage effects

our study site, and we visited the area twice, we do not have 32 experiments to analyse. In our second visit, time and weather conditions prevented us from returning to some of the streams. However, we do have several paired experiments (spring, summer) to present here and a large pool of experimental data for combined analysis.

4 Obtaining Hydraulic Parameters Computationally

The transport of a conservative tracer in a stream with transient storage regions may be described by the following mass conservation equations:

$$\frac{\partial c}{\partial t} + u\frac{\partial c}{\partial x} = D\frac{\partial^2 c}{\partial x^2} + k_1(s - c) - \frac{q}{A}c \tag{1}$$

$$\frac{\partial s}{\partial t} = -k_1\frac{A}{A_s}(s - c) \tag{2}$$

where c is the concentration of tracer in the main channel, s is the concentration of tracer in the transient storage zone, A is the stream channel cross-sectional area, A_s is the transient storage cross-sectional area, D is the longitudinal dispersion coefficient, k_1 is the solute exchange parameter between the main channel and the storage zones, q is the lateral inflow, x is the longitudinal spatial co-ordinate and t is time.

For the experiments described herein, we are concerned with solving these equations over some stream reach length, L, and over some time interval, T, during which the tracer experiment takes place. Thus any parameters derived are recognised as being reach average values. Solutions to these equations are functions of space and time: $c(x, t)$ and $s(x, t)$. The following boundary conditions are appropriate for the scenario in these experiments. At the upstream boundary, the tracer concentration entering the reach is specified for all time, $t = 0$ to T; at the downstream boundary, a zero diffusive flux is assumed which implies that solute is advected out of the domain unhindered. For the initial conditions all concentrations are assumed to be zero at $t = 0$. The data required to furnish the upstream boundary condition is supplied by the observed conductivity data from the upper end of the study reaches.

The model equations are discretised using a control (or finite) volume approach, evaluating the advection term explicitly in time using a semi-Lagrangian method (Manson and Wallis 1995) and by evaluating the dispersion and transient storage terms implicitly in time using the Crank–Nicolson method, which apportions equal weight to both present and future values of c and s (Hoffman 1992). The solution consists of estimates for c and s over some discretised spatial and temporal domain, i.e. (c_i^m, s_i^m) for $i = 1$ to n_x and $m = 1$ to n_t, where n_x is the number of points in the spatial domain and n_t is the number of points in the temporal domain. Note that the spatial domain is divided into $(n_x - 1)$ cells of size Δx and the temporal domain is divided into $(n_t - 1)$ time steps of size Δt. Equation (1) is an advection–diffusion–decay equation and represents a considerable challenge to existing numerical methods. In order to achieve a satisfactory solution, the DISCUS method (Manson and Wallis 1995, 2000) is adopted. This method employs a conservative semi-Lagrangian algorithm that combines a control volume discretisation, the method of characteristics and a flux-based interpolation scheme. The method and its accuracy portrait is explained in detail elsewhere (Manson et al. 2001), but note that in this work, in contrast to earlier applications of this model, the Crank–Nicolson method is adopted for the dispersion and transient storage terms. The discretised form of these equations is a coupled pair of equations linking c_i and s_i to their neighbouring cells for the whole computational domain at the future time level, $n + 1$,

$$\alpha_i c_{i-1}^{n+1} + \beta_i c_i^{n+1} + \gamma_i c_{i+1}^{n+1} + \delta_i s_i^{n+1} = \varepsilon_i \tag{3}$$

$$\beta_i^s c_i^{n+1} + \delta_i^s s_i^{n+1} = \varepsilon_i^s \tag{4}$$

in which $\alpha_i, \beta_i, \gamma_i, \delta_i, \varepsilon_i, \beta_i^s, \delta_i^s, \varepsilon_i^s$ are coefficients related to both physical and numerical parameters. Since there are $(n_x - 1)$ cells, the resulting $2(n_x - 1)$ equations are assembled into a matrix and solved to give c_i^{n+1} and s_i^{n+1} for each computational cell. Note that (4) may be used to eliminate the transient storage term from (3) before it is solved.

The model prediction for concentration versus time at the downstream end of the experimental reach was fitted to the observed data at the lower end of the study reach, which had been collected at n_t points in time. A fitting parameter may be defined as:

$$E = \sum_{m=1}^{n_t} \left(c_{n_x}^m(\text{observed}) - c_{n_x}^m(\text{predicted}) \right)^2 \tag{5}$$

The model fit to the observations was optimised by adjusting the parameters ($u, D, k_1, A_s/A$) strategically in order to minimise E. Note that lateral inflow (q) was obtained directly from solute dilution calculations and was therefore not adjusted as a fitting parameter. The parameter optimization was undertaken with a direct search method, this being a SIMPLEX method of the Nelder–Mead variety (Lagarias et al. 1998). The numerical solution was coded in FORTRAN and then compiled and linked with the MATLAB mex libraries to create a dynamic linked library, which is a callable function within MATLAB.

5 Results and Discussion

5.1 *Pooled Analysis*

The model was optimised to several tracer data sets from both the summer and spring conditions to obtain stream transport parameters ($u, D, k_1, A_s/A$) over a range of flow conditions (flows varied from 1 to 16 L/s in the experiments considered). Figure 2 shows a plot of the transient storage exchange parameter (k_1) versus flow velocity. We might expect that these parameters would correlate with each other; at larger velocities the shear gradient across the boundary layer will be larger with consequently larger mass transport. Admittedly the flow field in these small streams is much more complicated than the simple shear flow for which this exchange mechanism is envisaged and there is considerable scatter; however, we observe a fairly strong correlation ($R = 0.635, n = 16$) when a power law model is fitted; we

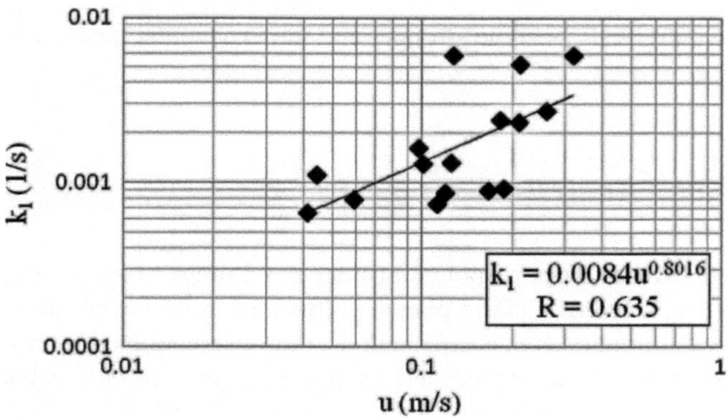

Fig. 2 Plot of transfer coefficient (k_1) versus velocity (u) for all studied streams pooled from summer and spring

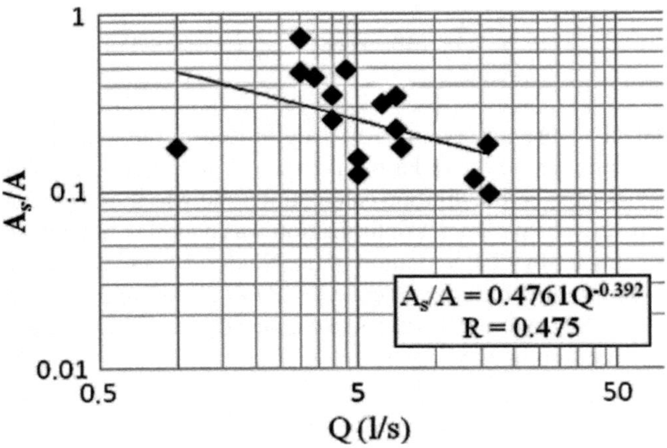

Fig. 3 Plot of fractional storage volume (A_s/A) versus discharge (Q) for all studied streams pooled from summer and spring

find that k_1 scales approximately with $u^{0.8}$. Note that these values of k_1 correspond with a range of transient storage residence times ranging from about 2 to 22 min. These values resonate with field observations.

Figure 3 shows the relationship between the ratio of transient storage volume to main channel volume, expressed as an area ratio on the basis that the reach length is common to both the stream channel and the transient storage zones (A_s/A) and discharge (Q). We observe an inverse power law relationship between A_s/A and discharge ($R = 0.475$, $n = 16$). In this watershed, we find A_s/A scaling approximately with $Q^{-0.4}$. We find for these streams that the transient storage volume varies from about 10% to 70% of the main channel volume. Once again there is some scatter, but these percentages seem reasonable when we consider in situ observations. The value of A_s/A obtained at the lowest flow appears to be an outlier, indeed it may reside in a different flow regime. If it were not included in the regression, the correlation would likely become stronger and the exponential decline steeper.

5.2 *Seasonal Analysis*

We observed several types of potential transient storage zones in these streams: seasonal development of aquatic plants, permanent moss cover, stones and deep pools. Table 1 summarises our expectations and observations. Generally, streams with large swathes of submerged and emergent aquatic vascular plants had less transient storage at the beginning of the spring (April 2009) than in late summer (August 2008), see Table 1 and Fig. 4, reflecting previous findings (Salehin et al. 2003; Ensign and Doyle 2005). The change in predicted transient storage in the

Table 1 Predictions and observations of seasonal changes in transient storage volume according to their nature

Sites	Nature of transient storage	Predictions (late summer to early spring)	Observations
2	Aquatic plants	−−	−
3	Aquatic plants	−−	−−
4	Stony bed	±	−
6	Stony bed	±	+[a]
7	Stony bed with moss	±	−
9	Stony bed (+ hyporheic zone?)	±	−
10	Stony bed with moss	±	+
11	Deep pools ($CaCO_3$ bed)	±	±
14	Stony bed	±	±

± = ±25% change, + or − = 25–50% change, ++ or −− = >50% change
[a]Extensive channel reworking and consequent geomorphological changes during the intervening winter.

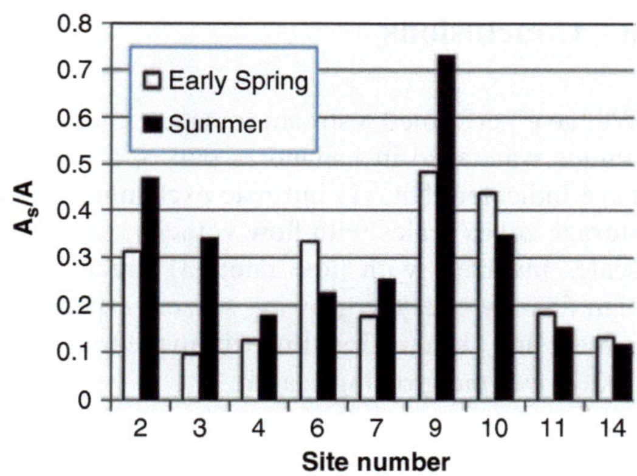

Fig. 4 Bar chart showing A_s/A for all sites for both early spring and summer

streams with more permanent structures such as moss cover and stones was generally slightly more variable than anticipated from the on-site observations of the stream characteristics, with no clear direction of change however.

Figure 5 shows some other interesting seasonal aspects of the system. First, velocities are higher in the early spring than in the previous summer. This is a direct consequence of the higher flows (the spring fed water being augmented by snow melt). We see a consistent increase in velocity (and flow) at virtually all sites and corresponding changes in the exchange coefficient.

We observe decreases in the exchange coefficient from early spring to late summer ranging from 35% up to 100%; these would impact the ecosystem with corresponding increases in the retention time in the transient storage zone. The increase in retention time would allow greater nutrient cycling in the transient storage zone over the growing season.

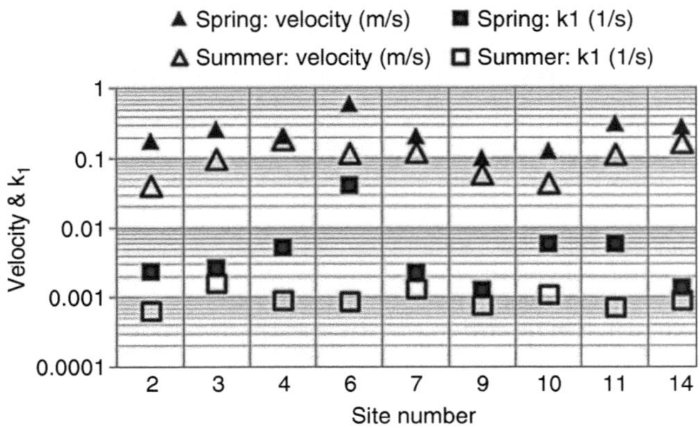

Fig. 5 Velocity and exchange coefficient for each site at both summer and early spring conditions

6 Conclusions

We have performed a stream transport characterisation of several streams within a unique watershed in Iceland as part of a larger stream metabolism study. Results have indicated that: (1) intrinsic exchange rate between main stream and transient storage zones scales with flow velocity; (2) the fractional transient storage volume scales inversely with flow rate; (3) stream transient storage volume can change significantly over the growing season, especially if submerged vascular plants are present and (4) retention time within the transient storage zone can increase by up to 100% over the growing season.

References

Arnason B, Theodorsson P, Björnsson S, Saemundsson K (1969) Hengill, a high temperature thermal area in Iceland. Bull Volcanol 33:245–259

Einarsson MA (1984) World survey of climatology. In: van Loon H (ed) Climate of the oceans, vol 15. Elsevier, Amsterdam, pp 673–697

Ensign SH, Doyle MW (2005) In-channel transient storage and associated nutrient retention: evidence from experimental manipulations. Limnol Oceanogr 50:1740–1751

Franzson H et al (2005) In: Proceedings world geothermal congress, Antalya, Turkey, pp 1–7

Friberg N, Dybkjaer JB, Olafsson JS, Gislason GM, Larsen SE, Lauridsen TL (2009) Relationships between structure and function in streams contrasting in temperature. Freshw Biol 54:2051–2068

Hoffman JD (1992) Numerical methods for engineers and scientists. McGraw-Hill, New York

Lagarias JC, Reeds JA, Wright MH, Wright PE (1998) Convergence properties of the Nelder–Mead simplex method in low dimensions. SIAM J Optim 9(1):112–147

Manson JR, Wallis SG (1995) An accurate numerical algorithm for advective transport. Commun Numer Methods Eng 11:1039–1045

Manson JR, Wallis SG (2000) A conservative, semi-Lagrangian fate and transport model for fluvial systems: Part 1 – theoretical development. Water Res 34(15):3769–3777

Manson JR, Wallis SG, Hope D (2001) Conservative semi-Lagrangian transport model for rivers with transient storage zones. Water Resour Res 37(12):3321–3329

Salehin M, Packman AI, Worman A (2003) Comparison of transient storage in vegetated and unvegetated reaches of a small agricultural stream in Sweden: seasonal variation and anthropogenic manipulation. Adv Water Resour 26:951–964

Simpson IA, Dugmore AJ, Thomson A, Vesteinsson O (2001) Crossing the thresholds: human ecology and historical patterns of landscape degradation. Catena 42:175–192

Using Experimental Research for an Analysis of Sorption Term in Groundwater Contaminant Transport Equation

Andrzej Aniszewski

1 Introduction

In the current literature on the determination of sorption parameter values, all authors use mathematical descriptions related either to kinetics or statics of this process (treated as separate processes) (Seidel-Morgenstern 2004; Chiang 2005). Thus, the literature lacks any attempts at determining relationships between the linear and nonlinear sorption kinetics rate constants (k_1, k_2) and the parameters of the adopted linear or nonlinear sorption isotherms (K_1, K_2, N) resulting from static equilibrium-controlled models. Such connections between kinetics and statics make it possible in practice to eliminate the dynamic complex experimental research with mobile groundwater (carried out on laboratory models – columns) and the very long-lasting and expensive field research in natural ground media by using considerably simpler and shorter static laboratory research with an immobile contaminant and groundwater.

In dynamic research of a sorption process for contaminant and groundwater movement through the ground, the determination of linear and nonlinear sorption kinetics rate constants (k_1, k_2) requires that the so-called breakthrough curves BTCs (named also transition curves through analyzed lengths of ground media) be defined every time (Letcher 2004; Bekhit and Hassan 2007). One has to remember, however, that the constant parameters of linear or nonlinear sorption isotherms (K_1, K_2, N) can be defined in a simple static laboratory research with immobile groundwater (Seidel-Morgenstern 2004; Aniszewski 2009).

A. Aniszewski
West Pomeranian University of Technology, Piastów Street 50a, Szczecin 70-310, Poland
e-mail: andrzej.aniszewski@ps.pl

P. Rowinski (ed.), *Experimental Methods in Hydraulic Research*,
Geoplanet: Earth and Planetary Sciences, DOI 10.1007/978-3-642-17475-9_7,

2 Approach and Methods

2.1 General Description of Contaminant Transport in Groundwater

In this chapter, the final 2D advection–dispersion equation resulting from the transport continuity equation has been used for describing the groundwater contaminants transport with consideration of the instantaneous nonlinear reversible sorption term (for inorganic chemicals) (Chiang 2005):

$$\frac{\partial C}{\partial t}\left(1+\frac{\rho}{m}\frac{\partial S}{\partial C}\right)+u_x\frac{\partial C}{\partial x}=D_x\frac{\partial^2 C}{\partial x^2}+D_y\frac{\partial^2 C}{\partial y^2} \tag{1}$$

where: C = the solute concentration in flowing groundwater in aqueous phase (in the local equilibrium conditions); S = the mass of the solute species adsorbed on the grounds per unit bulk dry mass of the porous medium (in the local equilibrium conditions); u_x = component of the so-called pore groundwater velocity in pore space along the x axis; D_x, D_y = components of the longitudinal and transverse dispersion coefficients along the x and y axes that depend on the longitudinal and transverse dispersivity (α_L, α_T); ρ = the bulk density of the porous medium; m = the effective porosity of the porous medium; t = coordinate of time; (x, y) = Cartesian coordinates of the assumed reference system; and $[1+(\rho/m)\cdot(\partial S/\partial C)]$ = the retardation factor, constant in time, resulting from sorption process (Seidel-Morgenstern 2004).

The reason for using the above-mentioned form of equation, along with detailed description of all the parameters being considered in it $(u_x, m, \rho, D_x(\alpha_L), D_y(\alpha_T))$ is given in Aniszewski (2009). In the nonequilibrium state analyzed further on $(\partial S/\partial t \neq 0)$, for the mathematical description of sorption models, the kinetic time-dependent models should be always used as the reversible first-order or N-order models of kinetics. Among many different kinetic models presented in the literature, the first-order kinetic sorption models (the nonlinear and linear kinetic equations) were chosen for further analysis. These kinetic models are also widely adopted in practice (Travis and Etnier 1981; Seidel-Morgenstern 2004; Chiang 2005).

The first-order kinetic sorption model (the nonlinear kinetic equation) can be written in the form:

$$\frac{\partial S}{\partial t}=\frac{m}{\rho}\cdot k_1\cdot C^N-k_2\cdot S \tag{2}$$

where: k_1 = the rate constant of adsorption process (the so-called forward adsorption rate constant); k_2 = the rate constant of desorption process (the so-called backward desorption rate constant); and N = adsorption constant characterizing geometric shape of adsorption isotherm (in the local equilibrium conditions).

However, (2) describes a fully reversible adsorption mechanism, in which the forward reaction is nonlinear and the backward desorption reaction is linear (Seidel-Morgenstern 2004; Chiang 2005).

For the linear kinetic sorption model, when the value of N is unity, (2) reduces to the reversible linear first-order kinetic sorption process described by the following equation that is frequently used in practice:

$$\frac{\partial S}{\partial t} = \frac{m}{\rho} \cdot k_1 \cdot C - k_2 \cdot S \tag{3}$$

Equation (3) assumes that the rate of solute sorption by the soil matrix is related to the difference between what can be adsorbed at some concentration and what has already been adsorbed.

In the case of a local equilibrium-controlled state ($\partial S/\partial t = 0$), (2) can be written in relation to the nonlinear sorption kinetics in the form (Aniszewski 2009):

$$S = \frac{k_1}{k_2} \cdot \frac{m}{\rho} \cdot C^N = K_2 \cdot C^N \tag{4}$$

where $\frac{k_1}{k_2} \cdot \frac{m}{\rho} = K_2$ is the constant nonlinear adsorption parameter.

Equation (4) is the Freundlich nonlinear isotherm, well known in literature and often used in practice; it has been presented in relation to static equilibrium-controlled sorption model for relatively large concentrations of moving contaminants in ground media (Seidel-Morgenstern 2004; Chiang 2005; Aniszewski 2009).

In turn, for the linear kinetic sorption model in the local equilibrium-controlled state ($\partial S/\partial t = 0$), (2) can be written in the form (Seidel-Morgenstern 2004; Chiang 2005):

$$S = \frac{k_1}{k_2} \cdot \frac{m}{\rho} \cdot C = k_1 \cdot C \tag{5}$$

where $\frac{k_1}{k_2} \cdot \frac{m}{\rho} = K_1$ is the constant linear adsorption parameter.

Equation (5) represents the Henry linear isotherm, well known in the literature and often put into practice; it has also been presented in relation to static equilibrium-controlled sorption model for relatively low concentrations of moving contaminants in ground media (Seidel-Morgenstern 2004; Chiang 2005; Aniszewski 2009).

At the same time, it should be noted that for the great sorption rates this process is an equilibrium-controlled one, and the concentration change of dissolved substance in liquid phase ($\partial C/\partial t$) is directly proportional to the concentration change of dissolved substance in solid phase ($\partial S/\partial t$) in the local equilibrium conditions (Seidel-Morgenstern 2004).

Hence, we can write the relationship:

$$-\frac{\partial C}{\partial t} = \frac{\rho}{m} \cdot \frac{\partial S}{\partial t} \tag{6}$$

Taking into consideration (6), (2) can be presented in the form:

$$-\frac{m}{\rho} \cdot \frac{\partial C}{\partial t} = k_1 \cdot \frac{m}{\rho} \cdot C^N - k_2 \cdot S \tag{7}$$

and after transformation and using the auxiliary expression ($a_0 = \rho/m$), in the form:

$$\frac{\partial C}{\partial t} = -k_1 \cdot C^N + k_2 \cdot a_0 \cdot S \tag{8}$$

One can say that the sorption kinetics as a first-order kinetically controlled process, being described by the reversible nonlinear (2) or linear (3) equations, defines the mechanism of gradually reaching the equilibrium-controlled state. In the final phase of this mechanism, the kinetic process changes into a static one (instantaneous equilibrium sorption models), taking advantage of the reversible Freundlich nonlinear or the Henry linear isotherms represented by (4) or (5), respectively, which are most often used in practice and are simple in mathematical description (Letcher 2004; Seidel-Morgenstern 2004; Aniszewski 2009).

The further part of this chapter addresses an attempt of practical determining of the rate constants of adsorption (k_1) and desorption (k_2), based on the nonlinear (2) and the linear (3) model of the sorption kinetics and using at the same time the static experimental research with immobile groundwater carried out by the author of this chapter. Generally, during the author's laboratory research (performed in the special closed vessels), the constant volumes (V) of the examined contaminant solutes (with the known initial concentrations C_0) were inundated and mixed with the ground samples with known, but varying masses m_0. After appropriate mixing time, in the local equilibrium conditions, the final stabilized concentrations C of the examined solutes in the particular samples were reached (in aqueous phase). Using the measured differences between these initial (C_0) and final stabilized (C) concentrations as well as the constant volumes of solutes (V), the masses of the examined solutes (S) adsorbed on the ground sample grains were calculated based on (9) as ratios to these ground sample masses (m_0) (in solid phase).

So, using in these static experimental research the final equilibrium state, the dimensionless mass balance of the moving contaminants (as adsorbate) can be written as:

$$S = \frac{V \cdot (C_0 - C)}{m_0} \tag{9}$$

where: V = the constant volume of examined contaminant solution with initial concentration C_0, used to inundate the ground samples of mass m_0 in laboratory research; m_0 = the mass of the examined ground samples; and C_0 = the initial concentration of the examined contaminant solution.

For the nonlinear model of sorption kinetics, taking into consideration (9) and using the auxiliary dimensionless expression $[a_1 = (a_0 \cdot V)/m_0]$, (8) can be written in the form:

$$\frac{\partial C}{\partial t} = -k_1 \cdot C^N + k_2 \cdot a_1 \cdot (C_0 - C) \tag{10}$$

Then, taking into consideration (4) and using also the auxiliary dimensionless expression $[a_2 = 1/(a_0 \cdot K_2)]$, (10) can be written for the nonlinear kinetics of sorption process as the final relationship:

$$\frac{\partial C}{\partial t} = -k_1 \cdot C^N + k_1 \cdot a_1^* \cdot (C_0 - C) \quad \left(a_1^* = a_1 \cdot a_2 = \frac{V}{m_0 \cdot K_2}\right) \tag{11}$$

For the linear kinetics of sorption process, taking into consideration (9) and using the auxiliary dimensionless expression $[a_1 = (a_0 \cdot V)/m_0]$, (3) can be written in the form:

$$\frac{\partial C}{\partial t} = -k_1 \cdot C + k_2 \cdot a_1 \cdot (C_0 - C) \tag{12}$$

Now, taking into consideration (5) and using auxiliary dimensionless expression $[a_3 = 1/(a_0 \cdot K_1)]$, (12) can be written as the final relationship as follows:

$$\frac{\partial C}{\partial t} = -k_1 \cdot C + k_1 \cdot a_2^* \cdot (C_0 - C) \quad \left(a_2^* = a_1 \cdot a_3 = \frac{V}{m_0 \cdot K_1}\right) \tag{13}$$

3 Results of Analytical Calculations

3.1 *Determination of the Rate Constants of Adsorption (k_1) and Desorption (k_2) for the Nonlinear and Linear Models of Sorption Kinetics*

In this section, two practical cases of determination of the rate constants of adsorption (k_1) and desorption (k_2) are presented for the nonlinear and linear models of sorption kinetics in relation to the nonlinear and linear sorption isotherms widely applied in practice (Seidel-Morgenstern 2004; Chiang 2005). So, the analytical solution of the equation describing the nonlinear sorption kinetics according to (11) was adopted, assuming exemplary numerical value of the constant Freundlich exponent ($N = 2$) in relation to the Freundlich nonlinear isotherm (4).

Taking into consideration the above assumptions, (11) can be written as:

$$\frac{\partial C}{\partial t} = -k_1 \cdot C^2 + k_1 \cdot a_1^* \cdot (C_0 - C) \tag{14}$$

After separation of the variables C and t and double-sided integration of (14), the final form of the analytical solution was defined as an exponential function; using appropriate transformations (Bronsztejn and Siemiendiajew 1990), we get:

$$C = \frac{B \cdot (A \cdot e^{k_1 \cdot t \cdot B} + 1) - a_1^* \cdot (A \cdot e^{k_1 \cdot t \cdot B} - 1)}{2 \cdot (A \cdot e^{k_1 \cdot t \cdot B} - 1)} \tag{15}$$

with the auxiliary expressions:

$$A = \frac{-2 \cdot k_1 \cdot C_0 - k_1 \cdot a_1^* - \sqrt{-\Delta}}{-2 \cdot k_1 \cdot C_0 - k_1 \cdot a_1^* + \sqrt{-\Delta}} \quad \text{and} \quad B = \sqrt{4 \cdot a_1^* \cdot C_0 + a_1^{*2}} \tag{16}$$

It should also be noticed that the value of the discriminant (Δ) of the expression $[-k_1 \cdot C^2 + k_1 \cdot a_1^* \cdot (C_0 - C)]$ in (14) is less than zero ($\Delta < 0$) and amounts to $\Delta = -4 \cdot k_1^2 \cdot a_1^* \cdot C_0 - k_1^2 \cdot a_1^{*2}$.

Having determined the rate constant of adsorption (k_1) from (15) and (16), the rate constant of desorption (k_2) can easily be determined, based on the parameter of the Freundlich nonlinear isotherm (K_2) occurring in (4) as the relationship $K_2 = k_1/(k_2 \cdot a_0)$, in the following form:

$$k_2 = \frac{k_1}{a_0 \cdot K_2} = \frac{k_1 \cdot m}{K_2 \cdot \rho} \tag{17}$$

However, for other empirical numerical values of nonlinear exponents (N) in the Freundlich nonlinear isotherms, the solution of (14) can be obtained using various more complex nonanalytical methods, for example, numerical ones, as presented in Aniszewski (2009).

In further analysis, the analytical solution of the equation describing the linear sorption kinetics according to (13) was accepted, assuming numerical value of the constant Freundlich exponent ($N = 1$) in relation to the Henry linear isotherm (5).

After separation of the variables C and t and double-sided integration of (13), the final form of analytical solution was defined as the exponential function; using the appropriate transformations (Bronsztejn and Siemiendiajew 1990), we get:

$$C = C_0 \left[\frac{e^{-k_1 \cdot t(1 + a_2^*)} - a_2^*}{1 - a_2^*} \right] \tag{18}$$

Having determined the rate constant of adsorption (k_1) from (18), the rate constant of desorption (k_2) can be defined based on the parameter of the Henry linear isotherm (K_1) occurring in (5) as the relationship $K_1 = k_1/(k_2 \cdot a_0)$ in the following form:

$$k_2 = \frac{k_1}{a_0 \cdot K_1} = \frac{k_1 \cdot m}{K_1 \cdot \rho} \tag{19}$$

4 Discussion and Conclusions

Based on simple analytical calculations, one can state that the main advantage of equations (15)–(19) presented here is connected with the possibility of their practical application involving relatively simple laboratory research, instead of dynamic and complex laboratory or field research with mobile groundwater in order to obtain the so-called breakthrough curves BTCs (Letcher 2004).

The attempts at finding practical relationships between kinetics and statics of sorption process in the groundwater contaminant transport equation (1) were presented in this chapter. The nonlinear (2) and linear (3) sorption kinetic models being applied most often in practice, along with the analytical solutions of these models in the form of (11)–(13), were also used. The current literature on contemporary research of other authors lacks such practical relationships between the rate constants of sorption process (k_1, k_2) and the constant linear and nonlinear isotherm parameters (K_1, K_2, N).

References

Aniszewski A (2009) Mathematical modeling and practical verification of groundwater and contaminant transport in a chosen natural aquifer. Acta Geophys 57:435–453

Bekhit HM, Hassan AE (2007) Subsurface contaminant transport in the presence of colloids: effect of nonlinear and nonequilibrium interactions. Water Resour Res 43:409–418

Bronsztejn IN, Siemiendiajew KA (1990) Matematyka. Poradnik encyklopedyczny [Mathematics. Encyclopaedic handbook]. Parts III and IV, 9 ed. Polish Scientific Publishers PWN, Warsaw (in Polish)

Chiang WH (2005) 3D-Groundwater modeling with PMWIN. A simulation system for modeling groundwater flow and pollution. Springer, Heidelberg, New York

Letcher TM (ed) (2004) Chemical thermodynamics for industry. Royal Society of Chemistry, Cambridge

Seidel-Morgenstern A (2004) Experimental determination of single solute and competitive adsorption isotherms. J Chromatogr A 1037:255–272

Travis CC, Etnier EL (1981) A survey of sorption relationships for reactive solutes in soil. J Environ Qual 10:8–17

2D Simulation of Discontinuous Shallow Flows

R. Canelas, J. Murillo, and R. Ferreira

1 Introduction

Open channels with mobile boundaries are subjected to entrainment, erosion, transport, and deposition of sediments. During intense floods, such as those originated by dam failure, bedload can be intense, leading to flow stratification: an uppermost layer of clean water and a lowermost layer of sheetflow can be observed. Such flows will herein be designated geomorphic flows. Since the physical system comprises a stratified mixture of fluid and granular matter, conservation equations must take into account such stratification along with the influence of sediment on inertia and pressure terms in the momentum balance. In the case of a purely IVP–Riemann problem, Ferreira (2005, 2008) showed that the system of conservation equations describing geomorphic flows is structurally similar to the clean water system. The main difference is that, considering equilibrium sediment transport, a new Riemann wave is introduced, expressing a genuinely nonlinear characteristic field, associated to the conservation of granular material in the transport layer. This new discontinuity is weak and unsusceptible to bring about major errors in the approximate Riemann solvers based on the clear-water conservation equations for conservative finite-volume discretization. Considering nonequilibrium sediment transport, the wave structure is identical to that of the clear-water problem as the extra equation; the mass conservation of the bed is formally uncoupled.

The purpose of this work is to present a 2DH mathematical model suited for potentially discontinuous geomorphic flows over complex geometries. The model will be based on the clear-water conservation equations and will later be adapted to mobile bed simulations.

R. Canelas (✉) and R. Ferreira
CEHIDRO, Instituto Superior Técnico, ULT, Av. Rovisco Pais, Lisbon 1049-001, Portugal
e-mail: ricardo.canelas@ist.utl.pt; ruif@civil.ist.utl.pt

J. Murillo
Fluid Mechanics Department, C.P.S. University of Zaragoza, Zaragoza, Spain

P. Rowinski (ed.), *Experimental Methods in Hydraulic Research*,
Geoplanet: Earth and Planetary Sciences, DOI 10.1007/978-3-642-17475-9_8,

Dam-break flows constitute an interesting case of geomorphic flow, featuring sediment transport at high shear stresses, discontinuities, and critical flow points. They share features with other flows of the same class, such as avalanches, debris flows, and intense floods. Hence, in this work, dam-break flows will be object of mathematical description, in the hope to build a more generalized model.

The physical model presented was developed within the granular-fluid flow paradigm (reviews in Armanini et al. 2005; Ferreira 2005), attending to the properties of the granular material, the properties of fluid, and the viscous interaction between both. The numerical model is developed within the finite-volume framework and based on the model presented Murillo and García-Navarro (2010), allowing for a fully conservative simulation. The implementation is on-going.

2 Physical System

2.1 Conservation Equations

Geomorphic flows are a class of free-surface, shallow, multiphase flows (water and granular solid material), exhibiting high shear rates and shear stresses. The micro-mechanical characteristics of the sediments are relevant for the definition of the constitutive equations of the system.

The flow structure is represented in Fig. 1, where three main layers can be pointed out: A is a clean water layer, with low concentration of suspended load, where turbulent stresses are dominant; B is a transport layer, where the sediment concentration decreases as we move upwards in the flow and most stresses are of

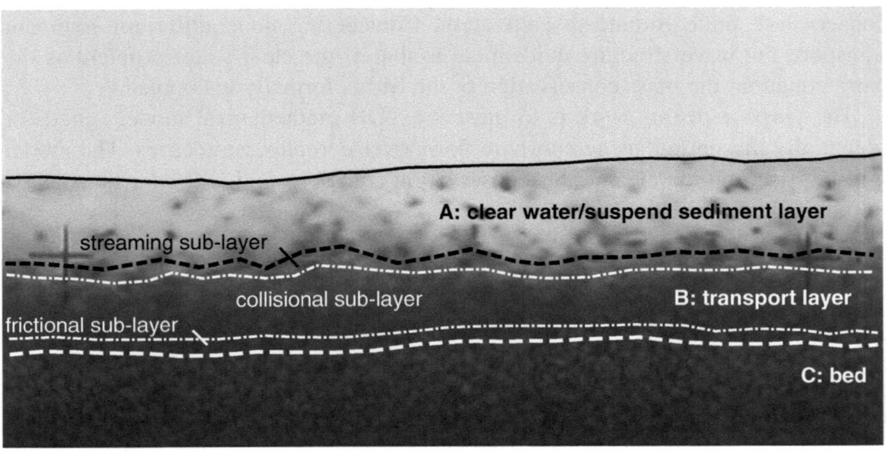

Fig. 1 Details in a stratified flow

collisional nature; and C is the bed, where horizontal movement of the sediments is not important.

Ferreira (2005, 2008) showed that the system of conservation equations describing geomorphic flows is structurally similar to the clean water system, where a new wave is introduced expressing the conservation of granular material in layer B. In this work, emphasis will be placed on showing applications of the model developed by Murillo and García-Navarro (2010). Hence, the conservation equations to be solved will be the clear-water shallow-water equations, i.e., no stratification or bedload will be considered and the influence of sediment density in the inertia and pressure terms will be disregarded. Additionally, boundary friction will be neglected as it is irrelevant to discuss the structure of the solutions. The total mass and momentum conservation equations in the x and y directions are thus:

$$\partial_t h + \partial_x(uh) + \partial_y(vh) = 0 \tag{1}$$

$$\partial_t(uh) + \partial_x\left(u^2h + g\frac{h^2}{2}\right) + \partial_y(uvh) = 0 \tag{2}$$

$$\partial_t(vh) + \partial_x(uvh) + \partial_y\left(v^2h + g\frac{h^2}{2}\right) = 0 \tag{3}$$

where h is the water depth and u and v are the flow velocities in x and y directions, respectively.

3 Domain Discretization: Meshing

3.1 Generation and Refinement

The numerical solution of the partial differential equations (1–3) requires the discretization of the computational domain in order to be able to replace the continuous differential equations with a system of simultaneous algebraic difference equations. In the finite-volume methods, the discretization is accomplished by dividing the computational domain into cells. The choice between a structured and an unstructured mesh nature follows from the nature of the work at hand; since the model must be prepared to deal with realistic, thus highly irregular domains, the unstructured type of mesh is chosen for this work.

The mesh generator employed was Gmsh, chosen for its built-in pre- and postprocessing capabilities. This generator provides three algorithms to generate unstructured surface meshes. The Frontal algorithm is preferred for the high quality of the output mesh.

This work uses conditions from two separate domains in order to locally control the mesh density, the spatial gradients from the hydrodynamic variables and from the bottom elevations, which are used to construct a background mesh for feedback to the mesh generator.

3.2 Merging of Altimetric Information with 2D Mesh

Since the 2D meshing procedures described are valid for the plane alone, the output elements from Gmsh have that coordinate set to 0 by default. If provided a Digital Terrain Model (DTM), which is a file with a set of points represented in 3D resulting of a topographical survey, it is possible to superimpose the mesh over the DTM and calculate altimetric coordinates for every point on the mesh. A simple algorithm, using linear interpolation on the DTM, was developed to efficiently estimate the elevation of the points in the mesh. Assuming that the DTM is regular (uniform spacing between orthogonal measured points), it is possible to devise a way to directly access the points in question in the altimetric matrix of the DTM that surrounds a point belonging to the mesh and interpolate information.

4 Numerical Discretization

4.1 Finite-Volume Discretization

The finite-volume method (FVM) was developed to represent conservation PDEs in the form of simple algebraic equations (reviews in LeVeque 1992; Hirch 2001; Toro 1999). In this method, volume integrals in a PDE that contains a divergence term are converted in surface integrals using Gauss's theorem. These terms can then be evaluated as boundary fluxes in each computational cell, and since the flux that enters one cell is the one that leaves the adjacent one, these methods are inherently conservative.

The hyperbolic, nonhomogeneous, first-order, quasi-linear system that expresses the conservation laws (1–3) can be written in vector notation as:

$$\partial_t(\boldsymbol{U}(\boldsymbol{V})) + \nabla.\boldsymbol{E}(\boldsymbol{U}) = \boldsymbol{H}(\boldsymbol{U}) \Leftrightarrow \partial_t(\boldsymbol{V}(\boldsymbol{U})) + \partial_x(\boldsymbol{F}(\boldsymbol{U})) + \partial_y(\boldsymbol{G}(\boldsymbol{U})) = \boldsymbol{H}(\boldsymbol{U}) \quad (4)$$

where $\boldsymbol{U} : \mathbb{R} \times]0, +\infty[\to \mathbb{R}^3$ is the independent conservatives variables vector; $\boldsymbol{V} : \mathbb{R}^3 \to \mathbb{R}^3$ is the primitive variables vector; $\boldsymbol{F} : \mathbb{R}^3 \to \mathbb{R}^3$ and $\boldsymbol{G} : \mathbb{R}^3 \to \mathbb{R}^3$ are the flux vectors in x and y, respectively; $\boldsymbol{H} : \mathbb{R}^3 \to \mathbb{R}^3$ is the source terms vector; and x, y, and t are space and time coordinates.

To obtain the discretization scheme, (4) is integrated in a computational cell, i, and Gauss's theorem is applied. This yields:

$$\partial_t A_i \langle \boldsymbol{V}_i \rangle + \int_{\Omega_i} \nabla \cdot \boldsymbol{E}(\boldsymbol{U}) dS = A_i \langle \boldsymbol{H}_i \rangle \quad (5)$$

$\boldsymbol{E} \cdot n = \boldsymbol{F}i + \boldsymbol{G}j$ with $n = (i, j)^T$. A_i is the cell area and the $\langle\rangle$ operator represents spatial average. Integrating along the n_i edges of the i cell and performing a first-order time integration, (5) becomes:

$$A_i \frac{\Delta\langle \boldsymbol{V}_i \rangle}{\Delta t} + \sum_{k=1}^{n_i} L_k \Delta_{ik} \langle \boldsymbol{E} \cdot n \rangle = A_i \langle \boldsymbol{H}_i \rangle \tag{6}$$

where L_k is the length of the k edge.

The fluxes through the k edge of cell i represent, in the simplest interpretation, the differences between the values of the independent variables on the adjacent cells i and j, separated by edge k. Δ_{ik} represents that difference, through edge k from cell i to j. The flux variations can be expressed as a function of the independent conservative variables using an approximate Jacobian matrix, $\underset{n,ik}{\widetilde{\boldsymbol{J}}}$, orthogonal to the edge in question:

$$\Delta_{ik} \langle \boldsymbol{E} \cdot n \rangle = \underset{n,ik}{\widetilde{\boldsymbol{J}}} \Delta_{ik} \langle \boldsymbol{V} \rangle \tag{7}$$

Such approximation arises from the fact that, for the shallow-water-type equations, even if the problem is hyperbolic, the flux vectors are not homogeneous functions of the dependent variables ($\boldsymbol{F} \neq \boldsymbol{J}(\boldsymbol{U})$). Therefore, it is not possible to use the natural Jacobian of the system, and an approximated matrix, built on the basis of a local linearization, must be used (Roe 1981).

The eigenvalues and the eigenvectors of the approximate Jacobian matrix are given by:

$$\tilde{\lambda}_{ik}^{(1)} = (\tilde{u} \cdot n + \tilde{c})_{ik}, \quad \tilde{\lambda}_{ik}^{(2)} = (\tilde{u} \cdot n)_{ik}, \quad \tilde{\lambda}_{ik}^{(3)} = (\tilde{u} \cdot n - \tilde{c})_{ik} \tag{8}$$

$$\tilde{\boldsymbol{e}}_{ik}^{(1)} = \begin{bmatrix} 1 \\ \tilde{u} + \tilde{c}|i| \\ \tilde{u} + \tilde{c}|j| \end{bmatrix}_{ik}, \quad \tilde{\boldsymbol{e}}_{ik}^{(2)} = \begin{bmatrix} 1 \\ -\tilde{c}|j| \\ \tilde{c}|i| \end{bmatrix}_{ik}, \quad \tilde{\boldsymbol{e}}_{ik}^{(3)} = \begin{bmatrix} 1 \\ \tilde{u} - \tilde{c}|i| \\ \tilde{u} - \tilde{c}|j| \end{bmatrix}_{ik} \tag{9}$$

The approximate variables are given by simple arithmetic means (Roe 1981).

$$\tilde{u}_{ik} = \frac{u_i\sqrt{h_i} + u_j\sqrt{h_j}}{\sqrt{h_i} + \sqrt{h_j}} \quad \tilde{v}_{ik} = \frac{v_i\sqrt{h_i} + v_j\sqrt{h_j}}{\sqrt{h_i} + \sqrt{h_j}} \quad \tilde{c}_{ik} = \sqrt{g\frac{h_i + h_l}{2}} \tag{10}$$

The dependant variables variations are projected on a new base, formed by the eigenvectors of the system:

$$\Delta\langle \boldsymbol{V} \rangle = \sum_{n=1}^{3} \alpha_{ik}^{(n)} \widetilde{\boldsymbol{e}}_{ik}^{(n)} \tag{11}$$

where the $\alpha_{ik}^{(n)}$ coefficients, the wave strengths, are:

$$\begin{aligned}
\alpha_{ik}^{(1)} &= \frac{\Delta_{ik}\langle h_i\rangle}{2} - \frac{1}{2\widetilde{c}_{ik}}\left(\Delta_{ik}\langle u_i h_i\rangle - \widetilde{u}_{ik}\Delta_{ik}\langle h_i\rangle\right) n_{ik} \\
\alpha_{ik}^{(2)} &= \frac{1}{\widetilde{c}_{ik}}\left(\Delta_{ik}\langle u_i h_i\rangle - \widetilde{u}_{ik}\Delta_{ik}\langle h_i\rangle\right) t_{ik} \\
\alpha_{ik}^{(1)} &= \frac{\Delta_{ik}\langle h_i\rangle}{2} + \frac{1}{2\widetilde{c}_{ik}}\left(\Delta_{ik}\langle u_i h_i\rangle - \widetilde{u}_{ik}\Delta_{ik}\langle h_i\rangle\right) n_{ik}
\end{aligned} \tag{12}$$

where $t = (-j, i)^{\mathrm{T}}$.

4.2 Flux Vector Splitting

FVS aims at generalizing upwinding schemes for nonlinear systems and uses the fact that the flux differences can be split explicitly has incoming and outgoing contributions for each cell, where $\lambda^{\pm} = \frac{1}{2}(\lambda \pm |\lambda|)$. If the fluxes are expressed in the local system eigenvector base, we have:

$$\Delta_{ik}\langle \boldsymbol{E}\cdot n\rangle = \sum_{n=1}^{3} \widetilde{\lambda}_{ik}^{(n)}\,\alpha_{ik}^{(n)}\,\widetilde{\boldsymbol{e}}_{ik}^{(n)} = \sum_{n=1}^{3}\left(\widetilde{\lambda}_{ik}^{(n)}\,\alpha_{ik}^{(n)}\,\widetilde{\boldsymbol{e}}_{ik}^{(n)}\right)^{+} + \sum_{n=1}^{3}\left(\widetilde{\lambda}_{ik}^{(n)}\,\alpha_{ik}^{(n)}\,\widetilde{\boldsymbol{e}}_{ik}^{(n)}\right)^{-} \tag{13}$$

In order to estimate a time step, Δt, that ensures stability, a CFL condition is used. In 1D, the imposition is that Δt is chosen small enough so that there is no interaction of waves from neighboring Riemann problems. In 2D, the framework is the same and a geometrical parameter that is the equivalent to distance in the 1D case must be computed. According to the work of Murillo et al. (2006),

$$\chi_i = \frac{A_i}{\max(L_{ki})}; \quad \Delta t \leq CFL\cdot\Delta t^{\tilde{\lambda}}; \quad \Delta t^{\tilde{\lambda}} = \frac{\min(\chi_i, \chi_j)}{\max\left(\left|\tilde{\lambda}^m\right|\right)} \tag{14}$$

It is however possible to find values of $h < 0$ for some region of the domain, which corresponds to a nonphysical solution.

Murillo (2006) defines two parameters that represent the intermediate state of the solution of the local Riemann problem, h_i^* and h_j^{***}:

$$h_i^* = h_i + \alpha_{ik}^{(1)} - \beta_{ik}^{(1)}/\tilde{\lambda}_{ik}^{(1)}; \quad h_j^{***} = h_j - \alpha_{ik}^{(3)} - \beta_{ik}^{(3)}/\tilde{\lambda}_{ik}^{(3)} \tag{15}$$

In order to control the occurrence of $h < 0$, the Δt can be locally defined, in order to contain the influence of the RP. Only in subcritical flow cases does this need to be computed.

$$\Delta t^{***} = \frac{\chi_j}{2\tilde{\lambda}_k^3}\frac{h_j^n}{h_j^n - h_j^{***}}; \quad \Delta t^{*} = \frac{\chi_i}{2\tilde{\lambda}_k^1}\frac{h_i^n}{h_i^n - h_i^{*}}; \quad \Delta t = \min(\Delta t^{***}, \Delta t^{*}, \Delta t^{\tilde{\lambda}}) \tag{16}$$

4.3 Bottom Source Terms

The bottom source terms are dealt with in the same conservative manner: they are projected in the local eigenvectors base and split into positive and negative contributions:

$$A_i\langle \boldsymbol{H}_i\rangle \approx \sum_{k=1}^{3} L_k \left(\sum_{n=1}^{3} \left(\beta_{ik}^{(n)} \tilde{\boldsymbol{e}}_{ik}^{(n)}\right)^{+} + \sum_{n=1}^{3} \left(\beta_{ik}^{(n)} \tilde{\boldsymbol{e}}_{ik}^{(n)}\right)^{-} \right) \tag{17}$$

where

$$\beta_{ik}^{(1)} = -\frac{1}{2\tilde{c}_{ik}} \left(\frac{p_b}{\rho_w}\right)_k; \quad \beta_{ik}^{(2)} = 0; \quad \beta_{ik}^{(3)} = -\beta_{ik}^{(1)} \tag{18}$$

Special attention must be given in the actual definition of $(p_{\mathrm{b}}/\rho_{\mathrm{w}})_k$, in order to optimize the pressure balance in various cases, where a difference of performance in the alternative formulations of pressure force terms is noted (Murillo and García-Navarro 2010). This discretization should follow a steady-state equilibrium criteria, i.e., it must allow the scheme to preserve steady states such as still-water equilibrium (well-balanced scheme, LeVeque 2002).

These requirements on the source terms of the numerical solution can lead to extremely small time steps, as Δt^{*} or Δt^{***} can be various orders of magnitude smaller than $\Delta t^{\tilde{\lambda}}$. Such a situation can be avoided by means of a reconstruction of the approximate solution, forcing positive values on h_i^{*} and h_j^{***} by reducing the numerical source term instead of the time step size (Murillo and García-Navarro 2010).

4.4 Wetting and Drying Algorithm

Regions with updated negative values of water depth near wet/dry interfaces, with or without discontinuous bed levels can occur. According to the Δt^{***} criteria in (16), the time step in those cases becomes nil. To ensure positivity and conservation in the solution in all cases, a redistribution of the fluxes is proposed. The flux in a general intercell edge k is computed as:

$$\text{if } h_j^n = 0 \text{ and } h_j^{***} < 0, \text{ set} \begin{cases} (\Delta\langle E\cdot n\rangle - \langle H\cdot n\rangle)_{i,k}^{-} = (\Delta\langle E\cdot n\rangle - \langle H\cdot n\rangle)_{i,k} \\ (\Delta\langle E\cdot n\rangle - \langle H\cdot n\rangle)_{j,k}^{-} = 0 \end{cases} \tag{19}$$

$$\text{if } h_i^n = 0 \text{ and } h_i^{*} < 0, \text{ set} \begin{cases} (\Delta\langle E\cdot n\rangle - \langle H\cdot n\rangle)_{j,k}^{-} = (\Delta\langle E\cdot n\rangle - \langle H\cdot n\rangle)_{j,k} \\ (\Delta\langle E\cdot n\rangle - \langle H\cdot n\rangle)_{i,k}^{-} = 0 \end{cases} \tag{20}$$

Otherwise, normal updating is used.

4.5 Conservative Entropy Fix

To avoid nonphysical results derived from the linearization, an entropy correction to filter sonic rarefactions is needed. A version of the Harten–Hyman entropy fix is applied (LeVeque 2002).

In the case of a left transonic rarefaction, characterized by $\lambda_i^1 < 0 < \lambda_j^1$, with $\lambda_i = \lambda(\boldsymbol{U}_i)$ and $\lambda_j = \lambda(U_j)$, the initial jump associated to $\tilde{\lambda}_k^{(1)}$ is decomposed in two new jumps:

$$\bar{\lambda}_k^1 = \lambda_i^1 \frac{\left(\lambda_j^1 - \tilde{\lambda}_k^1\right)}{\left(\lambda_j^1 - \lambda_i^1\right)}; \quad \hat{\lambda}_k^1 = \lambda_j^1 \frac{\left(\tilde{\lambda}_k^1 - \lambda_i^1\right)}{\left(\lambda_j^1 - \lambda_i^1\right)} \tag{21}$$

with $\bar{\lambda}_k^1 + \hat{\lambda}_k^1 = \tilde{\lambda}_k^1$, preserving the original value of the state U_i^* and, in consequence, the stability region. For a right transonic rarefaction, $\lambda_i^3 < 0 < \lambda_j^3$, the procedure is analogous, this time conserving the U_i^{***} state and its stability region.

$$\bar{\lambda}_k^3 = \lambda_i^3 \frac{\left(\tilde{\lambda}_k^3 - \lambda_i^3\right)}{\left(\lambda_j^3 - \lambda_i^3\right)}; \quad \hat{\lambda}_k^3 = \lambda_i^3 \frac{\left(\lambda_j^3 - \tilde{\lambda}_k^3\right)}{\left(\lambda_j^3 - \lambda_i^3\right)} \tag{22}$$

4.6 Numerical Scheme

The final formulation of the scheme is written as:

$$\langle \boldsymbol{V}_i \rangle^{m+1} = \langle \boldsymbol{V}_i \rangle^{m} - \frac{\Delta t}{A_i} \sum_{k=1}^{3} L_k \left\{ \sum_{n=1}^{3} \left(\left(\tilde{\lambda}_{ik}^{(n)} \alpha_{ik}^{(n)} - \beta_{ik}^{(n)} \right) \tilde{\boldsymbol{e}}_{ik}^{(n)} \right)^{-} \right\} \tag{23}$$

5 Results

Numerical solutions for the Riemann problem are presented. Four dam-break problem examples are analyzed, with the intention of demonstrating some properties of the scheme, namely the effects of being of first order in its diffusive behavior, its ability to cope with dry bed and wetting fronts, and the effectiveness of the entropy corrections.

The mesh was unstructured with an average cell side of 1.0 m. The value of the CFL number was 0.9.

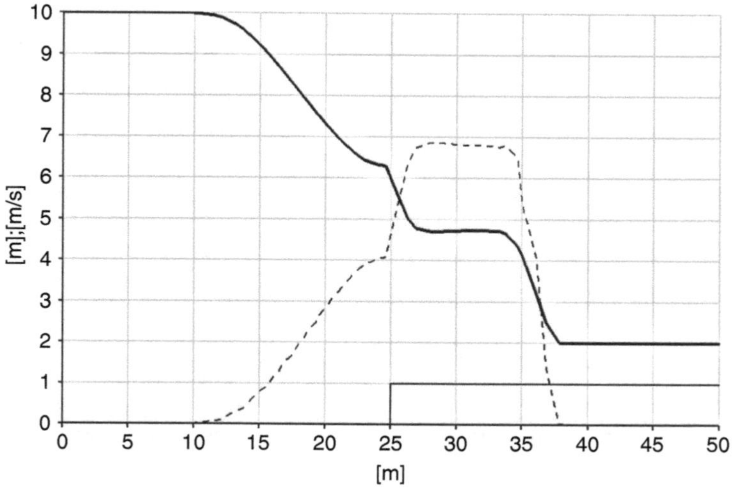

Fig. 2 Example 1 – Parameters: $h_L = 10.0$ m, $h_R = 1.0$ m, $Y_{bl} = -1.0$ m, $\alpha = 0.2$, and $\delta = 0$. *Solid line* is the height profile (m), *dashed line* the velocity (m/s), and *thin solid line* is the bottom profile; $t = 1$ s

The fundamental nondimensional parameters that describe the initial conditions are Ferreira et al. (2006):

$$\alpha \equiv \frac{h_R + \min(0, Y_{bl})}{h_L + \max(0, Y_{bl})}; \quad \delta \equiv \frac{Y_{bl}}{h_L + \max(0, Y_{bl})} \tag{24}$$

where h_R and h_L are the water heights of the initial right and left states, respectively, and Y_{bl} is the bed elevation on the left side considering the right side the reference horizontal plane.

Example 1 is a 1D result. The initial left state was a 10.0 m water column. The initial right state comprised a 1.0 m water column over a 1.0 m bed step. The solution, presented in Fig. 2, contains a left-moving rarefaction wave, a stationary shock at the step, and a right-moving shock wave. The presence of the step leads to a reduction of the discharge to the right side of the domain, induced by a head loss in the stationary shock. The position of the fan expansion and the shock are correct (cf. Alcrudo and Benkhaldoun 2001) and no numerical oscillations are produced.

Example 2 is a 1D simulation with dry bed. The initial left state had $h_i = 10.0$ m. The solution, presented in Fig. 3, contains a left-moving rarefaction wave and a right-moving wetting advancing front. It is comparable to an exact Ritter solution. The scheme being first order, the solution presents itself smoothed, without the typical discontinuity present in the exact solution. The velocity profile shows that the maximum speed is near 14 ms^{-1}, against the 19.7 ms^{-1} given by the Ritter solution. This can result from the use of a coarse mesh, together with the tendency of the scheme to present numerical diffusion and the design of the wetting–drying algorithm. A correct behavior is observed in the zone where a sonic rarefaction was expected, showing the effectiveness of the entropy corrections.

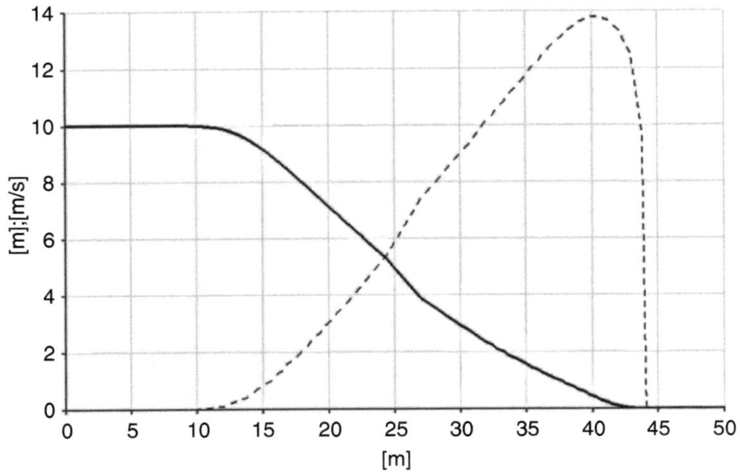

Fig. 3 Example 2 – Parameters: $h_L = 10.0$ m, $h_R = 0.0$ m, $Y_{bl} = 0.0$ m, $\alpha = 0$, and $\delta = 0$; $t = 1$ s

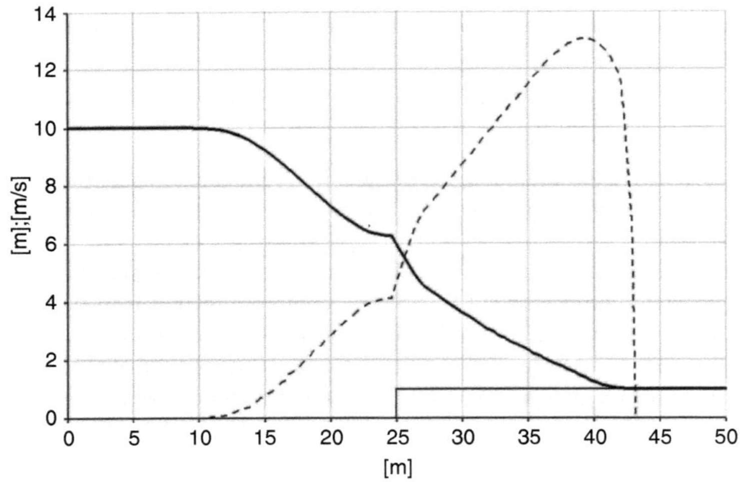

Fig. 4 Example 3 – Parameters: $h_L = 10.0$ m, $h_R = 0.0$ m, $Y_{bl} = -1.0$ m, $\alpha = 0.1$, and $\delta = 0$; $t = 1$ s

Example 3 in Fig. 4 is a 1D dry bed dam-break type problem, with a negative 1.0 m bed step. The solution presents a left-moving rarefaction wave, a stationary shock at the step and right-moving wetting advancing front. Again, the presence of the step leads to a reduction of the discharge to the right side of the domain, induced by the head loss in the stationary shock. The position of the fan expansion and the shock remains correct, and no numerical oscillations are produced.

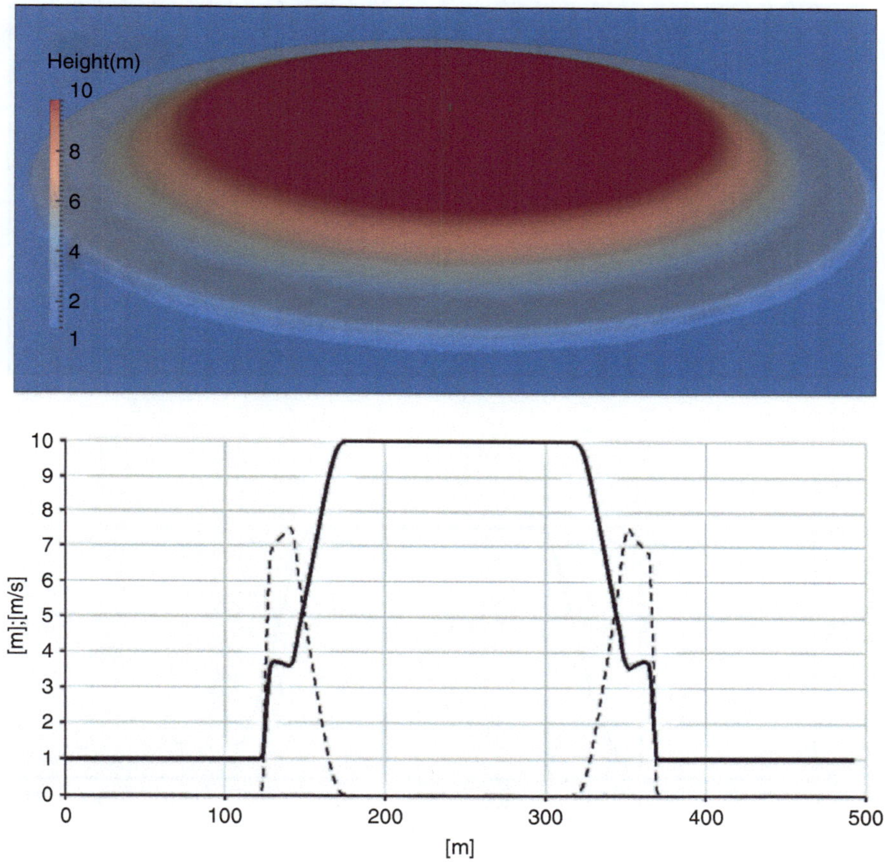

Fig. 5 Example 4(a) – Circular dam-break. 3D view and profile plot. $h_L = 10.0$ m, $h_R = 1.0$ m, $Y_{bl} = 0.0$ m, $\alpha = 0.1$, and $\delta = 0$; $t = 2$ s

Example 4, in Figs. 5 and 6 is a 2D dam-break case, a circular explosion. In case (a), the bed is wet, and a shock wave propagating in the bed layer can be seen; in (b), a dry bed and a wetting front is presented.

6 Conclusions

An approximate Riemann solver adapted to the structure of a system describing a highly unsteady discontinuous flow over complex geometries has been presented, where a new wave associated to the source terms is added.

Two modifications to the original Roe scheme are used: A reformulation of the stability condition that generalizes the classical CFL condition by including the

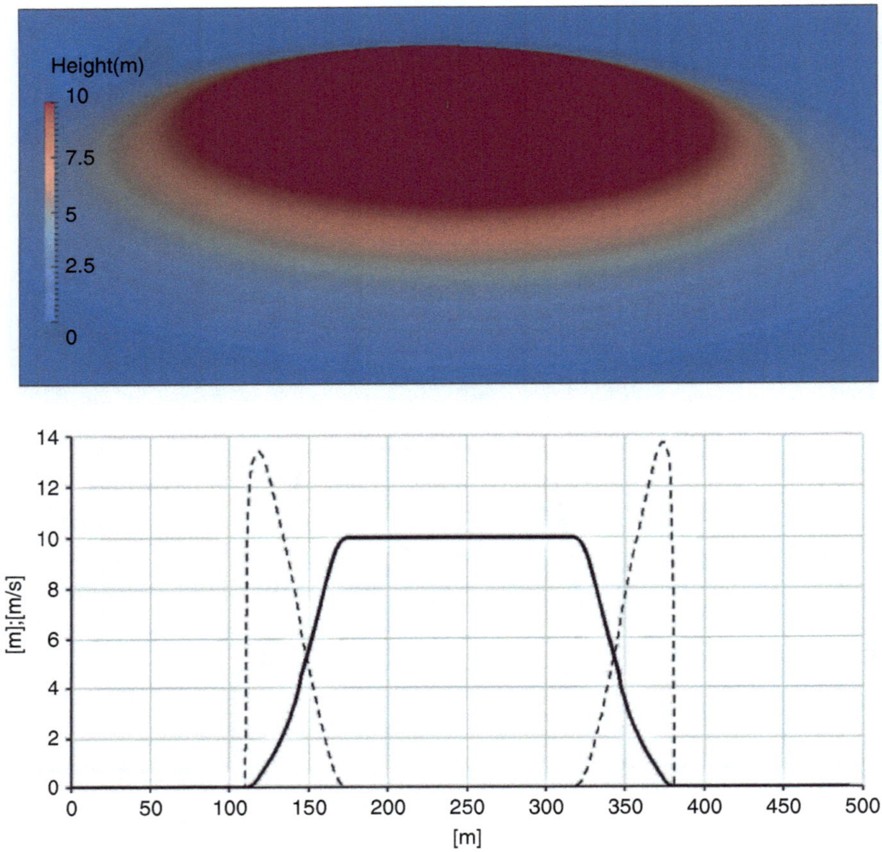

Fig. 6 Example 4(**b**) – Circular dam-break. $h_L = 10.0$ m, $h_R = 0.0$ m, $Y_{bl} = 0.0$ m, $\alpha = 0$, and $\delta = 0$. 3D view and profile plot; $t = 2$ s

influence of the source terms and the initial conditions in order to avoid nonphysical solutions, and a local redefinition of the numerical scheme in order to accurately overcome other nonphysical solutions with independence of the time step.

The model is expected to be fully compatible with debris flows with minor adaptations, due to the similar structure of the systems. Flows presenting simple quasi-Newtonian rheologies, like mud-flows, are also within the modeling capabilities of the scheme.

Further work is expected in order to implement proper inlet and outlet boundary conditions, as well as the development and implementation of an uncoupled mobile bed solver, where a weak discontinuity associated with the bedload will be introduced in the solution.

Acknowledgment This work has been partially supported by the FCT-funded project PTDC/ECM/65442/2006.

References

Alcrudo F, Benkhaldoun F (2001) Exact solutions to the Riemann problem of the shallow water equations with a bottom step. Comput Fluids 30:643–671

Armanini A, Capart H, Fraccarollo L, Larcher M (2005) Reological stratification in experimental free-surface flows of granular liquid mixtures. J Fluid Mech V532:269–319

Ferreira R (2005) River morphodynamics and sediment transport, conceptual models and solutions. Instituto Superior Técnico, Universidade Técnica de Lisboa, Lisbon

Ferreira R (2008) Fundamentals of mathematical modeling of morphodynamic processes. Application to geomorphic flows. Taylor & Fancis, London

Ferreira R, Amaral S, Leal JGB, Spinewine B (2006) Discontinuities in geomorphic dam-break flows. In: River flow 2006. Taylor & Fancis, London

Hirch C (2001) Numerical computation of internal and external flows, vol I and II. Wiley, New York

LeVeque RJ (1992) Numerical methods for conservation laws. Birkhauser, Basel

LeVeque RJ (2002) Finite volume methods for hyperbolic problems. Cambridge University Press, Basel

Murillo J (2006) Two-dimentional finite volume numerical models for unsteady free surface flows, solute transport, and erosion/deposition processes. University of Zaragoza, Fluid Mechanics. C.P.S., PHD thesis, February 2006

Murillo J, García-Navarro P (2010) Weak solutions for partial differential equations with source terms: application to the shallow water equations. J Comput Physics 229(11)

Murillo J, García-Navarro P, Burguete J, Brufau P (2006) A conservative 2D modelo f inundation flow with solute transport over dry bed. Int J Numer Methods Fluids 52(10):1059–1092

Roe PL (1981) Approximate Riemann solvers. J Comput Phys 43:357–372

Toro EF (1999) Riemann solvers and numerical methods for fluid dynamics. Springer, Berlin

Laboratory Modeling of Buoyant Jet in a Rotating Fluid

Natalya Demchenko

1 Introduction

It is known that the bottom topography significantly influences the dynamics and structure of thermal fronts in the freshwater basins and brackish seas. Field observations of the thermally induced fronts in the Baltic Sea during spring heating have revealed that large-scale fronts in the southern and central Baltic are aligned with the coasts and isobaths; significant correlation between the sea-surface temperature and topography was found (Bychkova et al. 1987). This was corroborated also by an analysis of 9-month time series of infrared imagery in the northern and central Baltic (Kahru et al. 1995) In some regions of the Ladoga Lake, nonuniform topography leads to the deceleration of thermal bar (frontal zone, associated with the temperature of maximum density, $Tmd = 3.98°C$) and formation of the "secondary" thermal front, marked by 5°C-isotherm at the surface, with very sharp temperature gradients in the upper layers (Naumenko and Karetnikov 1993). Also, there are natural observations and laboratory and theoretical studies of the effects of Earth rotation on the vertical and horizontal convection in the ocean and atmosphere (Boubnov and Golitsyn 1995) showing the importance of these effects for geophysical convective flows.

The main goal of this work is to investigate, in the rotating laboratory tank with a sloping and horizontal bottom, how the Coriolis force and bottom topography influence the propagation of the surface warm buoyant jet in fresh water layer – a qualitative manifestation of the "late" stage of the thermal bar phenomenon in lakes and marginal seas (Demchenko and Chubarenko 2007).

N. Demchenko
P.P. Shirshov Institute of Oceanology of RAS, Atlantic Branch, Prospect Mira 1, Kaliningrad 236022, Russia
e-mail: ndemchenko@mail.ru

P. Rowinski (ed.), *Experimental Methods in Hydraulic Research*,
Geoplanet: Earth and Planetary Sciences, DOI 10.1007/978-3-642-17475-9_9,

2 Methods

Experiments were performed in a cylindrical tank with radius $R_{tank} = 0.3$ m, mounted on a rotating table (Fig. 1). The whole depth of the tank was 0.24 m. In half of experimental runs, the bottom was horizontal but in other half of them the tank was equipped by a sloping bottom (slope angle $\alpha = 39°$, fluid depth was increasing toward the center of the tank). The tank was filled with fresh distilled water. Experiments were performed: (1) with rotation, for rotation periods of 5, 10, 15 s and Coriolis parameter $f = 2.5$, 1.25, 0.8 s^{-1}, respectively; and (2) without rotation. To reproduce the warm buoyant jet, the circular heating cable belt of 0.01 m thickness was used. It was mounted on the top of the water layer along the round wall of the tank. Series of experiments were performed using various specific power supplies q (25.7, 13.1, 6.37 Wt/m). The water temperature was controlled near the wall, at the middle of the radial distance, and at the center of the tank by three calibrated thermometers. For visualization of the buoyant jet and surface currents, the electrochemical thymol blue indicator technique and small paper pellets were used.

At the beginning of each experimental run, the tank was filled with fresh water that contained thymol blue indicator with concentration of about 25 g/65 L of water. To avoid vertical convection, the water temperature in the tank was 1–2°C below the air temperature. Before the beginning of the experiments with rotation, the rotating frame was switched on and the fluid in the tank was brought to solid-body rotation state. After that, the video record was switched on; the heating cable was initiated as a heater. Due to the electrochemical reaction, thymol blue solution near

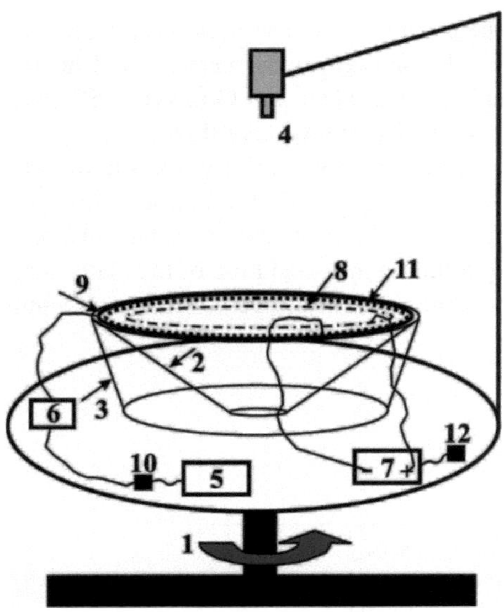

Fig. 1 Experimental setup. *1* – rotating table, *2* – slope, *3* – tank wall, *4* – video camera, *5* – autotransformator, *6* – ampermeter, *7* – voltage supply, *8* – copper plate (anode), *9* – heating resistance isolated cable, *10*, *12* – outlets, *11* – nichrome wire (cathode)

the heated cable became alkaline and changed its color, from maize yellow to dark blue. Radial propagation of warm buoyant jet from the heated cable toward the center of the tank was well marked by the blue color (Fig. 2a). The paper pellets were added to the free surface to visualize the azimuth flow and recalculate surface velocity fields. Data processing was performed using Tank Field Calculator, Excel, and Surfer.

3 Results

3.1 *Development of the Temperature Front in the Presence of Slope*

The laboratory experiments with rotation have shown that the radial velocity of the warm buoyant jet propagation toward the center of the tank depends on rotation rate (Coriolis parameter, *f*) and the rate of cable heating (specific power supply, *q*). In both cases (with and without rotation) the radial velocity of the jet increased with the rate of cable heating. It was revealed that in rotating fluid a strong along-wall cyclonic current is initiated (Fig. 2b).

The radial velocity of buoyant jet was one order less than in nonrotating fluid. This effect was produced by the Coriolis force that pressed the buoyant jet to the tank wall. In the presence of slope, along-shore currents were stable and kept

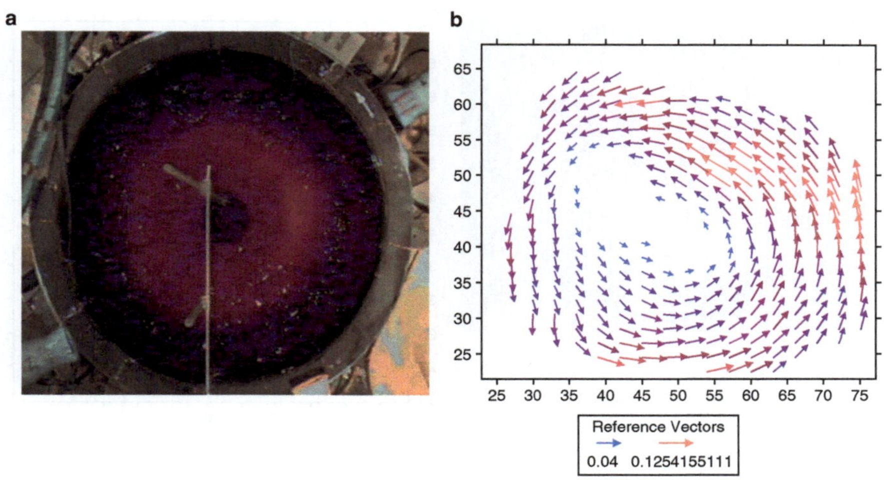

Fig. 2 Case with rotation, slopping bottom, view from above: (**a**) warm buoyant jet colored in dark blue propagating from the wall to the deep central part of the tank and (**b**) velocity field, calculated from paper pellets displacement using Tank Field Calculator program. Time from the start of the run is 35 min; specific power supply $q = 25.7$ Wt/m

axisymmetric shape during the entire experiment. Over the flat bottom, baroclinic eddies were formed, and the leading edge of the warm buoyant jet broke-up, becoming meandering. The radial propagation of the thermal front was related to lateral turbulent diffusion produced by eddies. In nonrotating fluid, the structure of the warm jet was threadlike due to the convective instability near the heater.

3.2 Development of the Temperature Front over Horizontal Bottom

The qualitative results of experimental runs with rotation have shown that over the horizontal bottom, the along-shore currents are unstable and meandering due to the baroclinic eddies (Fig. 3). Therefore, radial propagation velocity of the temperature front increases over horizontal bottom due to the lateral eddy diffusivity in comparison with the propagation velocity of axisymmetric buoyant jet over the slope. These effects are absent in nonrotating fluid in a tank with horizontal bottom. However, in the nonrotating fluid another type of buoyant jet instability was observed in both cases: with the horizontal and with sloping bottom. Due to that instability (of convective nature, as it was mentioned before) the propagating buoyant jet breaks down into the spatial periodic radial "strips"; so it becomes threadlike. Till that moment it is unknown how this instability affects the radial spreading of the buoyant jet.

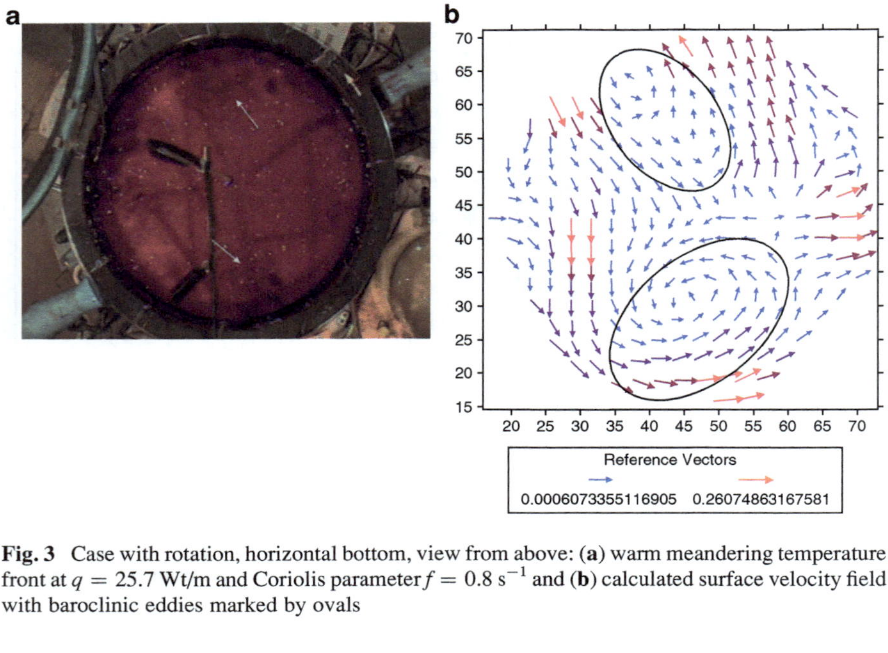

Fig. 3 Case with rotation, horizontal bottom, view from above: (**a**) warm meandering temperature front at $q = 25.7$ Wt/m and Coriolis parameter $f = 0.8\ \mathrm{s}^{-1}$ and (**b**) calculated surface velocity field with baroclinic eddies marked by ovals

4 Scaling Analysis

4.1 Rotating Case

The main goal of this section is to define the radial velocity of the jet propagation, U (m/s), in dimensionless form using the main external dimensional parameters: buoyancy flux (initiated by line source), B (m^3/s^3); Coriolis parameter, f (s^{-1}); depth of the tank, H (m); time, t (s); thermal diffusivity, k_T (m^2/s); and kinematical viscosity, ν (m^2/s).

Results of laboratory experiments have shown that the radial propagation velocity of the jet is proportional to the buoyancy flux, B, and Coriolis parameter, f, as $U \sim B^{1/2}$, $U \sim f^{-1}$, and that it does not depend on time. As a working fluid in all the experimental runs is fresh water, we can assume that the k_T and ν are nearly constant.

Thus, the main dimensionless parameters governing the process are the (flux) Rayleigh number, $\mathrm{Ra_F} = g\alpha F H^3/\rho_0 c_p \kappa_T^2 \nu$, and the Ekman number,

$$\mathrm{Ek} = \frac{\nu}{fH^2} \tag{1}$$

Here, g is the acceleration due to gravity (m/s^2); α the thermal expansion coefficient (°C^{-1}); F the heat flux (Wt/m); H the vertical length scale (m); ρ_0 the reference density (kg/m^3); c_p the water heat capacity (Wt/m kg °C); k_T the thermal diffusivity (m^2/s); ν the viscosity (m^2/s). The heat flux, F, causes the buoyancy fluxes, $B = F\alpha g/\rho c_p$ (m^3/s^3) into the surface layer. Thus, it is more convenient to use for the Rayleigh number the expression:

$$\mathrm{Ra}_B = \frac{BH^3}{\kappa_T^2 \nu} \tag{2}$$

Therefore, we may parameterize the velocity of the buoyant jet as follows:

$$\frac{U}{B^{1/3}} \sim \left(\mathrm{Ra}^{1/2}\mathrm{Ek}\right) \tag{3}$$

In the final form, the formula for the radial velocity of the jet is defined as:

$$\frac{U}{B^{1/3}} \sim \frac{B^{1/2}\nu^{1/2}}{k_T H^{1/2} f},$$

$$U \sim \left(\frac{B^{1/3}\nu}{k_T^2 H^{1/2} f}\right)^{1/2},$$

or, using the Pr number,

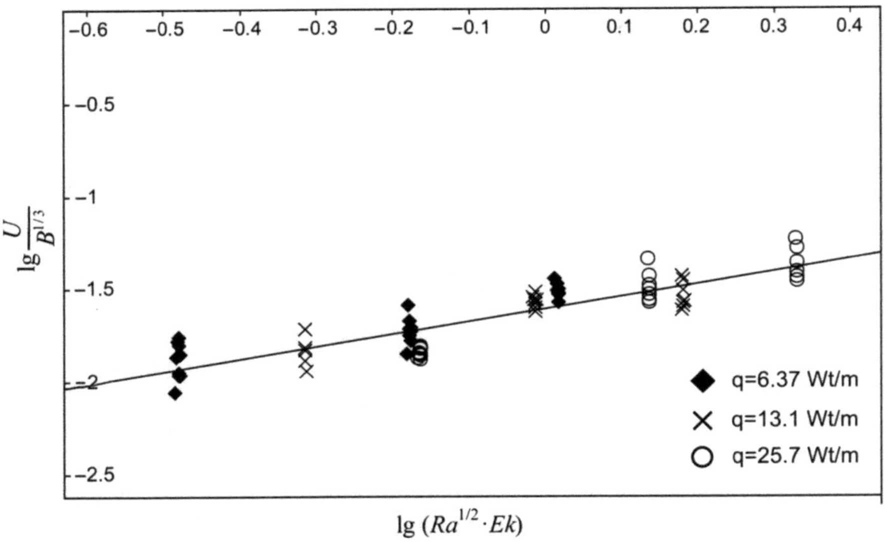

Fig. 4 Velocity of radial propagation of the warm buoyant jet, U, nondimensionalized by $U/B^{1/3}$ versus Rayleigh and Ekman numbers in double logarithmic scale. Data of the described experiments in the presence of slope configuration for rotation rates $f = 2.5, 1.25, 0.8\ \mathrm{s}^{-1}$

Table 1 Overview of the experimental parameters for the rotating case. The radius of the tank $R = 0.3$ m and the slope angle $\alpha = 39°$ is the same for all experiments

Type of markers (see Fig. 4)	q (Wt/m)	f (s^{-1})	B ($\mathrm{m}^3/\mathrm{s}^3$)
♦	6.37	2.5	$(3-3.3)\cdot 10^{-9}$
		1.25	$(3.13-3.35)\cdot 10^{-9}$
		0.8	$(3.1-3.4)\cdot 10^{-9}$
×	13.1	2.5	$(7.2-7.6)\cdot 10^{-9}$
		1.25	$(7.3-7.8)\cdot 10^{-9}$
		0.8	$(7.4-7.7)\cdot 10^{-9}$
○	25.7	2.5	$(1.5-1.6)\cdot 10^{-8}$
		1.25	$(1.52-1.6)\cdot 10^{-8}$
		0.8	$(1.5-1.56)\cdot 10^{-8}$

$$U \sim \left(\frac{B^{1/3}}{k_{\mathrm{T}} f^2 H}\right)^{1/2} \mathrm{Pr}^{1/2}, \tag{4}$$

where the Pr number is constant and for the experiments $\mathrm{Pr} \sim 7$.

The regularities for radial velocity of the warm buoyant jet in the rotating tank with sloping bottom are shown in Fig. 4. The solid line is the best fit of the predicted law (3) to the all-laboratory experiments with the high correlation coefficient $R^2 = 0.76$. The equation of the solid line $y = 0.76 \times -1.8$ gives clear dependence of the nondimensional velocity of the buoyant jet, $U/B^{1/3}$, versus Rayleigh and Ekman numbers as $U/B^{1/3} = C_1(\mathrm{Ra}^{1/2}\mathrm{Ek})$, where $C_1 = 1.2 \times 10^{-2}$. The high reliability of the linear approximation of the data shown in Fig. 4 proves the evident dependence of the radial propagation velocity of the buoyant jet on the buoyancy flux and Coriolis parameters (Table 1).

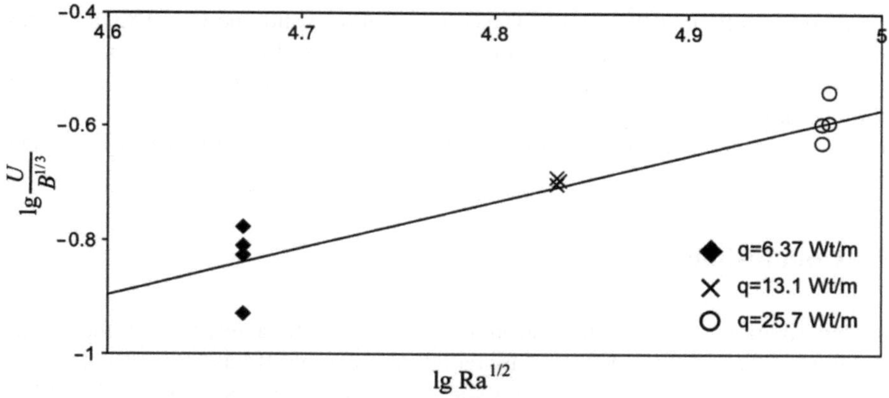

Fig. 5 Nondimensional radial velocity of the warm buoyant jet as a function of Rayleigh number in the double logarithmic scale

Table 2 Overview of the experimental parameters for nonrotating case. The radius of the tank $R = 0.3$ m and the slope angle $\alpha = 39°$ is the same for all experiments

Type of markers	q (Wt/m)	B (m^3/s^3)
♦	6.37	$(3.12 - 3.13) \cdot 10^{-9}$
×	13.1	$(6.4 - 6.6) \cdot 10^{-9}$
○	25.7	$(1.2 - 1.3) \cdot 10^{-8}$

4.2 Nonrotating Case

In the nonrotating case, the external dimensional parameters are the same as described in Sect. 4.1, excepting the Coriolis parameter (it is absent). Results of laboratory experiments have shown that the radial propagation velocity of the jet $U \sim B^{1/2}$ and that it is independent of time; k_T and v are nearly constant. Therefore, the main dimensionless parameter is (flux) Rayleigh number, $\mathrm{Ra}_B = BH^3/\kappa_T^2 v$, where B is the buoyancy flux (m^3/s^3); H the vertical length scale (m); k_T the thermal diffusivity (m^2/s); v the viscosity (m^2/s).

By analogy with Eq. 3 in Sect. 4.1, we should parameterize the velocity of the buoyant jet as follows:

$$\frac{U}{B^{1/3}} \sim \mathrm{Ra}^{1/2} \tag{5}$$

Thus, the formula for velocity jet is obtained in the final form:

$$U \sim \left(\frac{B^{1/3}H^3}{k_T^2 v}\right)^{1/2} \tag{6}$$

The results are presented in log–log scale in Fig. 5, which includes the data of all experimental runs in nonrotating case (Table 2). The solid line in Fig. 5 is the best fit of the predicted law (5) to the all-laboratory experiments with the high correlation coefficient $R^2 = 0.88$. The equation of the solid line $y = 0.83 \times -4.68$ gives clear dependence of the nondimensional velocity of the buoyant jet, $U/B^{1/3}$, versus Rayleigh number as $U/B^{1/3} = C_2\mathrm{Ra}^{1/2}$, where $C_2 = 2 \times 10^{-6}$. The high reliability of the linear approximation of the data shown in Fig. 5 proves an evident dependence of the radial propagation velocity of the warm jet on the buoyancy flux.

5 Conclusions

1. The rotation of the system strongly affects the characteristics of the warm buoyant jet and the velocity of its radial propagation. It has been revealed that in a rotating fluid a strong along-wall cyclonic current is formed. The radial velocity of buoyant jet propagation is about an order of magnitude less than in nonrotating fluid.
2. The bottom slope stabilizes the propagation of the temperature front of the buoyant jet, preventing for its breaking-up and formation of the baroclinic eddies. Upon the horizontal bottom, the radial propagation of the buoyant jet is effected and accelerated by the lateral eddy diffusivity.
3. Scaling analysis has shown that Rayleigh and Ekman numbers are the key nondimensional parameters that determine the regimes and regularities of the warm buoyant jet in the rotating case, but only the Rayleigh number is important for the nonrotating case. The processing of experimental data revealed that the nondimensional radial velocity of jet propagation depends on Rayleigh and Ekman numbers according to the formula $U/B^{1/3} = 1.2 \times 10^{-2}\mathrm{Ra}^{1/2}\mathrm{Ek}$ for the rotating case and the formula $U/B^{1/3} = 2 \times 10^{-6}\mathrm{Ra}^{1/2}$ for the nonrotating case.

Acknowledgments This work is supported by RFBR 10-05-90746_mob_st, 10-05-00472a. I would like to express great thanks to Drs. Irina Chubarenko and Andrei Zatsepin for stimulation of my work and valuable discussions, and to leading engineer of the Laboratory of Experimental Physics of the Ocean (P.P. Shirshov Institute of Oceanology, Moscow) Dmitrij Elkin for his technical assistance.

References

Boubnov BM, Golitsyn GS (1995) Convection in rotating fluids. Kluwer Academic, London, 224 pp

Bychkova IA, Viktorov SV, Losinsky VN (1987) Structure of coastal fronts of the Baltic Sea from remote sensing data of infra-red range. Proceedings of the 3rd congress of Soviet oceanographers, III, pp 64–65 (in Russian)

Demchenko N, Chubarenko I (2007) Coastal cooling/heating events based on laboratory experiments. Acta Geophys 55(1):56–64

Kahru M, Hakansson B, Rud O (1995) Distributions of the sea-surface temperature fronts in the Baltic sea as derived from satellite imagery. Cont Shelf Res 15(6):663–679

Naumenko MA, Karetnikov SG (1993) Using of the IR-information for the study of the Ladoga Lake thermal regime. Earth Investigation from Space, pp 69–78 (in Russian)

Analysis of Turbulent Flow Measurements in a Flume with Induced Upward Seepage

Oscar Herrera-Granados and Stanisław W. Kostecki

1 Introduction

Turbulence is considered one of the most interesting and not completely understandable phenomena in environmental sciences. Due to the fact that almost all the natural streams flow under turbulent regime, it is a necessity in hydraulic research to analyze many of the roles that turbulence plays in hydrodynamics. Thus, this chapter is focused on analyzing the influence of seepage flow in the free-surface hydrodynamics at the laboratory scale. These experimental works were carried out at the open-air hydraulic laboratory of the Wrocław University of Technology.

River flow is complicated because it depends on many different local conditions and on the interaction with the environment. Therefore, river flow is directly associated with turbulence, seepage, and sediment transport among other phenomena. The first part of this chapter presents at a glance the theoretical background of turbulence and its association with seepage (and an example that involves both seepage and river hydrodynamics). Afterwards, the experimental setup is described and the statistical analysis of several turbulent flow measurements is presented. This demonstrates that the presence of this small inflow (around 0.01% of the main flow) changes the open-channel hydrodynamics. Additionally, the statistical analysis of the flow measurements with and without seepage is compared.

1.1 Seepage and River Hydrodynamics

In rivers and natural streams, where the material of the hyporheic zone is constituted by coarse soils (such as mixtures of gravel and sand), the uppermost layer of the bed constitutes a porous medium and seepage flow can take place.

O. Herrera-Granados (✉) and S.W. Kostecki
Institute of Geotechnics and Hydro-engineering, Wrocław University of Technology, Wrocław, Poland
e-mail: Oscar.herrera-granados@pwr.wroc.pl; Stanislaw.kostecki@pwr.wroc.pl

P. Rowinski (ed.), *Experimental Methods in Hydraulic Research*, 163
Geoplanet: Earth and Planetary Sciences, DOI 10.1007/978-3-642-17475-9_10,

The hydrodynamic processes that occur in this zone are often neglected in river engineering. Nevertheless there are cases where the groundwater velocity field influences the open channel hydrodynamics (Herrera-Granados 2008b), especially in shallow waters with flow velocities that close to the bottom tend to be laminar.

In the majority of the European heavily modified rivers (according to the WFD classification), several hydraulics structures were built for river regulation, flood control, or for different socioeconomic purposes (Herrera-Granados 2008a). The water level, upstream these constructions, is higher than the water level downstream. Therefore, the difference between the downstream and tailwater levels represents a hydraulic head that can provoke seepage under or beside the structure (if the soil of the hyporheic zone is permeable enough). Hence, seepage cannot be neglected for such conditions because it changes not only the velocity field, but also the sediment transport rate and the bed forms (Herrera-Granados 2008b). Downstream, in the vicinity of this kind of structures, the direction of the seepage is practically upward. Hence, this chapter if focused on analyzing the influence of the seepage in the free-surface flow at the laboratory scale where upward groundwater flow was induced.

1.2 Theoretical Background about Turbulent Incompressible Flows

The Navier–Stokes Equations (NSE) give an accurate description of a great variety of fluid flows including turbulent flow with unordered seemingly chaotic fluid dynamics. For incompressible turbulent flows, the momentum (1) and the continuity (2) equations that describe the fluid's motion are:

$$\frac{\partial \boldsymbol{u}_i}{\partial t} + \boldsymbol{u}_j \frac{\partial \boldsymbol{u}_i}{\partial \boldsymbol{x}_j} = \upsilon \frac{\partial^2 \boldsymbol{u}_i}{\partial \boldsymbol{x}_j^2} - \frac{1}{\rho}\frac{\partial \boldsymbol{p}}{\partial \boldsymbol{x}_i} + \boldsymbol{g}_i \tag{1}$$

$$\frac{\partial \boldsymbol{u}_i}{\partial \boldsymbol{x}_i} = 0 \tag{2}$$

where $\boldsymbol{u}_i$ is the velocity component in the i-direction, $\boldsymbol{p}$ is the pressure, ρ is the fluid's density, υ is the kinematic viscosity, and $\boldsymbol{g}_i$ are the external body forces acting on the system in the i-direction. A full analysis of all the turbulence structures within the fluid is practically impossible in engineering and out of the scope of this chapter. Thus, the Reynolds averaging form of the NSE or RANS (Reynolds' Average Navier Stokes) equation (3) and its parameters are briefly mentioned.

RANS is the basis of the most popular numerical approaches for analyzing turbulent flows in engineering because not all the turbulent scales are resolved but modeled. This fact considerably reduces the computational costs and time. The RANS equation can be expressed in the following way (Herrera-Granados 2009):

$$\frac{\partial \boldsymbol{U}_i}{\partial t} + \boldsymbol{U}_j \frac{\partial \boldsymbol{U}_i}{\partial \boldsymbol{x}_j} = \frac{1}{\rho} \frac{\partial}{\partial \boldsymbol{x}_j} \left(-\boldsymbol{P}\delta_{ij} - \rho \overline{\boldsymbol{u}_i' \boldsymbol{u}_j'} \right) \tag{3}$$

where δ_{ij} is the Kronecker delta; the velocity is divided in two terms: its time averaged value $\boldsymbol{U}$ and its velocity fluctuation $\boldsymbol{u}'$. The last term of (3) represents the Reynolds stresses, which are modeled using the Boussinesq approximation (Herrera-Granados 2009):

$$-\rho \overline{\boldsymbol{u}_i' \boldsymbol{u}_j'} = \rho \, \nu_T \left(\frac{\partial \boldsymbol{U}_i}{\partial \boldsymbol{x}_j} + \frac{\partial \boldsymbol{U}_j}{\partial \boldsymbol{x}_i} \right) - \frac{2}{3} \rho \boldsymbol{k} \delta_{ij} \tag{4}$$

where $\boldsymbol{k}$ is the *turbulent kinetic energy* (TKE) and ν_{T} is the eddy viscosity. These are another two unknowns of the RANS equation that are to be solved based on the model closure (e.g., the Re-normalized group *RNG*, the standard *k*-ε or the *k*-ω model). Some of these turbulence parameters can be estimated in the laboratory.

2 Experimental Works

2.1 Experimental Setup

The laboratorial research was carried out at the open air laboratory of the Wrocław University of Technology. This laboratory is provided with two rectangular channels and the biggest of these flumes was used for the experiments (Fig. 1).

Fig. 1 The flume and the tank that provokes the induced seepage through the sandy layer

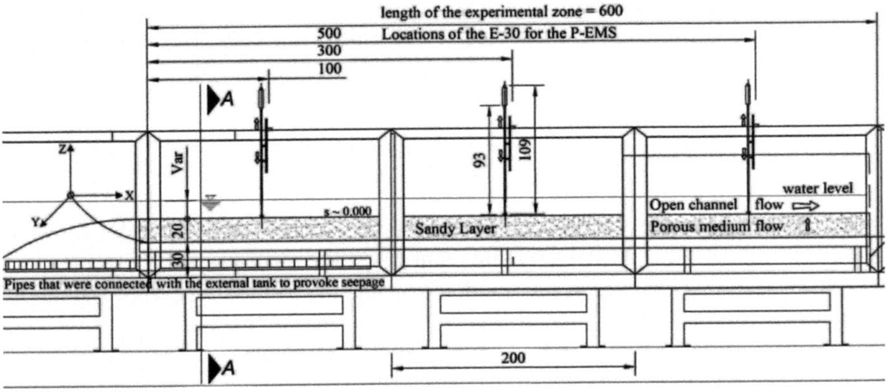

Fig. 2 General scheme of the flume at the laboratory (dimensions in centimeters)

The experimental zone was 8 m long, 0.5 m width, and 1.0 m high. The first 2 m of this experimental zone were used to stabilize the flow in order to address it into subcritical regime to the region where the measurements were taken (Herrera-Granados 2008a). Figure 1 shows the flume and the tank that provoked upward seepage. In order to allocate a 0.2-m porous bed (constituted by a sandy fine soil), a special structure was built 0.30 m over the lowest part of the flume to protect the small pipes at the bottom of the channel (Herrera-Granados 2008a). These pipes played an important role because the water from the external tank flowed through them and provoked distributed hydraulic pressures acting on the lowest part of the channel (Herrera-Granados 2008b).

The small lines under the sandy layer (shown in Fig. 2) represent the pipes that were connected to the external tank. The function of the tank was to feed with water the 0.3 m region under the sandy layer in order to gain a constant hydraulic head higher than the water level of the open-channel flow. The tank was provided with three hoses, one flowmeter, and a small spillway to control the induced hydraulic pressures provoking upward seepage through the porous sandy layer.

Figure 2 depicts the parts of the flume and the location of the velocimeter P-EMS in three cross sections of the flume ($X = 1.0$, 3.0, and 5.0 m, according to the established reference frame, which is depicted in the same figure). The P-EMS (with its accessory E-30) is able to measure the instantaneous velocities in two directions (x, y) with an accuracy of 0.001 m/s. The turbulent flow measurements were taken in 35 different points at each cross section. The general scheme of the external tank, its main components and the representation of the inducted hydraulic head (ΔH) as well as the view of the flume from the cross section A–A (see Fig. 2) with its geometry and the locations of the above-mentioned measurement points are depicted in Fig. 3. A thin layer of geotextile was allocated in between the sandy layer and the metal base that was symmetrically drilled to allow upward flow.

The geotextile let the pressure and consequently the seepage flow to be distributed uniformly in the whole porous medium.

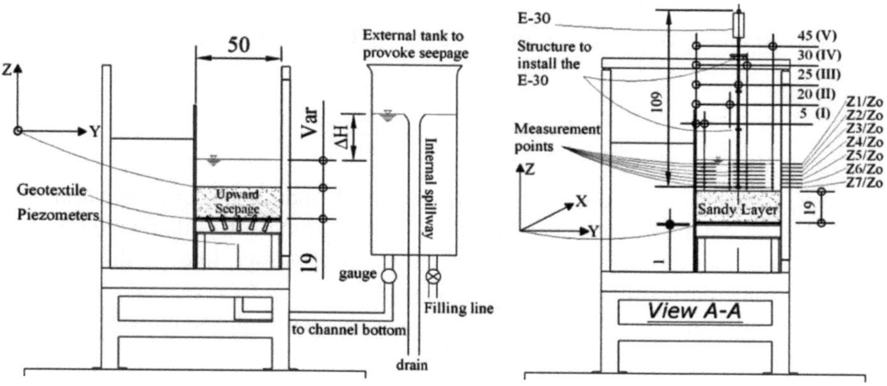

Fig. 3 General tank scheme and view of the flume from A–A (dimensions in centimeters)

2.2 *Statistical Analysis*

The influence of the seepage flow on the flow velocity components and velocity fluctuations of the open-channel flow was analyzed by estimating the statistical moments of the flow measurements; namely, the arithmetic mean, covariance, skewness, and kurtosis. Other turbulence parameters that were compared with and without seepage are the *Reynolds stresses* in the *xy*-plane $-\rho\overline{u'v'}$ and TKE.

In statistics, any random variable can be characterized by statistical moments of different orders. A statistical moment of *n*th order for data collected at one point is defined according to the expectation theory (Czernuszenko and Holley 2007):

$$\boldsymbol{E}[\xi^n] = \int\limits_{-\infty}^{\infty} \xi^n \boldsymbol{p}(\xi)\, d\xi \tag{5}$$

where $\boldsymbol{p}(\xi)$ is the probability density function (PDF) of the variable ξ. For discrete variables, the first statistical moment (expected value) is called the mean of the random variable (Czernuszenko and Holley 2007).

From this part of the chapter and to facilitate the identification of the flow velocity measurements, the velocity components in the *x*-, *y*-, and *z*-directions are denoted as $\boldsymbol{V}_x$, $\boldsymbol{V}_y$, and $\boldsymbol{V}_z$, respectively. Hence, the first statistical moment of the *i*-velocity component ($\boldsymbol{V}_i$) of an stationary ergodic event with an established time period T is calculated with the formula:

$$\boldsymbol{E}[\boldsymbol{V}_i] = \mu = \lim_{T\to\infty} \frac{1}{T} \int_{t_o}^{t_o+T} \boldsymbol{V}_i \, dt \tag{6}$$

Thus, considering the velocity component in the x-direction ($\boldsymbol{V}_x$) as a discrete random variable of $\boldsymbol{N}$ number of measurements, its expected value is estimated as:

$$\boldsymbol{E}[\boldsymbol{V}_x] = \mu_{V_x} = \frac{1}{N}\sum_{\alpha=1}^{N}\boldsymbol{V}_\alpha \tag{7}$$

where $\boldsymbol{V}_\alpha$ is the instantaneous velocity measurement, and μ represents the variable's mean. Figure 4 depicts one of many of the $\boldsymbol{V}_x$ and $\boldsymbol{V}_y$ time-series that were taken at the laboratory. In the same figure, the mean (red dot-dashed line) $\pm$ the standard deviation (blue dashed lines) are shown as well.

The second statistical moment or *variance* is defined in the same way based on the expectation theory as shown below (Reeve 2010):

$$variance = \boldsymbol{E}[(\boldsymbol{V}-\mu)^2] = \boldsymbol{s}^2 \tag{8}$$

The first statistical moment is probably the most popular measurement of central tendency; the second statistical moment is useful to know the dispersion of the random variable while the third (skewness) and fourth (kurtosis) moments describe the shape of the density function (Reeve 2010). The *kurtosis* (9) and *skewness* (10) are defined as well based on the expectation theory as follows:

$$skewness = \frac{\boldsymbol{E}\left[(\boldsymbol{V}-\mu)^3\right]}{\boldsymbol{s}^3} \tag{9}$$

$$kurtosis = \frac{\boldsymbol{E}\left[(\boldsymbol{V}-\mu)^4\right]}{\boldsymbol{s}^4} - 3 \tag{10}$$

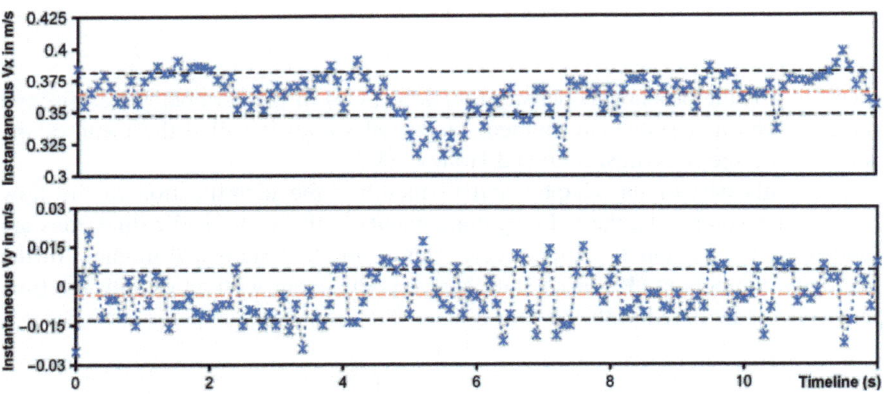

Fig. 4 $\boldsymbol{V}_x$ and $\boldsymbol{V}_y$ time-series at $X = 1.00$ m, $Y = 0.25$ m, and $Z/H = 0.5$ ($Q = 10\ \mathrm{dm^3s^{-1}}$)

Equation 10 is corrected by subtracting 3 in order to make the kurtosis of the normal distribution equal to zero (Czernuszenko and Holley 2007). In the case of turbulence parameters, the Reynolds stresses in the *xy*-plane $-\rho\overline{u'v'}$ are directly calculated from the velocity fluctuations while the TKE is defined in the (11a), but at the laboratory is estimated by the relation (11b) (Garcia et al. 2005):

$$\boldsymbol{k} \equiv \frac{1}{2}\overline{\boldsymbol{u}_i'\boldsymbol{u}_i'} \tag{11a}$$

$$\boldsymbol{k} = \frac{1}{2}\left(\boldsymbol{V}_x'^2 + \boldsymbol{V}_y'^2 + \boldsymbol{V}_z'^2\right) \tag{11b}$$

where $\boldsymbol{V}_x'^2$, $\boldsymbol{V}_y'^2$, and $\boldsymbol{V}_z'^2$ are the variance of the flow velocity components in *x*, *y*, and *z*, respectively. Nevertheless, a correction of the TKE is to be applied because the velocimeter used for the experiments only provided the velocity components in two directions. Hence, formula (11b) is corrected (Liiv and Lagemaa 2008):

$$\boldsymbol{k} \approx \frac{1.33}{2}\left(\boldsymbol{V}_x'^2 + \boldsymbol{V}_y'^2\right) \tag{12}$$

In the next section, a detailed statistical analysis of the turbulent flow measurements, for the flow rate $Q = 10\ \mathrm{dm^3s^{-1}}$ at the cross section $X = 1.0$ m without seepage and with two hydrostatic pressures that induced seepage through the flume's bed ($\Delta H = 10$ and 30 cm), is presented. The magnitude of the groundwater flow velocities measured at the laboratory oscillated between 3.3 and $4.6 \times E^{-5}\ \mathrm{ms^{-1}}$. Thus, this small seepage flow was around 0.011–0.015% of the open-channel flow.

2.3 Experimental Results

Figure 5 depicts the mean velocity profiles (five profiles according to Fig. 3) as a function of the relative depth at the cross section $X = 1.0$ m without seepage (Fig. 5a) and with induced hydraulic heads corresponding to 10 cm (Fig. 5b)

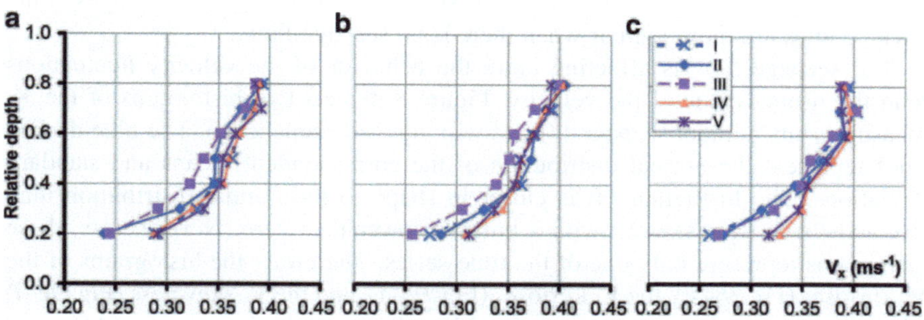

Fig. 5 V_x at $X = 1.0$ m, $Q = 10\ \mathrm{dm^3s^{-1}}$, and $\Delta H = 0$ (**a**), $\Delta H = 10$ (**b**) and $\Delta H = 30$ cm (**c**)

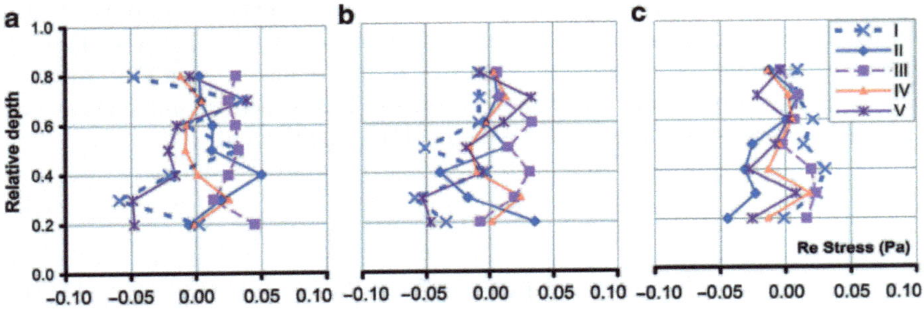

Fig. 6 Reynolds stress at $X = 1.0$ m, $Q = 10$ dm^{3}s^{-1}, and $\Delta H = 0$ (**a**), $\Delta H = 10$ (**b**), and $\Delta H = 30$ cm (**c**)

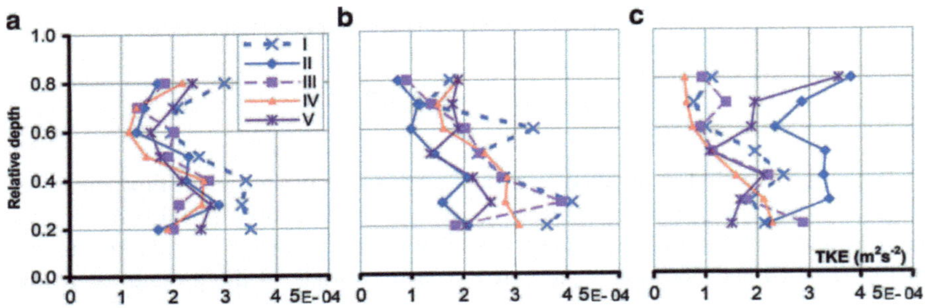

Fig. 7 TKE at $X = 1.0$ m, $Q = 10$ dm^{3}s^{-1}, and $\Delta H = 0$ (**a**), $\Delta H = 10$ (**b**), and $\Delta H = 30$ cm (**c**)

and 30 cm (Fig. 5c) acting in the bottom. Figure 6 depicts the profiles of the Reynolds stresses and Fig. 7 depicts the TKE as a function of the relative water depth.

Based on Fig. 5, there is no clear evidence that the seepage influences the open-channel hydrodynamics. Nevertheless, seepage is influencing the turbulence parameters as depicted in Figs. 6 and 7. In Fig. 6 it is possible to observe that the seepage is smoothing the gradient of the Reynolds stresses while in Fig. 7, the shapes of the TKE profiles are more regular when there is no seepage flow.

The seepage flow is affecting more the behavior of the velocity fluctuations than the mean values of the velocity. Figure 8 depicts two histograms of the $\boldsymbol{V}_x$ measurements without seepage (a) and with upward seepage (b). The blue dashed lines represent the normal distribution of the correspondent means and standard deviations. The histogram (a) is closer in shape to the normal distribution than (b), and (b) is right skewed with a kurtosis lower than zero. Nevertheless, these histograms represent only one of the time-series. Therefore, the histograms of the $\boldsymbol{V}_x$ kurtoses (Fig. 9a–c), the $\boldsymbol{V}_y$ kurtoses (Fig. 9d–f), and the $\boldsymbol{V}_x$ skewness (Fig. 9g–i) of all the flow measurements without ($\Delta H = 0$ cm) and with seepage ($\Delta H = 10$ and 30 cm) were categorized in 16 data classes and analyzed.

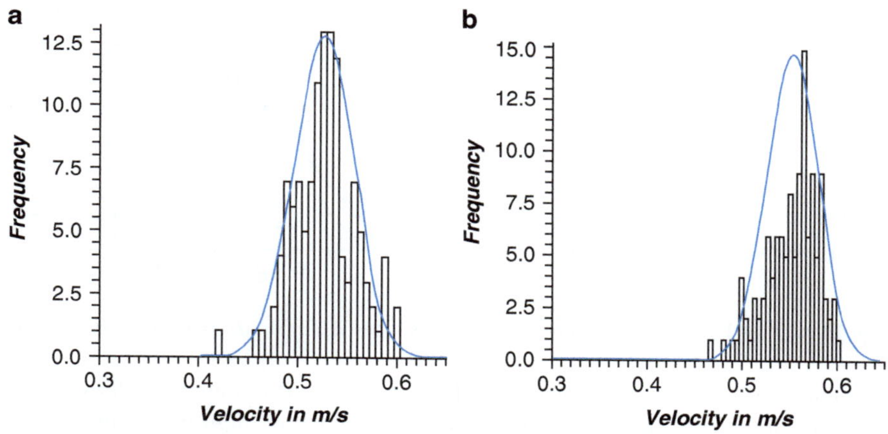

Fig. 8 Comparison of a histogram of V_x time-series without (**a**) and with (**b**) seepage

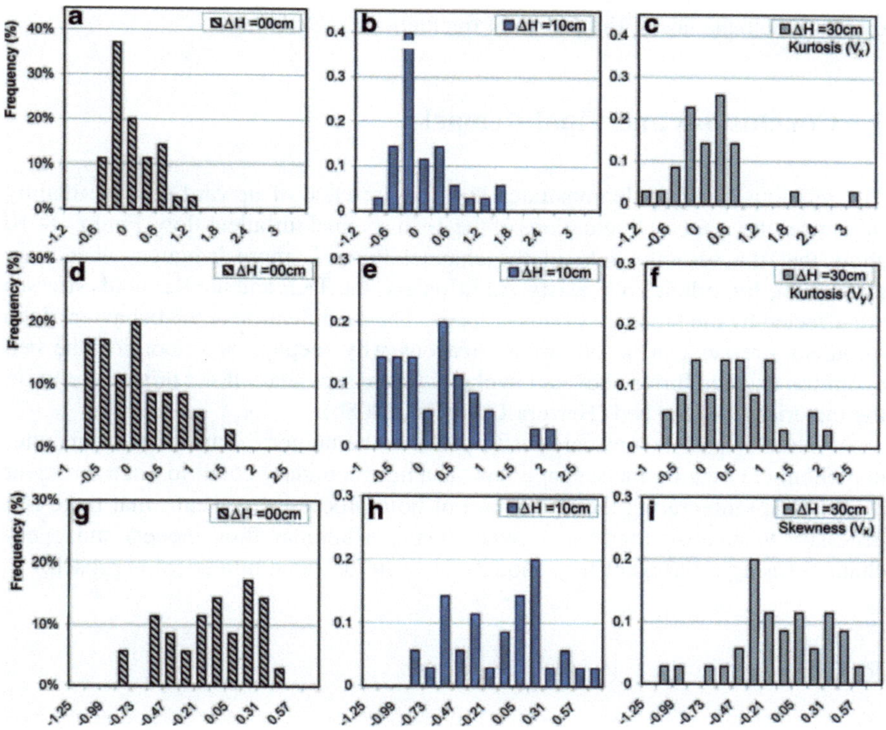

Fig. 9 Histogram of V_x kurtosis (**a–c**), V_y kurtosis (**d–f**), and V_x skewness (**g–i**) for $Q = 10\ \mathrm{dm^3 s^{-1}}$ and $\Delta H = 0$, $\Delta H = 10$, and $\Delta H = 30$ cm

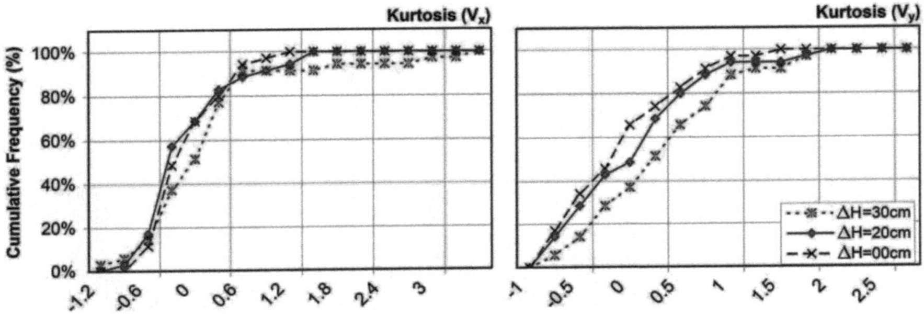

Fig. 10 CDF of V_x and V_y kurtosis for $Q = 10\ \mathrm{dm^3s^{-1}}$ and $\Delta H = 0$, $\Delta H = 10$, and $\Delta H = 30$ cm

Figure 9a–f shows that the values of the V_x and V_y kurtoses are spreader with the presence of seepage; hence the density functions are flatter. Figure 10 confirms this fact (looking at the kurtoses *cumulative density function*, CDF). The last three histograms of Fig. 9g–i show that the values of the V_x skewness (grouped in 10 data classes without seepage and in 13 with seepage) are spreader when seepage flows through the bed. Therefore, the density functions of the flow measurements are skewer when seepage exists. Hence, the histograms depicted in Fig 8 are typical cases of the behavior of the turbulent fluctuations.

3 Conclusions and Final Remarks

The previous analysis demonstrates that the presence of upward seepage mainly influences the velocity fluctuations of the open channel turbulent flow. Figures 8–10 show that the velocity field of the channel flows is more irregular when flow through the hyporheic zone exists. Additionally, the TKE and the Reynolds stresses are affected by the presence of this seepage. The modification in the behavior of the turbulent stresses can be one of the reasons why seepage is smoothing the bed evolution and bed forms above all in places where secondary flows arise and entrain the material from the bed (Herrera-Granados 2008b).

A future step of this research is to perform the numerical analysis of this phenomenon: (1) treating the seepage flow as a new boundary condition in a turbulent model or (2) integrating the interaction of both processes; it means that there is a necessity to analyze together seepage (using a laminar flow model) and open-channel (using a turbulent flow model) flows in the same numerical approach.

References

Czernuszenko W, Holley E (2007) Open-channel turbulence measurement with a three-component acoustic Doppler velocimeter. In: Transport phenomena in hydraulics. Polish Academy of Science, Warsaw, pp 49–79

Garcia CM, Jackson PR, Garcia MH (2005) Confidence intervals in the determination of turbulence parameters. Exp Fluids 40(4):514–522

Herrera-Granados O (2008a) The variation of the water's density due to sedimentation processes and its role in modeling open channel flow. In: Dohmen-Janssen M, Hulscher SJMH (eds) RCEM 2007. Taylor and Francis, London, pp 477–484

Herrera-Granados O (2008b) Effect of seepage flow on channel bed evolution at the laboratory scale. In: Altinakar MS et al (eds) River flow. Kubaba Congress Department, Izmir, pp 1117–1126

Herrera-Granados O (2009) Computational fluid dynamics in river engineering: a general overview. In: InterTech 2009. Poznań University of Technology, Poland, pp 259–263

Liiv T, Lagemaa P (2008) The variation of the velocity and turbulent kinetic energy field in the wave in the vicinity of the breaking point. Est J Eng 14(1):42–64

Reeve D (2010) Risk and reliability: coastal and hydraulic engineering. Spon Press, London

Changes of Sediment Distribution in a Channel Bifurcation – 3D Modeling

Leszek Książek and Douwe G. Meijer

1 Introduction

There are many factors involved in how suspended sediment and bed-load sediment are distributed over the branches of a bifurcation. Some of these factors are:

- Discharge Q_0 and its distributions Q_1 and Q_2 over the branches.
- Geometry of the bifurcation: cross sections A_0, A_1, and A_2 (area and shape); depths h_0, h_1, and h_2; widths B_0, B_1, and B_2; and angles α_1 and α_2, slope angles of the embankments.
- Streamlining of bifurcation.
- Conditions in the approaching channel: straight channel/bend flow, turbulence intensity (caused by bed roughness/structures) sediment characteristics, particle size and fall velocity, bed load/suspended transport.
- Sediment management measures: sills, guide vanes, sediment extracting measures.

Physical model studies were performed by Bulle (1926), Riad (1961), Den Dekker and Voorthuizen (1994), Islam (2000), and others. In general, it was found that sediment distribution is generally more than linearly related to the discharge distribution, a bigger diverting angle attracts more (bed load) sediment, and highly suspended sediment (wash load) is distributed more or less proportionally to the discharge distribution. Richardson and Thorne (2001), on the basis of field measurements point out that the circumstances leading to the bifurcation of a single channel are a topic of fundamental importance for understanding the physical processes responsible for braiding. The main channel downstream the bifurcation can carry less sediment load which may cause a degradation (Garde and Raju 2000).

L. Książek (✉)
Department of Hydraulic Engineering and Geotechnics, University of Agriculture in Krakow, Al. Mickiewicza 24/28, Kraków 30-059, Poland
e-mail: rmksiaze@cyf-kr.edu.pl

D.G. Meijer
Arcadis Netherlands, Piet Mondriaanlaan 26, Amersfoort 3812 GV, The Netherlands

P. Rowinski (ed.), *Experimental Methods in Hydraulic Research*,
Geoplanet: Earth and Planetary Sciences, DOI 10.1007/978-3-642-17475-9_11,

In the case the diversion takes out only water, the main channel will carry the same load with reduced discharge, which may result in aggradation.

In river morphology models, the bifurcation problem is usually approached by defining nodal-point relations in which the sediment distribution is a function of the discharge distribution (Fokkink and Wang 1993). However, the sediment distribution in a bifurcation is locally defined and strongly influenced by the local flow field. For this reason, nodal-point relations can only be found using prototype measurements, physical model studies, or higher order (three-dimensional) numerical computations.

Some of the nodal-point relations used in 1D morphological computations are given below.

$$\frac{S_1}{S_2} = \frac{Q_1}{Q_2} \tag{1}$$

with S_1, S_2 = sediment inflow into branch 1 and branch 2 ($m^3\ s^{-1}$), Q_1, Q_2 = discharges in branches 1 and 2 ($m^3\ s^{-1}$).

Equation (1) assumes proportionality between the sediment transport and the discharge in the bifurcation. This relation is quite plausible for highly suspended material that is nearly uniformly distributed over the channel cross section, such as fine sediment and silt. If there are significant vertical concentration gradients, for example, in channels with mainly bed-load transport, the sediment distribution is highly influenced by the local flow field and not necessarily proportional to the discharge distribution.

$$\frac{S_1}{S_2} = \alpha \frac{Q_1}{Q_2} + \beta \tag{2}$$

The formulation (2) is used in the 1D-modeling system (Vermeer 1993), with user-defined values for α and β.

Wang et al. (1993) propose the following configuration of nodal-point relations:

$$\frac{S_1}{S_2} = \left(\frac{Q_1}{Q_2}\right)^m \left(\frac{B_1}{B_2}\right)^n \tag{3}$$

with theoretically $n = 1 - m$.

For practical use, the exponents m and n in (3) are to be determined empirically and are dependent on local conditions. In alluvial channels, the inflow of sediment into a diverting branch can be in disharmony with the transport capacity further downstream, causing a change of the channel bed. This change can, as a second-order effect, again influence the sediment distribution. Alluvial bifurcations can be stable or unstable. Both types can be recognized in nature. If a bifurcation is unstable, one of the diverting branches eventually disappears. Examples of unstable bifurcations are branches in breeding rivers that mainly carry suspended load.

The experimental results of the distribution of sediment at channel bifurcations suggest that this distribution can be expressed by the general nodal-point relation

(Islam et al. 2006). A good knowledge of nodal-point relations is a basic condition for any kind of morphological study in which bifurcations are involved. Therefore, this study makes an attempt to find nodal-point relations by 3D flow modeling for some basic geometries. The relations will be formulated in terms of (2) and (3), in which α, β and m, n are analyzed as functions of the previously mentioned factors.

This study is limited to a few cases of the main channel with a diverting branch. The influence of the discharge distribution, the diverting angle on the distribution of the sediment is investigated (Meijer and Ksiazek 1994).

2 Material and Methods

2.1 Mathematical Model

FLUENT is a general-purpose package for 2D or 3D modeling of fluid flow. The model consists of two parts: preprocessor that allows grid generation and a main module for defining the physical models and fluid and material properties, and the boundary conditions (Fluent 1993).

The model simulates a range of physical phenomena by solving conservation equations for mass, momentum, and energy using a control volume-based finite volume method. It solves the Navier–Stokes equations as momentum conservations and describes turbulence using several turbulence models. The governing equations are discretized on a curvilinear grid to enable computations in complex geometries. A non-staggered system is used for storage of discrete velocities and pressures. The equations are solved using an algorithm with an iterative line-by-line matrix solver and multigrid acceleration or with the full-field iterative solver.

The model computes trajectories of particles in the flow. The trajectory of a particle is determined by its diameter, its Reynolds-dependent drag coefficient C_D, and the external forces caused by the flow and gravity. The fluctuating component of this force is related to the turbulence intensity. In this way, several identical particle releases can yield different trajectories.

2.2 Channel Geometries

Three bifurcation geometries are distinguished, in which the angle of diversion is varied. In addition, a basic geometry without a diverting channel is added for preliminary computations in order to judge the computational behavior of the grid, the numerical convergence, and the adaptation lengths of the inlet conditions. This results in the following geometries: channel 1 – without bifurcation, and channels 2, 3, and 4 – with bifurcation of 90°, 45,° and 135°, respectively (Fig. 1). The

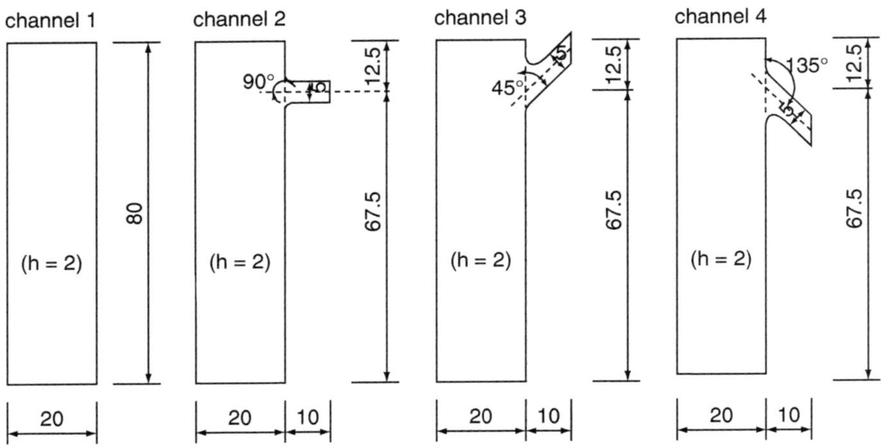

Fig. 1 Channel geometries (dimensions in meters)

geometrical parameters for all the channels are as follows: main channel width of 20 m and length of 80 m, diverting channel width of 5 m, and the water depth of 2 m.

The grid (86,428 cells) is made in curvilinear body-fitted coordinates in which the grid lines are determined by a coordinate system that conforms to the geometric boundary of model. The largest variations in the fluid flow, sediment transport, and turbulence occur in the region of the bifurcation and in the boundary layer, so grid refinement is applied in these regions.

For all computations, the following flow parameters are set: density of water $\rho = 1{,}000\ \mathrm{kgm}^{-3;}$ wall roughness $k_s = 0.003$ m; and sediment density $\rho_s = 2{,}650$ kgs^{-1}. The inflow discharge is 30 $\mathrm{m^3s^{-1}}$. At the inlet, a uniform flow profile is defined: $u(z) = 0.75\ \mathrm{ms}^{-1}$. The turbulent kinetic energy profile at the inlet is approximated by a linear function: $k(z) = k_0(1 - z/h)$, where $k_0 = u_*{}^2 c_\mu{}^{-0.5}$ and $u_* = (\tau_w \rho^{-1})^{0.5}$, in which c_μ = empirical constant in k–ε model ($c_\mu = 0.09$), h = flow depth (m), k_0 = turbulent kinetic energy above the wall boundary layer ($\mathrm{m^2/s^2}$), u_* = wall shear stress velocity (ms^{-1}), z = vertical coordinate (m), and τ_w = wall shear stress ($\mathrm{N/m^2}$) (Rodi 1993). The dissipation rate profile is approximated by the equation $\varepsilon(z) = \varepsilon(z_b) z_b / z(1 - z/h)$, where $\varepsilon(z) = u_*{}^3/\kappa z_b$, in which κ = von Karman's constant, z_b = reference level at the bed (equal to the vertical coordinate of the first computational cell; $z_b = 0.046$ m).

For channel 1, one flow computation is carried out, whereas for each of the channels with bifurcation, two discharge ratios are chosen with the following discharges and flow velocities in the branches:

$Q = 30\ \mathrm{m^3s^{-1}}$	$Q_1 = 5\ \mathrm{m^3s^{-1}}$	=>	$u_1 = 0.500\ \mathrm{ms^{-1}}$	(branch)
	$Q_2 = 25\ \mathrm{m^3s^{-1}}$	=>	$u_2 = 0.625\ \mathrm{ms^{-1}}$	(main)
$Q = 30\ \mathrm{m^3s^{-1}}$	$Q_1 = 10\ \mathrm{m^3s^{-1}}$	=>	$u_1 = 1.000\ \mathrm{ms^{-1}}$	(branch)
	$Q_2 = 20\ \mathrm{m^3s^{-1}}$	=>	$u_2 = 0.500\ \mathrm{ms^{-1}}$	(main)

In preliminary computations using channel 1, various particle tracking experiments were carried out (particle diameters: 0.01, 0.05, 0.1, 0.15, and 0.2 mm). Two diameters are chosen that characterize bed load and suspended sediment, respectively: $D_1 = 0.15$ mm ($w_1 = 20.2$ mms^{-1}) and $D_2 = 0.01$ mm ($w_2 = 0.09$ mms^{-1}).

Van Rijn (1993) defined, as a criterion for the extent of suspension of the sediment, the suspension parameter $Z = w/\kappa u_*$. For the two chosen particle sizes, this parameter has the values: $z_1 = 2.0$ (bed load and suspended sediment in near-bed region), $z_2 = 0.009$ (suspended sediment). The Shields parameter is defined as $\theta = \tau_w/(\rho_s - \rho)gD$, whereas the beginning of motion occurs if $\theta > \theta_{cr}$ with $\theta = 0.03 \sim 0.06$, depending on the Reynolds number. For the chosen particle sizes, the Shields numbers are $\theta_1 = 0.41$ and $\theta_2 = 5.98$. According to Shields diagram (Simons and Sentürk 1992), these particles will be moved by the flow.

The larger particles ($D_1 = 0.15$ mm) are released in the lowest 0.10 m of the inlet. The smaller ($D_2 = 0.01$ mm) are released uniformly over the inlet. The particle motion on the bottom is defined as saltation (jumping); sediment bounces along the stream bed by the energy and turbulence of the flow. Interaction between the particles is not modeled.

After each flow computation, two particle tracking computations are executed using different particle sizes (0.15 and 0.01 mm, respectively). This makes a total of 14 runs, each particle tracking being repeated three times using 1,000 particles so as to check the replicability of the results.

3 Results

The channel without bifurcation is established as a basic geometry to investigate the fluid flow and the movement of sediment particles. It allows to check the inlet conditions at the inlet and the development of the profiles of the flow velocity, the turbulent kinetic energy, the eddy dissipation, and the particle concentration.

A discharge of $Q = 30$ m^{3}s^{-1} shows a Reynolds number of $Re = 1.5 \times 10^6$, which indicates turbulent flow. The uniform velocity profile ($u = 0.75$ ms^{-1}) develops over approximately half the length of the channel to an equilibrium profile, which is in agreement with the theoretical log-profile.

At the inlet, the turbulent kinetic energy and dissipation rate profiles are set in accordance with function $k(z)$ and $\varepsilon(z)$, respectively. The adaptation to equilibrium profiles takes place mainly in the first 40 m, so the channel length is large enough for the adjustment of these profiles to equilibrium contours.

The turbulent kinetic energy has the highest value in the boundary layer at the bottom. At the water surface, the value is near zero, because the kinetic energy of turbulence is very small. For both the turbulent kinetic energy and the eddy dissipation, the maximum value appears in the first computational cell above the bottom.

The motion of sediment particles in the basic channel is shown in Fig. 2. The particle trajectories in Fig. 2a represent bed-load transport ($D_1 = 0.15$ mm). These

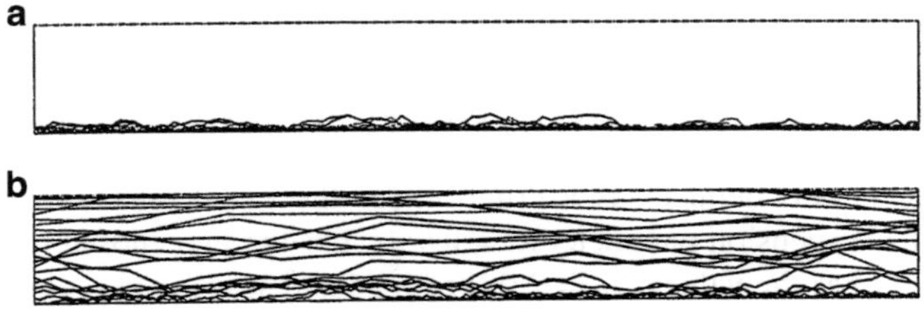

Fig. 2 Side view of the particle trajectories for channel without bifurcation: (**a**) as bed-load transport and (**b**) as suspended transport

particles are spread over the lowest 0.10 m of the cross section at the inlet. Particles fall onto the bottom, saltate, get into suspension again by the turbulence, and are transported further. However, the plot shows that the particles all stay in the near-bed region. The fall velocity is apparently too high for the particles to get into full suspension. Fig. 2b shows particle trajectories for suspended load ($D_1 = 0.01$ mm) which are spread uniformly over the channel cross section at the inlet. The crossing lines represent particle trackings at different y-coordinates. Some particles go up and others go down. The driving force on that motion is turbulence.

For both particle diameters, the concentration profiles of sediment at the inlet and the outlet are the same or nearly the same, which implies that those diameters represent well bed-load and suspended transport.

3.1 Bifurcation 90°: Runs 2a/b ($Q_1/Q_2 = 0.2$) and 2c/d ($Q_1/Q_2 = 0.5$)

The direction of the fluid motion in the bifurcation is shown by streaklines in Fig. 3. Streaklines visualize flow patterns by drawing continuous lines joining all fluid particles which originate from the same point in the fluid. Figure 3a shows streaklines that originate 0.01 m above the bottom at the inlet, and are released 1.93 m above the bottom (i.e., 0.07 m under the surface) for discharge $Q_1/Q_2 = 0.2$.

The fluid motion into the branch is different for both layers. At the bottom, eight streaklines are banded into the branch, whereas at the surface only four. This is caused by the difference in flow velocity between the bed region and the surface region. The driving force on the fluid motion into the branch, caused by the negative pressure gradient, bends the fluid particles at the bottom more easily into the branch because of their lower momentum (Fig. 4).

A more detailed analysis of the shape of the streaklines in the branch shows that the fluid moves up from the bottom in the left part of the channel, and down from the surface in the right part of the channel, which indicates a spiral motion of the fluid in the branch.

The behavior of the turbulence in the bifurcation region represented by kinetic energy of turbulence differs in the lowest and the highest cell layer (on the bottom

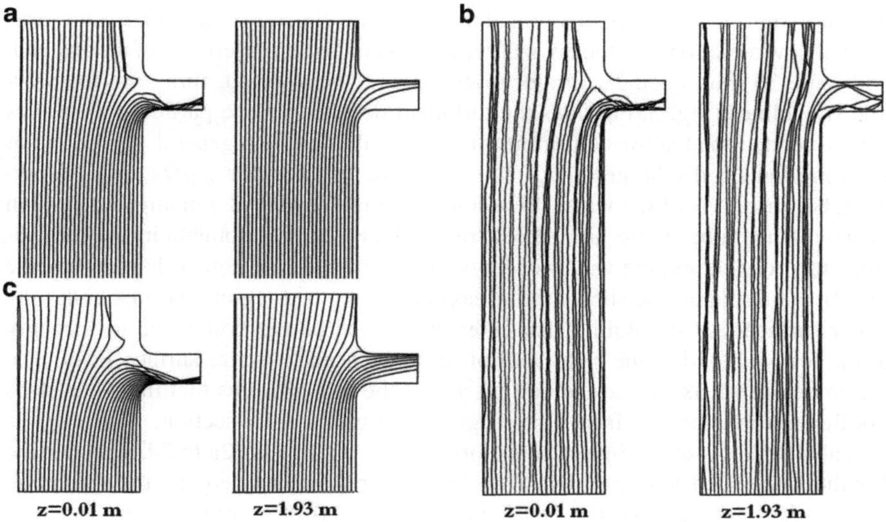

Fig. 3 Flow conditions at height $z = 0.01$ m and $z = 1.93$ m: (**a**) streaklines for discharge $Q_1/Q_2 = 0.2$, (**b**) particle trajectories for $Q_1/Q_2 = 0.2$, and (**c**) streaklines for $Q_1/Q_2 = 0.5$

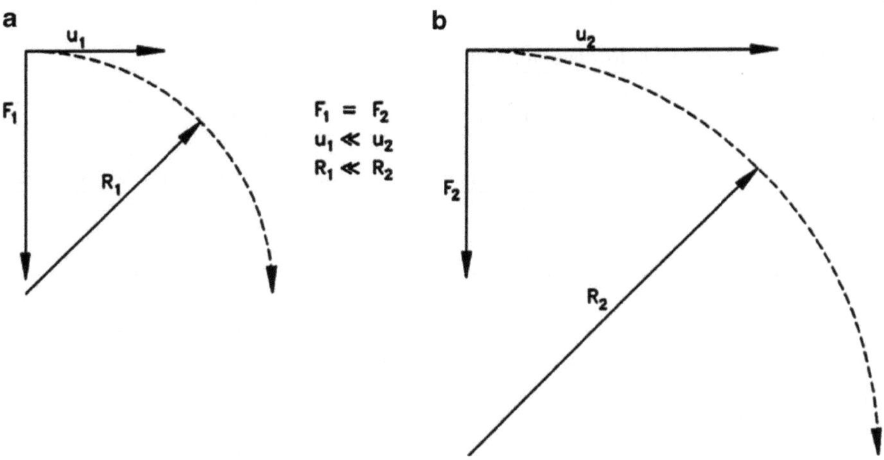

Fig. 4 Resultant trajectories of fluid particles under centripetal force: (**a**) at the bottom, and (**b**) at the surface

$z = 0.046$ m and under the water surface $z = 1.870$ m, respectively). The maximum value of the kinetic energy of turbulence occurs at the bottom at the right wall of the branch. At the surface, the maximum turbulence intensity appears on the left wall of the branch. These results confirm that turbulence arises where flow lines diverge. The locations of diverging flow lines near the bed and near the surface are at opposite sides, causing a spiral motion. In addition, the velocity vectors in the outlet of the branch evidently represent a circular motion of the fluid. The direction of motion is anticlockwise in flow direction, and the pivot curve of the spiral is on the left side of the channel.

Figure 3b shows bed-load trajectories that are, as expected, similar to the near-bed streaklines (Fig. 3a) besides the turbulent deviations. The particle trajectories for suspended load show conformity to the streaklines. The general conditions of fluid motion for discharges $Q_1/Q_2 = 0.5$ are the same as for $Q_1/Q_2 = 0.2$. However, because 33% of the water flows into the branch and 67% remains in the main channel (run 2a/b: 16.7% and 83.3%, respectively) the phenomena in the bifurcation are stronger, and the values of physical magnitudes are higher than in the case $Q_1/Q_2 = 0.2$. Figure 3c shows the streaklines in the bed region and just under the water surface. All streaklines that enter the branch on the bottom go up strongly at the left wall and at the right wall of the branch under water surface level. The difference in streaklines at the bottom and at the top indicates that the water does not flow into the branch from a rectangular part of the cross section.

Table 1 shows the sediment transport ratio S_1/S_2 for runs 2a to 2d. Apparently, for the suspended load problem (run 2b), this ratio nearly equals the discharge ratio Q_1/Q_2, whereas for the bed-load problem (2a) this ratio is about twice as high. The ratios S_1/S_2 for run 2c and 2d again indicate the sediment is practically proportionally distributed in the suspended load case (run 2d), in contrast to the bed-load case (run 2c). In run 2c, more than half of the bed-load material flows into the branch, while only a third of the water is extracted. Figure 5 shows an

Table 1 Ratio of sediment distribution for different conditions

Channel run	Angle of diverting branch (°)	S_1/S_2			
		$Q_1/Q_2 = 0.2$		$Q_1/Q_2 = 0.5$	
		(a) $D = 0.15$ mm	(b) $D = 0.01$ mm	(c) $D = 0.15$ mm	(d) $D = 0.01$ mm
2	90	0.425	0.219	1.213	0.542
3	45	0.280	0.205	0.781	0.541
4	135	0.378	0.214	0.932	0.531

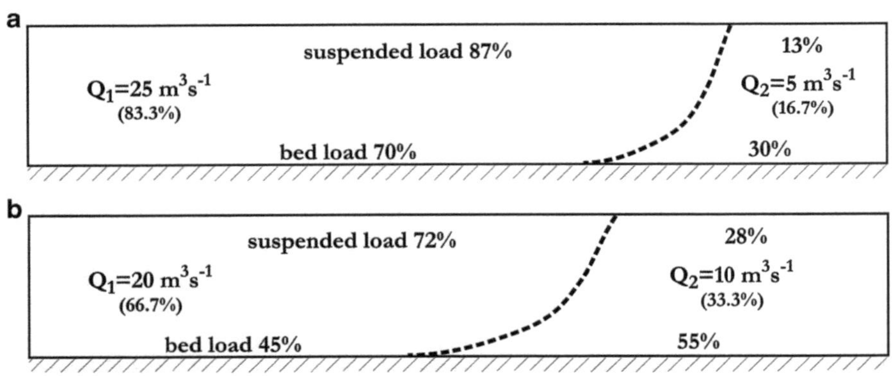

Fig. 5 Part of upstream cross section that flows into the branch for channel 2 for discharges: (**a**) $Q_1/Q_2 = 0.2$ and (**b**) $Q_1/Q_2 = 0.5$

approximation of the part of the upstream cross section that flows into the branch for both flow cases.

3.2 Bifurcation 45°: Run 3a/b ($Q_1/Q_2 = 0.2$), Run 3c/d ($Q_1/Q_2 = 0.5$)

In Fig. 6a streaklines are shown which originate 0.01 m above the bottom. In the diverting branch a small recirculation zone appears, which has a local character, and disperses a few centimeters above the bottom. An analysis of the shape of the streaklines on the bottom and near the water surface indicates a spiral motion of the fluid in the branch. With the higher velocity in the branch – discharge $Q_1/Q_2 = 0.5$, Fig. 6b – the vortex observed in the case $Q_1/Q_2 = 0.2$ seems to disappear. Probably, the higher velocity dissipates that kind of disturbance.

Table 1 shows a proportional sediment distribution for the suspended sediment (run 3b). The amount of bed load in the branch (run 3a) is higher, but much lower than for channel 2 (run 2a). Apparently, the lower diverting angle of channel 3 attracts less bed load. The sediment distribution in the bifurcation for suspended load (run 3d) coincides with the discharge ratio. The amount of bed-load sediment attracted into the branch (run 3c) is 40% higher, which is less than for channel 2 (run 2c).

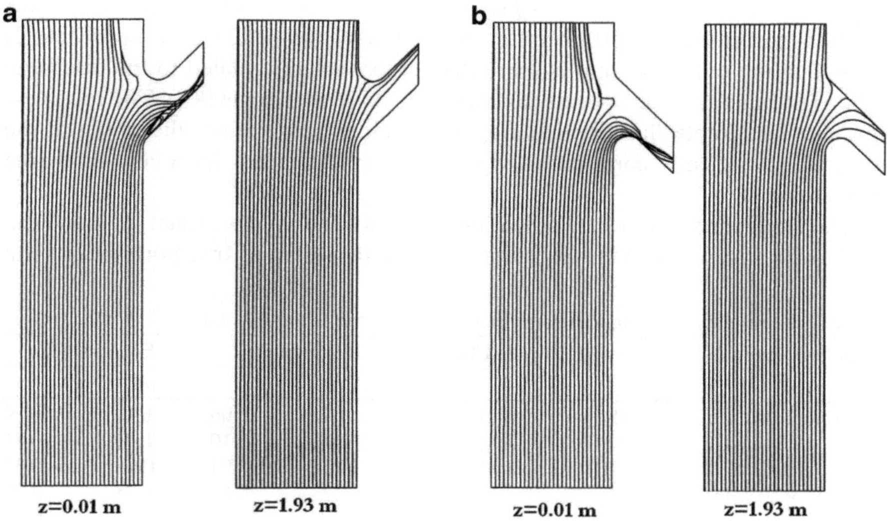

Fig. 6 Streaklines for channels 3 and 4 for discharge $Q_1/Q_2 = 0.2$

3.3 Bifurcation 135°: Run 4a/b/c/d ($Q_1/Q_2 = 0.2$ and 0.5)

The streaklines show the fluid motion in both sides of the branch and indicate a spiral fluid motion (run 4a/b). The same phenomena can be observed for the discharge ratio $Q_1/Q_2 = 0.5$ (run 4c/d). An analysis of the sediment distribution for suspended load (Table 1, run 4b/d) yields $S_1/S_2 = Q_1/Q_2$. Again, the bed load is attracted into the branch more than proportionally (run 4a/c). However, this ratio is higher than in channel 3 and less than in channel 2.

4 Analysis of Sediment Distribution

For every simulation, the 1,000 released particles are divided into three groups: escaped, trapped, and aborted. Particles are defined to distinguish the results, as escaped when they reach the outflow boundary of the main channel, or trapped when they reach the outflow boundary of the diverting branch. Particles that reach neither of the outflow boundaries, because they get into near-stagnant zones or recirculation areas, are classified as aborted. The results show a small amount of aborted particles (up to 3%).

Table 1 gives the sediment transport ratios S_1/S_2 based on the results of the particle tracking computations (runs 2a up to 4d).

Based on the numerical results described, the nodal-point relations for the bifurcations are sought. The coefficients α, β and m, n are found by fitting the nodal-point relations, (1) and (3), respectively, to the data found by the numerical computations. Table 2 shows the determined coefficients.

The sediment transport ratio S_1/S_2 is a function of the discharge ratio Q_1/Q_2 for the bed-load cases ($D = 0.15$ mm). The high α-values (i.e., $\alpha > 1$, see Table 2) show that the sediment ratio is more than proportionally related to the discharge ratio. The highest value appears for channel 2 (diverting angle 90°). The relatively low β-values imply that the linear approximations nearly cross the origin of the axis, which logically corresponds to a zero-sediment intake for a zero-discharge extraction.

The suspended sediment distribution is a practically proportional relation (i.e., $S_1/S_2 = Q_1/Q_2$) between the discharge ratio and the sediment transport ratio, as can

Table 2 Coefficients α, β, m, and n for fitted nodal-point relations (2) and (3)

Particle D (mm)	Angle (°)	Channel	Equation (2)		Equation (3)	
			α	β	m	n
0.15 (bed load)	45	3	1.68	−0.06	1.13	−0.39
	90	2	2.63	−0.10	1.15	−0.71
	135	4	1.85	0.01	0.98	−0.44
0.01 (suspended load)	45	3	1.12	−0.02	1.05	−0.09
	90	2	1.07	0.00	0.99	−0.05
	135	4	1.05	0.00	0.99	−0.04

be expected with highly suspended material. The α-values approximate unity, which indicates a nearly uniform sediment concentration profile. Together with the zero-values of β, the nodal-point relation (2) approximates the proportional relation (1).

In the exponential nodal-point relation (3), the m-values are slightly higher than unity for channel 2 and 3, indicating a weak nonlinearity. For channel 4, the value $m = 1$ shows a fully linear dependence. The negative n-values suggest a dependence on the widths of the branches, which is quite expectable for bed-load problems. However, conclusions cannot be drawn because the influence of the channel widths has not been investigated. The theoretical relation $n = 1 - \text{m}$ (see (3)) is not satisfied. It should be noticed that Wang et al. (1993) recommend an empirical determination of n.

For the suspended load problems, the values $m \approx 1$ and $n \approx 0$ show a nearly full proportionality of the nodal-point relation (3) approximating ($S_1/S_2 = Q_1/Q_2$), and an independence of the width ratio B_1/B_2. This result was to be expected and satisfies the theoretical relation $n = 1 - m$.

Resuming the foregoing, it can be concluded that for each of the bed-load problems, relatively more sediment is attracted to the diverting channel, whereas the sediment distribution over the branches for suspended load is practically proportional to the discharge distribution.

4.1 *Influence of Diverting Angle*

The relation between the sediment transport ratio S_1/S_2 and the angle of diversion is presented in Fig. 7. It shows that for bed-load problems, a critical angle exists, causing a maximum sediment load in the diverting branch. With an increasing angle, the sediment load decreases again behind this point for both discharge ratios. A possible explanation could be that the particles cannot follow the sharply bending

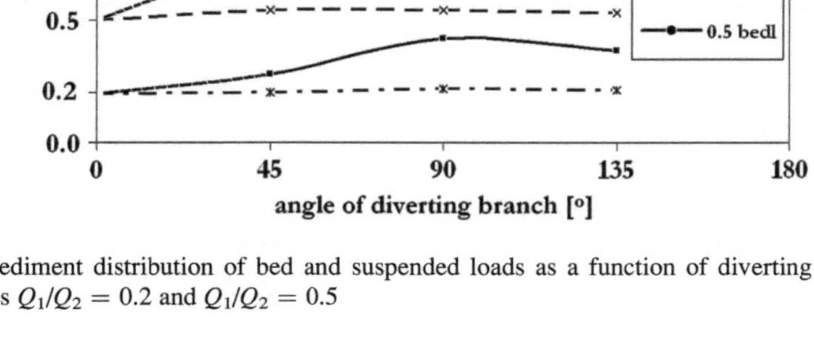

Fig. 7 Sediment distribution of bed and suspended loads as a function of diverting angle for discharges $Q_1/Q_2 = 0.2$ and $Q_1/Q_2 = 0.5$

flow lines for these large angles. Because of the limited number of experiments, this critical angle cannot be determined accurately.

The figure shows clearly the sensitivity of the sediment ratio for the angle of diversion. It is interesting to mention that Bulle (1926) found a similar relation in model experiments with critical angles between 90° and 120°.

For suspended load, the correlation between the diverting angle and the sediment transport ratio is ultimately weak, which again shows that the nodal-point relation is practically proportional for highly suspended sediment, independent of geometrical conditions.

5 Conclusions

In this study, fluid flow and sediment transport in a channel bifurcation have been simulated by means of the 3D flow modeling software FLUENT, which proves to be a useful computation fluid dynamics tool for optimization purposes in the design of hydraulic structures.

For suspended load, a practically entire conformity of sediment transport with fluid motion is found, independent of geometrical conditions as the angle of the diverting branch. The distribution of bed load is dependent on the shape of the bifurcation. The amount of attracted sediment into the diverting branch is smallest for the channel with a diverting angle of 45°, and highest for the channel with an angle of 90°.

The results of sediment distribution are considered to be satisfying because both nodal-point relations, (2) and (3) are fit to the cases of this study. Nevertheless, for engineering purposes, the nodal-point relation (3) is preferred for the following reasons:

- For strongly nonlinear relations, (3) will behave better than a linear approximation like (2).
- The relation (3) crosses the origin by definition (which is a physically logical point in the relation).
- Relation (3) shows the influence of the width ratio B_1/B_2.
- Stability analysis in bifurcation equation (2) is useless because of its unconditional instability.

Acknowledgments The authors thank A.C.H. Kruisbrink for technical assistance and valuable suggestions.

References

Bulle H (1926) Investigations on the trapping of bed-load in branching rivers. VDI-Verlag, Forschungsarbeit auf dem Gebiet des Ing.wesens, Berlin, Heft 283

den Dekker P, van Voorthuizen JM (1994) Research on the morphological behaviour of bifurcations in rivers. Delft University of Technology, Delft, the Netherlands

Fluent (1993) Fluent user's guide, version 4.2. Fluent Incorporated, New Hampshire
Fokkink RI, Wang ZB (1993) Study on fundamental aspects of 1D-network morphodynamic models. Delft Hydraulics, Delft, the Netherlands, Report Z654
Garde RJ, Raju KGR (2000) Mechanics of sediment transportation and alluvial stream problems. Taylor & Francis, London
Islam GMT (2000) On the distribution of sediments at channel bifurcation. ASCE Conf Proc 104, 160. doi:10.1061/40517(2000)160
Islam GMT, Kabir MR, Nishat A (2006) Nodal point relation for the distribution of sediments at channel bifurcation. J Hydraul Eng 132:1105. doi:10.1061/(ASCE)0733-9429(2006) 132:10(1105)
Meijer DG, Ksiazek L (1994) Sediment distribution in a channel bifurcation, Fluent simulation. Delft Hydraulics, Delft, the Netherlands, Reports Q1941
Riad K (1961) Analytical and experimental study of bed load distribution at alluvial diversions. Waltman, Delft, the Nertherlands
Richardson WR, Thorne CR (2001) Multiple thread flow and channel bifurcation in a braided river: Brahmaputra–Jamuna River, Bangladesh. Geomorphology 38(3–4):185–196. doi:10.1016/S0169-555X(00)00080-5
Rodi W (1993) Turbulence models and their application in hydraulics – a state-of-the-art review. Balkema, Rotterdam, the Netherlands
Simons DB, Sentürk F (1992) Sediment transport technology – water and sediment dynamics. Water Resources Publications, Littleton
van Rijn LC (1993) Principles of sediment transport in rivers, estuaries and costal seas. Aqua Publications, Amsterdam, the Netherlands
Vermeer (1993) SOBEK user's manual. Delft Hydraulics, Delft, the Netherlands
Wang ZB, Fokkink RI, Karssen B (1993) Theoretical analysis on nodal point relations in 1D network morphodynamic models. Delft Hydraulics, Delft, the Netherlands, Report Z473

Model Investigations of Side Channel Spillway of the Złotniki Storage Reservoir on the Kwisa River

Jerzy Machajski and Dorota Olearczyk

1 Introduction

The equivalent tasks of the Złotniki storage reservoir are flood protection of the downstream Kwisa River Valley and water storing for energetic purposes (Machajski 2009; Machajski and Olearczyk 2009). Installed electric power output is equal to 4.42 MW, for three turbines with the capacity of 20.29 $m^3\ s^{-1}$. During the period of most frequent occurrence of flood waves, that is, July–October, a permanent flood storage capacity of 1.60 hm^3 is kept in the reservoir. Water surplus flows through outlet installations, consisting of a diverse channel, two bottom outlets with diameters of 1,400 and 1,000 mm, and a side channel spillway. Bottom outlets are built-in dam body from left-sided abutment, whereas two conduits with 1,400 mm diameter each are built-in diverse channel cross section, dividing it into two sections: pressure section and free flow section. From the basic device of outlet installation, that is, spillway consisting of two sections of 22.50 m length each, divided by concrete pillar of 2.0 m width, water outflows downstream through flume and multistage cascade hewn in the rock massif. At present, the maximum discharge of outlet installations is equal to 484 $m^3\ s^{-1}$, from this 380 $m^3\ s^{-1}$ flows through side channel and cascade. The dam of 36 m height makes a storage reservoir with the maximum capacity of 12 hm^3. The dam was built in 1924–1927 and its arch body was made of broken stone in concrete. A functional plan of the Złotniki dam is shown in Fig. 1.

J. Machajski (✉)
Wrocław University of Technology, Institute of Geotechnics and Hydrotechnics, Wroclaw, Poland
e-mail: jerzy.machajski@pwr.wroc.pl

D. Olearczyk
Wrocław University of Environmental and Life Sciences, Institute of Environmental Engineering, Wroclaw, Poland
e-mail: dorota.olearczyk@up.wroc.pl

P. Rowinski (ed.), *Experimental Methods in Hydraulic Research*,
Geoplanet: Earth and Planetary Sciences, DOI 10.1007/978-3-642-17475-9_12,

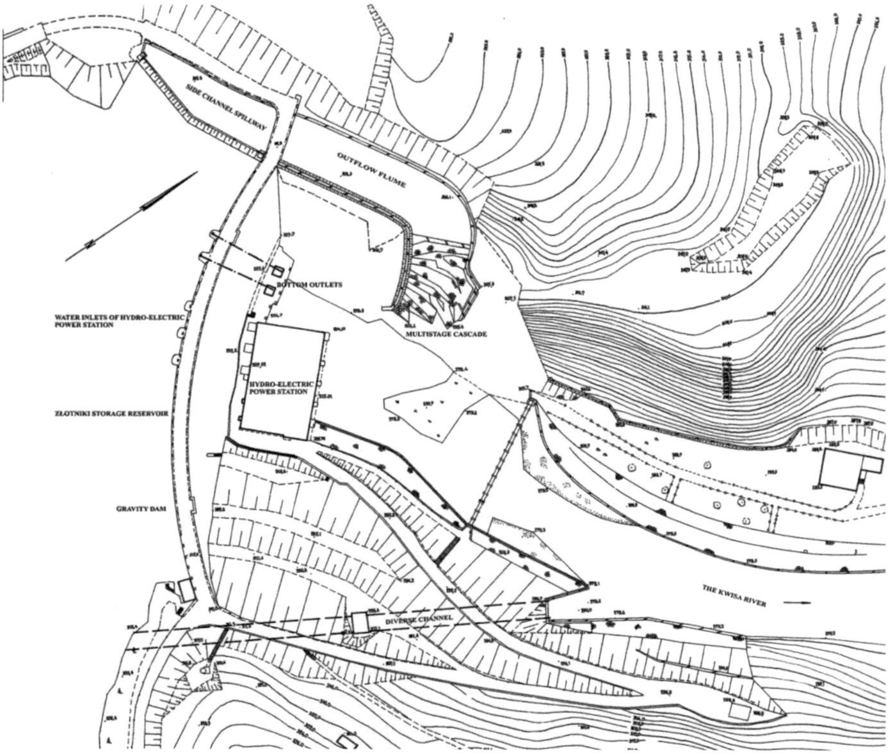

Fig. 1 Functional plan of the Złotniki storage reservoir

2 The Kwisa River

The Kwisa River is a left-sided tributary of the Bóbr River, with its mouth in 87.01 km. The headspring of the Kwisa River is on the hillside of the Wysoki Grzbiet, at the Przednia Kopa slope foot, at a height of about 900 m asl. The Kwisa River up to Leśna cross section flows through two basic geographical regions: the Izery Mountains and the Izery Foothills. The catchment area to the Złotniki dam cross section equals 288.14 km^2. The river length on that section is equal to about 36.2 km, and the average slope of catchment is about 3.5% in mountain and about 1% on foothills. Rivers and watercourses in the upper Kwisa River catchment are supplied in waters mainly from rain in summer season and from snow in winter season. The nature of water supply causes two maximum discharges from the catchment, summer (mainly July–September) and winter (mainly February–March), within a hydrological year. Summer freshets are the biggest in size and magnitude in comparison with winter freshets (Radczuk and Olearczyk 2006). A hydrographical network in the upper Kwisa River catchment is well developed. A drainage density is about 0.74 1 km^{-1}.

On the Kwisa River, upstream of the Złotniki reservoir, there is only one gauging station located in Mirsk. For the analysis of the Złotniki reservoir operation, the hydrological data from Mirsk gauging station were transferred, using recommended hydrological methods, to dam cross section. Characteristic discharges in that cross section are as follows (Polish abbreviations): minimum NNQ $= 0.214$ m^3 s^{-1}, average low flow SNQ $= 0.817$ m^3 s^{-1}, mean SSQ $= 4.449$ m^3 s^{-1}, average maximum flow SWQ $= 94.07$ m^3 s^{-1}, and maximum WWQ $= 323$ m^3 s^{-1} (1981). Maximum discharges with a given probability of exceedance are $Q_{50\%} = 85$ m^3 s^{-1}, $Q_{10\%} = 198$ m^3 s^{-1}, $Q_{5\%} = 245$ m^3 s^{-1}, $Q_{1\%} = 351$ m^3 s^{-1}, $Q_{0.5\%} = 397$ m^3 s^{-1}, and $Q^{\alpha}_{0.1\%} = 600$ m^3 s^{-1}, respectively.

According to the regulations (Rozporządzenie 2007) valid in Poland, Złotniki dam is classified into hydrotechnical structures of the first class of importance. For such structures, not subjected to destruction due to their overflow, a probability of exceedance for computational discharges should be set at 0.5% and 0.1%, but for the higher discharge – with the upper extension at $t_{\alpha} = 1$ and confidence level 0.84. For Złotniki dam cross section, the discharges are $Q_{0.5\%} = 397$ m^3 s^{-1} and $Q^{\alpha}_{0.1\%} =$ 600 m^3 s^{-1}, respectively. Computational discharges significantly exceed the outlet installation capacity ability of the reservoir, so model tests were carried out, preceded by analytical computations for the reconstruction of outlet installation. First, a side channel spillway was analyzed due to its possible significant over-dimensioning (Machajski 2009; Machajski and Olearczyk 2009).

3 Analytical Calculations

Side channel spillway fulfills two parallel functions, that is, that of overflow part and that of flume which is conducted parallel to overflow edge. This flume takes over the water overflowing spillway and conveys it to installations that link the side channel spillway with the downstream river section. Analysis of water movement conditions within the reach of side channel spillway has changed gradually. At present (Khatsuria 2005; Novak et al. 2007; Vischer and Hager 1998; Tančev 2005; Şentürk 1994), the analysis is available based on the law of linear momentum conservation assuming that the only force ensuring flow of water in flume results from slope of water level in the direction of water routing. Next, the energy of water overflowing a spillway is dissipated as a result of its mixing with water in flume, with simultaneous lack of influence on flow conditions in flume.

The same purpose of hydraulic computations of side channel spillway is the determination of necessary length of spillway edge assuming exploitation conditions on reservoir, and the determination of pattern of water-table lines in flume at the assumed parameters of flume cross section, slope of its bottom, and steady growth of discharge along the length of flume.

The spillway discharge was determined assuming that it will work in the whole range of expected discharges as not-submerged. Calculations were carried out using formula for free spillway, with straight insert on crest (Rogala et al. 1991; Vischer and Hager 1998):

$$Q = \varepsilon\, 2/3\, \mu\, B\, \sqrt{2\, g}\, H_o^{3/2} \tag{1}$$

where ε is the coefficient of side contraction weir, $\varepsilon = 1.0$; μ is the discharge coefficient, $\mu = 0.654$; B is the spillway crest length, $B = 45.0$ m; H_o is the energy height calculated with relation to spillway crest.

For calculations, two characteristic damming levels in reservoir were taken. They result from water level with regard to weir crest (309.30 m asl) and dam crest, determining water layer thickness and the resultant energy of stream. The calculation results are as follows:

$$312.12(\text{dam crest from downstream}) - 309.30 = 2.82\ \text{m} \rightarrow Q = 411.55\,\text{m}^3\text{s}^{-1}$$

$$312.00(\text{dam crest from upstream}) - 309.30 = 2.70\ \text{m} \rightarrow Q = 385.58\ \text{m}^3\text{s}^{-1}$$

In the further analysis for flume parameters determination, the discharge equal to 385.58 $\text{m}^3\ \text{s}^{-1}$ was assumed.

In the calculations for the determination of flume parameters, an impact of external forces, in compliance with Newton's second dynamic principle, should be considered in two cross sections, 1-1 and 2-2, at distance dx, including gravity force, material roughness of bed cross section, and resultant force of hydrostatic pressures in the two cross sections mentioned above.

A gravity force is represented by horizontal component of water weight W between two cross sections, 1-1 and 2-2, given in the form:

$$W\ \sin\Theta = \gamma\, S_o\, A\, \mathrm{d}x \tag{2}$$

where S_o is the longitudinal slope of bottom line of flume, A is the wetted area between cross sections 1-1 and 2-2, γ is the weight density of water, and Θ is the inclination angle of bottom line of flume.

The force that results from an impact of roughness of the flume material is as follows:

$$F_s = \gamma\, A\, S_f\, \mathrm{d}x \tag{3}$$

where S_f is the slope of energy line.

The force that results from differences of hydrostatic pressures in cross sections 1-1 and 2-2 can be written in the form:

$$P_1 - P_2 = -\gamma\, A\, \mathrm{d}y \tag{4}$$

where P_1 is the force of hydrostatic pressure in cross section 1-1, P_2 is the force of hydrostatic pressure in cross section 2-2, and dy is the depth difference in cross sections 1-1 and 2-2.

The resultant change of momentum in time unit between cross sections 1-1 and 2-2 is equal to the sum of external forces and is described by the equation:

$$\frac{\mathrm{d}M}{\mathrm{d}t} = P_1 - P_2 + W \sin \Theta - F_s \tag{5}$$

After substitution of $A = Q/v$ and $q = \mathrm{d}Q/\mathrm{d}x$ and transformations, the following equation is derived (Khatsuria 2005; Şentürk 1994):

$$\frac{\mathrm{d}y}{\mathrm{d}x} = \frac{S_o - S_f - 2\,\alpha\, Q\, q/g\, A^2}{1 - \alpha\, Q^2/g\, A^2\, D} \tag{6}$$

where α is the correction coefficient of kinetic energy (Saint Venant), $\alpha = 1.10$; q is the rise of flow per length of flume, $\mathrm{d}Q/\mathrm{d}x$; D is the hydraulic depth, $D = A/B$; and B is the width of flume, $B = 15$ m.

The above differential equation can be solved in two ways: either by applying numerical methods available for equations of this type or by applying simplified methods, which in general consist in introduction of finite differences in place of differentials. For example, the equation determining a change of water levels Δy for the length Δx is (Khatsuria 2005; Novak et al. 2007; Şentürk 1994):

$$\Delta y = -\frac{\alpha\, Q_1\,(v_1 + v_2)}{g\,(Q_1 + Q_2)} \left[(v_2 - v_1) + \frac{v_2\,(Q_2 - Q_1)}{Q_1} \right] + S_o \Delta x - S_f\, \Delta x \tag{7}$$

The analytical calculations applying the above formula confirmed a possibility of improving the capacity ability of spillway to about 400 $\mathrm{m^3\,s^{-1}}$, but only for a situation when it works as not-submerged. In the existing solutions, during water inflow into a flume, a quick overfilling of flume and changes of weir work conditions from not-submerged to submerged are observed. Hence, the results of calculations of capacity ability showed that for damming water in a reservoir of 312.00 m asl, the maximum attainable capacity ability is equal only to 250 $\mathrm{m^3\,s^{-1}}$, that is, about 130 $\mathrm{m^3\,s^{-1}}$ less than it results from the discharge curve that is currently valid for the weir. The corrected discharge curve of the Złotniki reservoir's spillway, for its present solutions, is shown in Fig. 2.

4 Concept of Outlet Installation Reconstruction

In the concept of reconstructing the Złotniki reservoir's outlet installations, a safety of reservoir was taken into account due to computational discharge values and capacity ability of existing outlet installations (Machajski and Olearczyk 2009). The consequences of computational discharges passage through the reservoir were determined, analyzing possibilities of their reduction to safe discharges both for downstream area and for reservoir (Jain 2001).

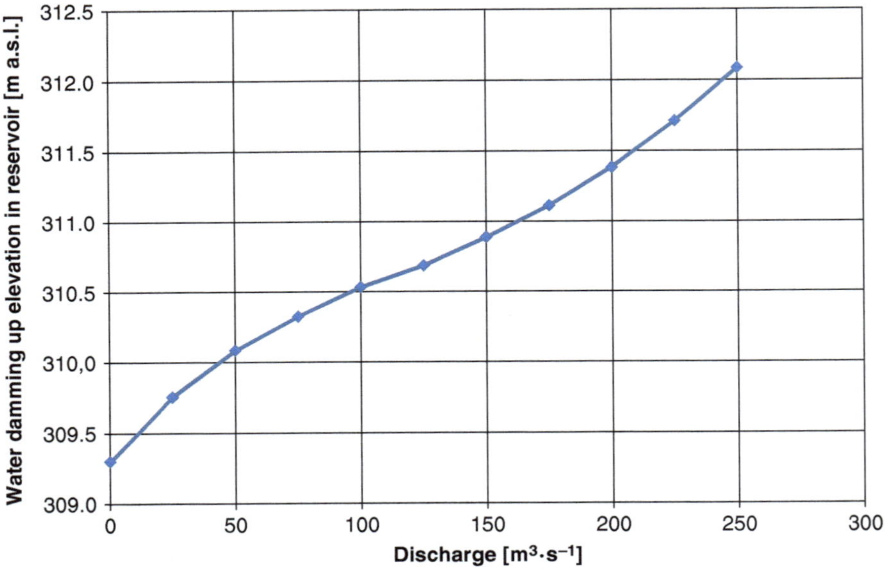

Fig. 2 Hydraulic characteristic of side channel spillway of the Złotniki reservoir for the existing state (analytical solution)

Calculations of flood waves transformation (Machajski and Olearczyk 2009) showed that for the existing outlet installations there is no possibility to preserve a safe dam crest height above the maximum water level in reservoir equal to 1.0 m (Rozporządzenie 2007). So, for flood wave $Q_{0.5\%}$ passage, in an initial state of reservoir fulfillment of 303.03 m asl, the water level in the reservoir reaches 312.35 m asl, that is, exceeds by 0.35 m the dam crest, and in an initial state of reservoir fulfillment of 309.30 m asl, the water level in the reservoir reaches 312.50 m asl, that is, exceeds by 0.50 m the dam crest. For the flood wave $Q^{\alpha}_{0.1\%}$ passage, a water level in the reservoir will exceed by about 0.50 m the dam crest in an initial state of water level equal to 303.03 m asl, and by about 0.80 m in an initial state of 309.30 m asl, which means a dam crest overflow in each computational case. Results of transformation calculations are shown in Figs. 3 and 4.

The calculations reveal that a free flow of computational discharge through the existing spillway is not possible, due to too shallow and too narrow flume in the upper part (Fig. 5), hence the flume should be deepened by about 2.50 m with simultaneous set of its width at 15.0 m. Under such conditions, the assumed discharge of spillway can be obtained for damming water level in the reservoir for 312.00 m asl. These requirements offer a possibility of the rebuilding in question, mainly in economic aspect. The reconstruction would force a necessity of constructional changes of the weir – new static conditions of its work, and a necessity of constructional changes of wall on the opposite side. However, the essential advantages of rebuilding convince for such realization.

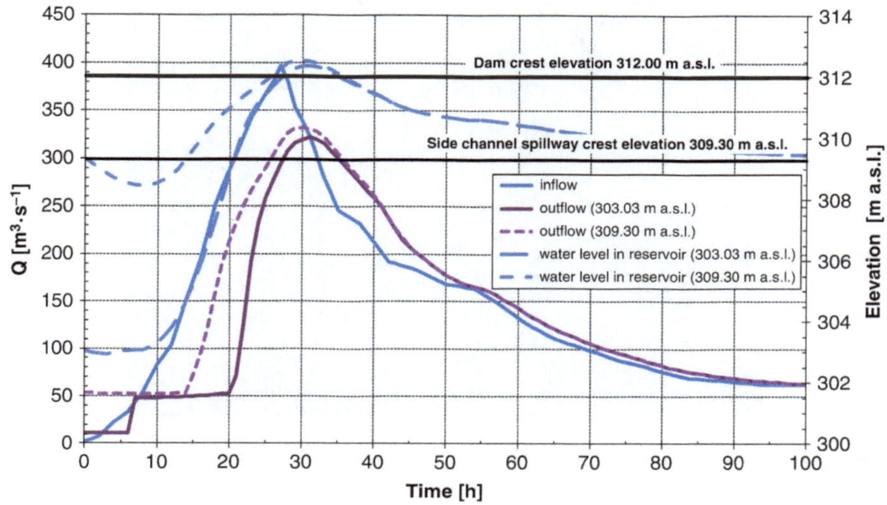

Fig. 3 Transformation of flood wave $Q_{0.5\%}$ for the existing outlet installations (corrected spillway discharge), initial levels of water dammed in reservoir 303.03 and 309.30 m asl

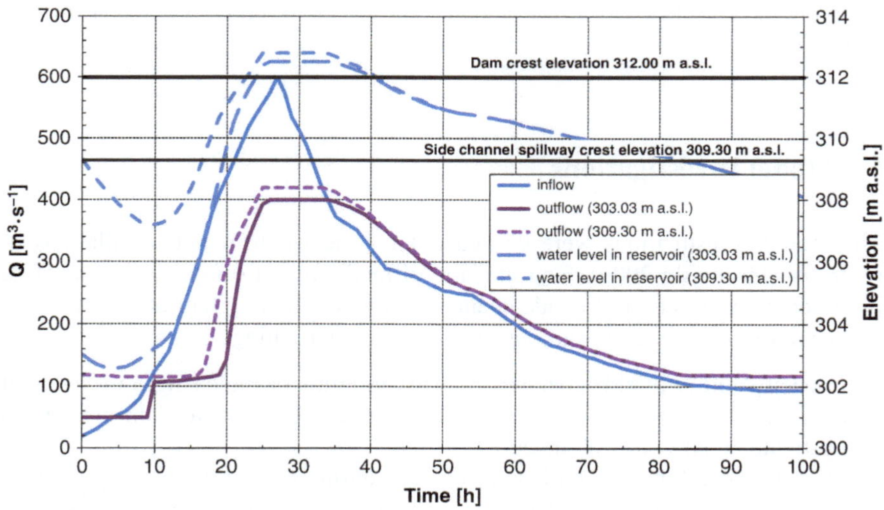

Fig. 4 Transformation of flood wave $Q^{\alpha}_{0.1\%}$ for the existing outlet installations (corrected spillway discharge), initial levels of water dammed in reservoir 303.03 and 309.30 m asl

The spillway can be reconstructed without a necessity of reservoir emptying; also its safety is ensured because flood waves can pass through spillway without any negative consequences. A significant drawback of spillway reconstruction is the lack of possibility of improving the operating conditions of freshet wave routing.

Fig. 5 View of side channel spillway model for existing state (Change I)

At present, the operation of outlet installation is carried out until a damming level in reservoir will reach dam crest, that is, the outflow will be somewhat over 100 $m^3 s^{-1}$.

In view of the above-mentioned requirements for exploitation safety of the reservoir, the side channel spillway applying the proposed solutions of reconstruction was modeled.

5 Model Investigations

The model investigations were carried out for the whole outlet installations of reservoir (Machajski 2009). First, the aim of model investigations was to verify a proposed reconstruction of a side channel spillway. The program of investigation consisted of measurements and observations for determining:

- Capacity ability of side channel spillway, in terms of the changes of levels of damming up in the reservoir ranging from 309.30 m asl to dam crest from upstream equal to 312.00 m asl
- Operating conditions of multistage cascade during greater discharge passage
- Operating conditions of arch bridge at passage section from flume to multistage cascade
- Impact of spillway constructional changes, introduced in minimal range – possible to be comparatively easily realized

The investigations, carried out on a spatial model at scale 1:40, included all component elements important for water passage in this side channel spillway with multistage cascade (Machajski 2009). For the studied problem, the water over weir crest and in a flume flows predominantly due to gravity force; therefore, the Froude criterion of similarity was applied to calculate the appropriate forces occurring both in nature and in the model (Cunge and Ackers 1987; Novak and Čábelka 1981; Vischer and Hager 1998). For the studied phenomenon, also a certain contribution comes from a viscous force. To limit the influence of viscous force to minimum, the same kind of flow in the model as that in nature had to be assured. According to input data, under the conditions of computational discharges passage, the conditions of turbulent flow occur; therefore, it is required that the turbulent flow, with relatively great Reynolds number, should also occur in the model (Novak and Čábelka 1981). Under model conditions and at variable discharges, the Reynolds numbers from $Re = 15{,}000$ for discharge $Q_{0.5\%}$ to $Re = 35{,}000$ for discharge $Q^{\alpha}_{0.1\%}$ were obtained. Hence, it was inferred that the phenomenon modeled was significantly affected by the gravity force.

In maintaining the similarity of phenomena observed both in model and in nature, the design of model with appropriate roughness seems to be of prime importance. According to input data, all elements of external outlet installation are made of concrete, hence roughness coefficient n is equal to 0.020. Because the model scale is $\alpha_l = 40$, therefore, after counting $\alpha_n = 1.8493$ (Machajski 2009), the resultant roughness coefficient of the material that should be used for essential details of the model outlet installations should be equal to:

$$n_M = \frac{n_N}{\alpha_n} = \frac{0.020}{1.8493} = 0.01081$$

This equation proves that elements of outlet installation should be made of a very smooth material, for example, Plexiglass® or Vinidur®.

The aim of changes introduced into the model was to check a possibility of capacity ability improvement of spillway, that is, to reconstruct or to build some elements that were not present on the object earlier and to correct the height configuration of flume bottom. First, the investigations were carried out for the existing state to confirm a consistence of a discharge curve currently valid for spillway with the calculated one – Change I (Fig. 5).

A significant change introduced into the model consisted in setting a constant width of flume – Change II (Fig. 6). The aim of Change III was to correct the height configuration of flume bottom, that is, its lowering by about 1.50 m (Fig. 6). Change IV concerned the checking of an impact of dividing pillar and steering walls on the conditions of water flow through side channel spillway (Fig. 6). The last Change V consisted in the deepening of flume bottom by about 2.50 m (Fig. 7).

6 Hydraulic Characteristics of Spillway

The spillway characteristic was determined for discharges ranging from 0.0 to 400 $m^3\ s^{-1}$ and for damming up level varying from 309.30 m asl (spillway crest elevation) to 312.00 m asl (elevation of dam crest from upstream). During

Fig. 6 View of side channel spillway model after introducing Changes II, III, and IV

Fig. 7 View of side channel spillway model after introducing Change V

investigations on spillway, the bottom outlets, hydroelectric plant, and diverse channel were closed. Taking measurements for this characteristic determination, the operating conditions of spillway were observed, mainly in respect of disturbances at spillway entry cross section at the side of narrowed flume (Fig. 5). Moreover, observations of conditions of great discharges entry into bridge cross section and conditions of their passage through the first stage of cascade were carried out.

The spillway hydraulic characteristic as a function of water dammed up in the reservoir is shown in Figs. 8 and 10 in order to compare changes introduced into the model. In Fig. 9, a discharge coefficient is presented for the proposed solutions of spillway.

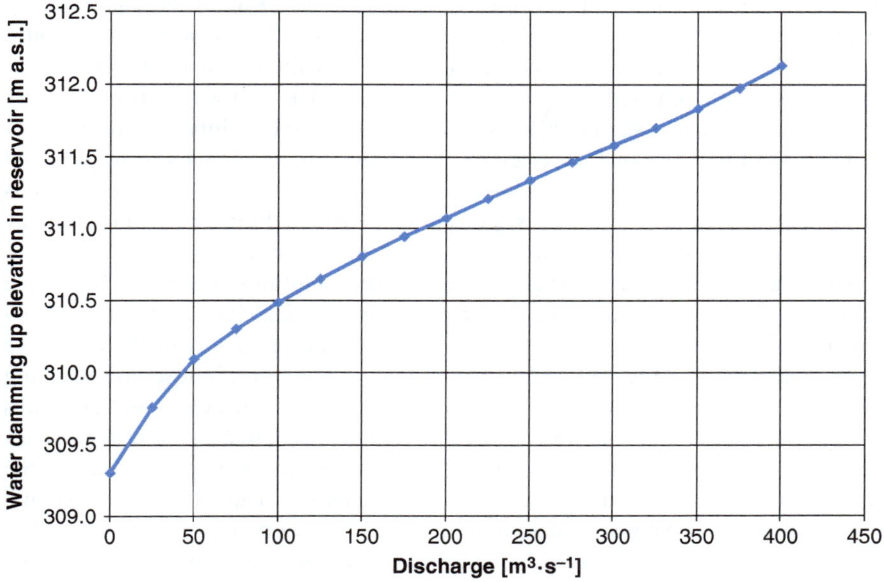

Fig. 8 Hydraulic characteristic of side channel spillway for the proposed solution

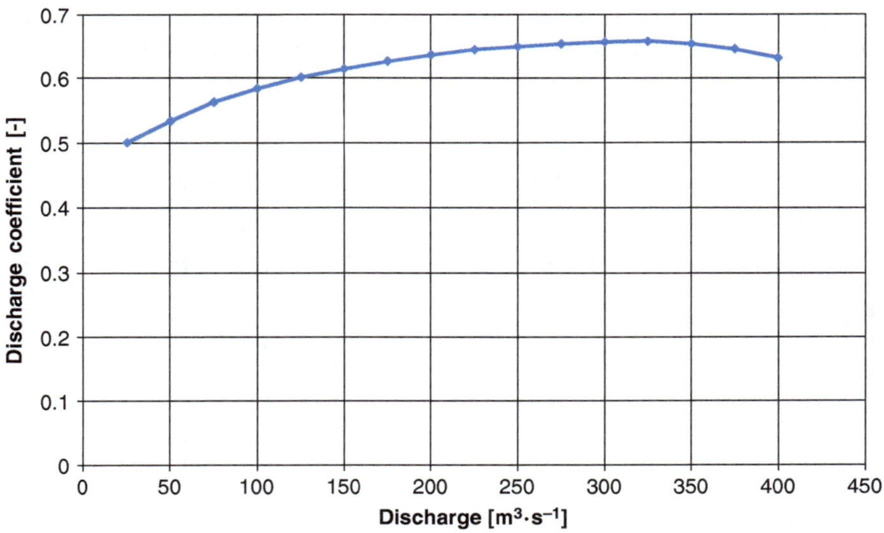

Fig. 9 Changes of discharge coefficient of side channel spillway for the proposed solution

The rise of discharge curve and a curve of discharge coefficient, visible in Figs. 8 and 9, show a moment of change in the character of weir work from not-submerged to submerged. This is a situation that, for this type of installation, should never develop; in every condition determined by weir parameters and water damming up levels in the reservoir, this installation should work as not-submerged. Because investigations concerned an existing object, it was important to determine the conditions for the changes in weir work character. The spillway capacity ability obtained based on investigations is close to that calculated, mainly from the point of view of water quantity that under given conditions can overflow by weir, provided that in the whole range of discharges it works as not-submerged. This corresponds to discharge of about 385 $m^3\ s^{-1}$, which could be obtained for the existing spillway layout in plan, its crest length, weir shape, and flume parameters.

Under the conditions of Change I, for the present solutions of weir and flume, the spillway capacity ability is lower, which results from quick changes of its operating conditions, from not-submerged to submerged. This takes place for the discharge equal to 125 $m^3\ s^{-1}$ (Fig. 10) and the water damming up level in reservoir reaching 310.70 m asl. The maximum spillway capacity ability under these conditions was 245 $m^3\ s^{-1}$ at the water damming up level of 312.00 m asl, that is, the water reaches the dam crest. After introducing Change II, a minimum improvement in the spillway operating conditions was obtained, whereas its capacity ability was not improved and no changes of damming up levels in reservoir were obtained. The introduced Change III allowed the spillway operating conditions to be improved and caused the expected changes of flow conditions in weir, which occurred at discharge of 175 $m^3\ s^{-1}$ (Fig. 10), but did not increase the expected spillway capacity ability. It is only upon introducing Change V that the spillway capacity ability increased to the expected 400 $m^3\ s^{-1}$ (Fig. 11). This took place for water damming up in reservoir to 312.13 m asl, which slightly exceeded the dam crest from downstream.

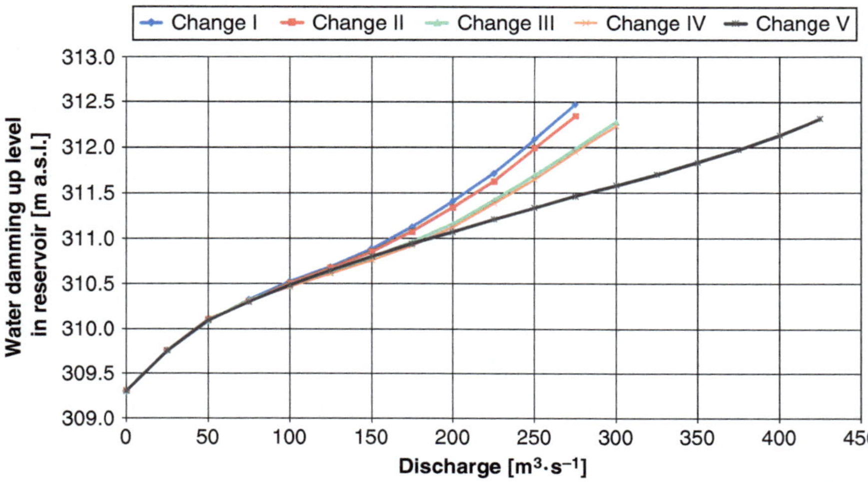

Fig. 10 Comparative hydraulic characteristics of side channel spillway for changes introduced into the model

Fig. 11 Spillway work under conditions of discharge passage of about 400 $m^3 s^{-1}$ (Change V)

The spillway work conditions after introducing Change V are evaluated as correct. Any intensive disturbances at the input cross section, both from the left and right sides of the bridge abutment, are not observed; any disturbances at the weir output cross section are not observed either: the stream closely adheres to the weir contour and gently flows in flume cross section in the cascade direction.

Also an improvement of the conditions of water flow into a bridge cross section is observed. During discharge passage in the amount of 400 $m^3 s^{-1}$, the bridge still works as free; only from the left side of its abutment, a small water damming up is observed.

The investigation confirmed the comments published in technical literature. They reveal that side channel spillways should be designed in such an elevation configuration of flume that in the whole range of computational discharges they work as not-submerged. Their discharge will be then a function of weir crest length, thickness of overflowing water, and weir shape – crest shape and its contour (Khatsuria 2005; Novak et al. 2007; Vischer and Hager 1998; Tančev 2005; Şentürk 1994).

7 Summary

Recapitulating the model investigations of side channel spillway of Złotniki reservoir connected mainly with hydraulic characteristic, the authors discuss a possibility of improving the capacity ability of this installation, hence the improvement of exploitation safety of the whole object. Proposed solutions that correct the configuration of the flume of side channel spillway are verified during model tests (Machajski 2009). This allows to draw the following conclusions:

1. In order to improve the exploitation safety of Złotniki reservoir, it is necessary to introduce the proposed changes in existing state of flume of side channel spillway.

2. Changes in side channel spillway solution consist in flume bottom deepening by about 2.50 m in relation to the present depth.
3. A flume deepening increases the capacity ability of this installation by about 135 $m^3 s^{-1}$.
4. An improvement of conditions of computational discharges passage through bridge cross section is obtained.

References

Cunge JA, Ackers P (1987) Topics in hydraulics modeling. In: Proceedings of XXII congress international association IAHR for hydraulic research, Lausanne

Jain SC (2001) Open-channel flow. Wiley, New York

Khatsuria RM (2005) Hydraulics of spillways and energy dissipators. Marcel Dekker, New York

Machajski J (2009) EW Złotniki. Przebudowa stopnia wodnego. Badania modelowe – Etap II. [Złotniki Storage Reservoir. Reconstuction. Stage II] Raport serii SPR nr 6/2009. Politechnika Wrocławska, Instytut Geotechniki i Hydrotechniki, Wrocław (in Polish)

Machajski J, Olearczyk D (2009) EW Złotniki. Przebudowa stopnia wodnego. Koncepcja – Etap I. [Złotniki Storage Reservoir. Reconstuction. Stage I] Raport serii SPR nr 5/2009. Politechnika Wrocławska, Instytut Geotechniki i Hydrotechniki, Wrocław (in Polish)

Novak P, Čábelka J (1981) Models in hydraulic engineering – physical principles and design applications. Pitman Publishers, London

Novak P, Moffat AIB, Nalluri C, Narayanan R (2007) Hydraulic structures. Taylor & Francis, New York

Radczuk L, Olearczyk D (2006) Studium ochrony przed powodzią zlewni rzeki Kwisy. Hydrologia wielkich wód. [Protection of Kwisa River basin against flooding. Hydrology of large waters] Podstawy hydrologiczne dla wyznaczenia zasięgu zalewów powodziowych. Wrocławska Agencja Rozwoju Regionalnego, Wrocław (in Polish)

Rogala R, Machajski J, Rędowicz W (1991) Hydraulika stosowana. [Applied hydraulics] Przykłady obliczeń. Wydawnictwo Politechniki Wrocławskiej, Wrocław (in Polish)

Rozporządzenie (2007) Rozporządzenie Ministra Środowiska z dnia 20 kwietnia 2007 w sprawie warunków technicznych, jakim powinny odpowiadać budowle hydrotechniczne i ich usytuowanie [Directive of the Poland's Minister of Environment of April 20, 2007, on technological conditions of hydrotechnical structures]. Dz.U. Nr 86/2007, poz. 579. (in Polish)

Şentürk F (1994) Hydraulics of dams and reservoirs. Water Resources Publications, Littleton, CO

Tančev L (2005) Dams and appurtenant hydraulic structures. A.A. Balkema, London

Vischer DL, Hager WH (1998) Dam hydraulics. Wiley, Chichester

Sediment Investigation at the 30° Water Intake

Mehdi Karami-Moghaddam, Mahmood Shafai-Bejestan, and Hossein Sedghi

1 Introduction

The study of flow division in open channels, which has been under consideration by hydraulic engineers for many years, is much used in designing the irrigation and drainage networks. Water intakes are used to divert flow from a main channel into irrigation systems and from a river into irrigation channels, at installation of water treatment and entrance for hydropower. The configurations of diversion flows are either natural, as braiding that causes a cutoff in meander rivers, or artificial diversion flows, as intakes from rivers for agricultural, industrial, and urban usages. One problem occurring in most intakes is the change of direction of the thalweg path toward the opposite bank which arises from gathering and influence of sediments into the entrance. The sediments delivery, if not restrained, may transmit into the channels and installations, thereupon carrying and depositing them in different parts. Small ting particles suspended in water render damage to the installations in case of high flow speed, especially when mechanical tools such as turbines and pumps are applied. Among the afore-said problems are the following: (1) Reduction of the flow discharge capacity in the channels as a consequence of sedimentation. (2) Rough materials could lead to erosion of the channel walls. (3) Interruption of water source for dredging of the channels may cut the water supplies providing to the farms. (4) Channel dredging is expensive and not economical. (5) Sedimentation facilitates the conditions for growing weeds that are harmful to the covers and result in leakage from the channel walls.

Figure 1 shows the great intake of Ohio River, in which the gathering of sediments in the entrance causes a decrease in the flow width, and so in the efficiency (Neary et al. 1999).

M. Karami-Moghaddam and H. Sedghi
Department of Water Engineering, Science and Research Branch, Islamic Azad University (IAU), Tehran, Iran

M. Shafai-Bejestan (✉)
College of Water Science and Engineering, Shahid Chamran University, Ahwaz, Iran
e-mail: m-shafai@scu.ac.ir

P. Rowinski (ed.), *Experimental Methods in Hydraulic Research*,
Geoplanet: Earth and Planetary Sciences, DOI 10.1007/978-3-642-17475-9_13,

Fig. 1 The great intake of Ohio River, where the gathering and sediments entry causes efficiency decline (Neary et al. 1999)

Therefore, considering the sediment problem in the intake channels is of great importance. Extensive methods are tested and applied during the recent years to control the sediments; the most common is the periodic dredging. The chief difficulty in these methods is that they take high costs and time. Other measures could be taken to the effect of improving the flow pattern, such as exerting optimal conditions for the intake, for example, changing the deviation angle and geometric shape of the intake. Full knowledge of the diversion flow pattern is a necessary condition to study the intake sediments. The diversion flows are essentially three dimensional. Some of their features are represented in Fig. 2 (Neary et al. 1999). These include a separation zone in the inside wall of the branch channel (Zone A), a contracted flow zone in the branch channel, a secondary circulation beside the outside wall of the branch channel, and a stagnation point near the junction of the downstream edge and the main channel (Zone C). The recirculation flow at the center of the separation zone is completely slow. The width of separation in the surface is greater than that in the bed. At the junction downstream in the opposite wall, there may occur a separation due to flow expansion (Zone B). The vertical velocity profile in open channels is nonuniform. According to the no-slip conditions, the velocity at the bed is necessarily zero, close to the water surface is high, and in between these two surfaces is logarithmic.

As the flow comes close to the intake, it is accelerated laterally due to the suction pressure at the end of the intake. The acceleration divides the flow into two parts: one entering the inside of the intake, and the other continuing downstream of the channel. The former is shown in Fig. 2 by a surface called the Dividing Stream

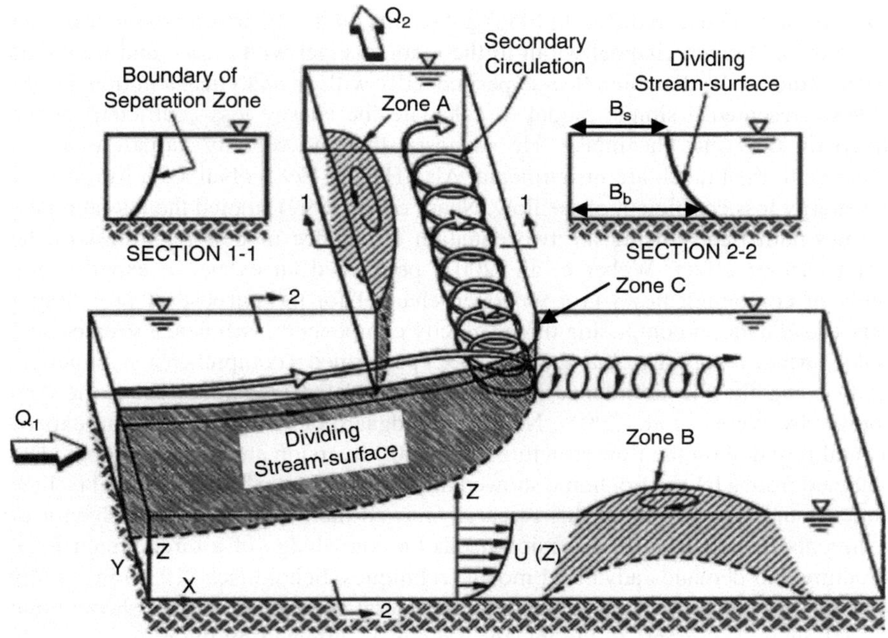

Fig. 2 The flow pattern in the intake entrance (Neary et al. 1999)

Surface (DSS) or stream tube. As seen in Fig. 2 (Sect. 2.2), in the main channels with rectangular section, the diversion flow width at the bed (B_b) is greater than that in the surface (B_s), which causes the sediments entry into the intake, resulting from their high density in the bed. The streamwise curvature in DSS yields imbalance among the transverse pressure gradient, the centrifugal force, and shear force, as a consequence of which a secondary current in clockwise direction is formed. Such a secondary vortex is also formed along the main channel wall. The more this current advances toward downstream, the more reduces its strength primarily due to the fluid viscosity. The secondary current, besides the separation zone along the inner wall of the branch channel (Zone A), gives rise to a complex three-dimensional flow.

The extent of DSS in the main channel determines the rate of discharge to the branch channel. The diversion flow width or stream tube at each surface (plan) is defined as the distance from the main channel bank at the intake side to the stream line ending in the stagnation point near the corner of downstream junction of the intake and the main channel.

Taylor (1944) studied the flow in the 90° intake and proposed a graphical method for determining the flow pattern. The method was used also by Thomson (1949) for an analytical solution to the intake, though his assumptions based on the flow depth to be constant is not practicable. Also Tanaka (1957) and Murota (1958), assuming that the water depth in all channels is constant, analytically solved the flow problem. Grace and Priest (1958) got results on the diversion flow in different ratios of the branch channel width to the main channel width. Law and Reynolds (1966) studied the diversion flow experimentally with an analytical solution. Hager (1984) presented a simple model to calculate the energy loss coefficient of the diversion flow into the intake. He supposed that the velocity variations at the entrance to the branch are insignificant. Also Hager (1992) obtained a formula for the energy loss coefficient of the flow. Neary et al. (1999) studied the lateral intake inflows numerically using the two equation turbulence models regardless of the water surface effect. Weber et al. (2001) performed an extensive experimental study of combining flows in a 90° open channel for the purpose of providing a very broad data set comprising three velocity components, turbulence stresses, and water surface mappings. Huang et al. (2002) performed a comprehensive numerical study using the 3D turbulence models, and validated the model using the data applied by Weber et al. (2001). Neary and Odgaard (1993) carried out an experimental research on the flow structure with a 90° diversion angle. The velocity data obtained from a laboratory flume showed that the flow in the branch channel is three dimensional. The results of this research showed that to describe the behavior of sediments transmission in diversion needs the knowledge of a three-dimensional structure and demands advanced model techniques. Schoklitsch (1937) in a study with the goal of a comparison between the lateral and frontal intakes showed that the inflowing sediments to the intakes are always affected by the roughness ratio (K_s/y_0) and the Reynolds number $Re_* = u_* d_{50}/v$, where K_s is the bed roughness of the main channel, y_0 is the depth of water in the main channel, D_{50} the size of sedimentary particles, u_* the shear velocity, and v the kinematic viscosity. Raudkivi (1993) investigated the effect of bed roughness on the sediments delivery into the intake. According to his study, the sediments delivery to the lateral intake decreases along with reducing the secondary currents strength, and this happens when the bed roughness coefficient increases. For intakes in bends, the decrease in the secondary currents strength leads to the increase of the sediments delivery as the bed roughness coefficient becomes greater. Studies are done by Razvan (1989) about the impression of the sill height in the lateral intakes. Suggestions are proposed by Novak et al. (1990) concerning the intake angle. Barkdoll et al. (1999) showed in his researches on the lateral intake, which are carried out in straight path with 90° intake angle, that the diversion flow ratio has the greatest effect on the sediment delivery ratio. The results obtained by Abassi et al. (2002) on the intake in the straight path of river showed that the existence of sill reduces the vortex width at the entrance, as a consequence of which, the sediments entry decreases. The sill affects more strongly with high diversion flow ratios than with lower ratios. Thirty-four experiments by Shafai Bajestan and Nazari (1999) carried out on an intake at a 90° bend with a 60° position showed that among the five different intake angles of 15°, 45°, 60°, 75°, and 90° with mobile bed, the 60° transmits more flow with the least rate of sediments. Using experimental data and comparing it with a numerical model which solves the standard three-dimensional equations RANS for unsteady turbulent flows, Ramamurthy et al. (2007) have shown that at the dividing flows, the mean exit angle of the streamlines for flow entering the branch is larger at the surface compared to the exit angles of the streamlines located at the bottom.

As it was said before, the stream tube dimensions affect the rate of the suspended sediments delivery to the intake. The length and width of the stream tube change along with variations of the diversion flow ratio. With the help of experimental data as well as the three-dimensional model SSIIM2, Karami Moghadam et al. (2010) studied the stream tube cases of the main channel, with inclined and vertical bank, and drew conclusions about the stream tube width. They inferred that slopping the main channel bank improves the flow pattern and the stream tube width in an inclined bank case, in contrast to the vertical case, increases in the surface and decreases in the bed much to the reduction of the sediments delivery. Also it was found out that as the flow diversion ratio increases, the stream tube width increases in the surface more vigorously. So, when the discharge ratio grows, more excessive discharge is provided from the surface than from the bed, and, consequently, in case the main channel flow contains sediments, much less of them is delivered into the intake.

Although many researches are done on the flow pattern and the sediments in intakes, most of them are directed toward the transmission of the bed load and to the lateral intakes installed on rectangular channels, and none is carried out yet on the suspended load delivery and into the intakes installed on trapezoidal ones. So, in the present research, the case is treated with the 30° water intake installed on trapezoidal channels. The angles of intake recommended by researchers such as Novak et al. (1990) and Yang et al. (2009) are, respectively, 30° and 30°–45°.

2 Materials and Methods

To study the flow and sediments in rivers and channels with inclined bank, some experiments are carried out in a non-recirculating long flume with a 30° branch channel. The experimental model was built in the hydraulic laboratory of Chamran University, Ahwaz, Iran. Figure 3 shows the setting of the laboratory equipment. The main channel and lateral channel were 8- and 5-m long, with bed widths of

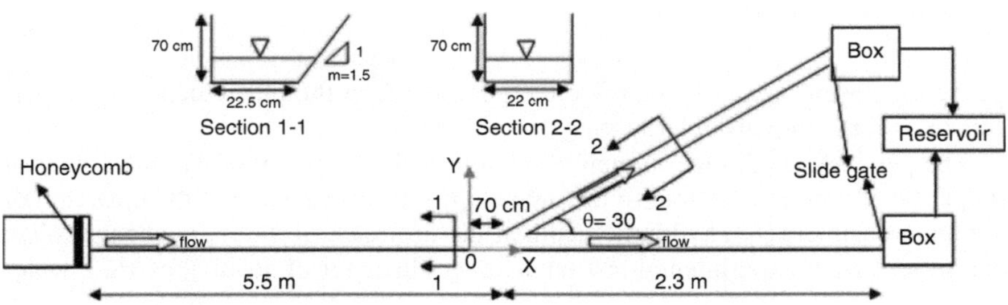

Fig. 3 The experimental equipment plan of the present study

22.5 and 20 cm, respectively. The main channel section was trapezoidal, and the branch channels were rectangular. The slope of the inclined bank was set at 1.5:1 ($m = 1.5$). The heights of both channels were chosen at 70 cm, and 10-cm thick Plexiglas is used as the channel walls. The branch channel was set at a distance of 5.5 m from the entrance of the channel. To adjust the discharge as well as the water depth in the channels, two sluice gates are installed at their ends. The water flow was issued from an underground source. To assure the flow expansion as well as low turbulence, a honeycomb was set up at the entrances of the main channel. The discharges from the main and branch channels were measured by means of two V-shaped weirs of 56° and 90°, respectively. The water depths in the upstream of the main channel were 10, 20, and 25 cm, and the chosen Froude numbers for the upstream of the main channel were 0.25, 0.30, 0.35, 0.40, and 0.45.

To perform any case of the experiment, first of all, the discharge of the main channel for the corresponding depth and Froude number was calculated, then this discharge was established in the main channel where both gates were completely down (free state), and after the flow became steady, the diversion flow ratio was measured using the V-shaped weirs. Afterward, the gates were brought up to the extent that both the diversion flow ratio and the desired depth were safeguarded. At the initial part of the main channel, there is a sediment injective source besides an electromotor with variable revolution which makes it possible that with different discharges, the sediment with the same concentration is injected. The applied sediments are from colored crystal with $\rho = 1.05$. At the end part of the main and branch channels, there exists a basket to trap the sediments. In the experiments concerning suspended sediments, using the electromotor with variable revolution, the rate of the injected sediments in the main channel upstream was so adjusted that in all the cases of the experiment the concentration would be the same and equal to $1\ \mathrm{g\ L^{-1}\ s^{-1}}$.

The sampling of the entered sediments in the branch channel was regularly performed and measured by trapping, and after reaching the steady state the main sampling has begun. Each sediment test took 90 min to be carried out. After the termination of the experiments, the trapped sediments were gathered, dried, and then weighed.

3 Dimension Analysis

Using dimension analysis, the effective parameters in the phenomenon were perceived, and the dimensionless ratios determined.

The parameters are the sediment discharge at the upstream of the main channel (Q_{su}), the sediment discharge in the intake (Q_{si}), flow discharge at the upstream of the main channel (Q_u), hydraulic depth at the upstream of the main channel (D_u), duration of each experimental test (t), average diameter of sediments (d_{50}), water density (ρ), sediment density (ρ_s), gravitational acceleration (g), bed roughness (k_s), fluid viscosity (v), shear velocity (u_*), the intake angle (θ), the stream tube width (B), and the stream width at a certain distance from the bed (T). Applying the Buckingham theory and eliminating the constant parameters, the dimensionless equation may be written as follows:

$$\frac{Q_{si}}{Q_{su}} = f\left(\frac{Q_i}{Q_u}, \frac{k_s}{D_u}, Fr, Re_*\right) \tag{1}$$

where Fr is the approaching Froude number at the upstream, $G_r = Q_{si}/Q_{su}$ is the ratio of entered sediment into the intake, $Q_r = Q_i/Q_u$ the diversion flow ratio, and Re_* is the shear Reynolds number of the particle. Since in all experiments the shear Reynolds number is greater than the supposed least value, it is abandoned. The ratio k_s/D_u is the roughness ratio, a parameter which is used in the analysis too.

4 Results and Discussion

Table 1 shows what is done in the laboratory and gives the corresponding explanations. As it is said before, the choice of the diversion flow is based on the free flow condition. Figure 4 displays the reverse proportion between the Froude number and the diversion flow ratio. The reason is that in the free state (the end gates are completely open), at a constant depth, the greater the Froude numbers are, the higher the flow velocity would be, as a consequence of which, in the intake extent the momentum force is not sufficient to divert the stream, which leads to the reduction of the diversion flow. The ratio k_s/D_u equal to 18.75E-6, 10.50E-6, and 8.73E-6 corresponds to, respectively, 10, 20 and 25 cm.

Because in the experiments the diversion flow ratios (Q_r) are different, the dimensionless parameter G_r/Q_r is taken into use to judge the suspended load ratio (G_r).

Table 1 The flow condition in this study

Side slope	Fr	Q_r	d_u (m)
1.5	0.25	0.445	0.1
	0.3	0.423	0.1
	0.35	0.37	0.1
	0.4	0.383	0.1
	0.45	0.363	0.1
	0.25	0.312	0.2
	0.3	0.297	0.2
	0.35	0.258	0.2
	0.4	0.3	0.2
	0.45	0.27	0.2
	0.25	0.312	0.25
	0.3	0.297	0.25
	0.35	0.258	0.25
	0.4	0.3	0.25
	0.45	0.27	0.25

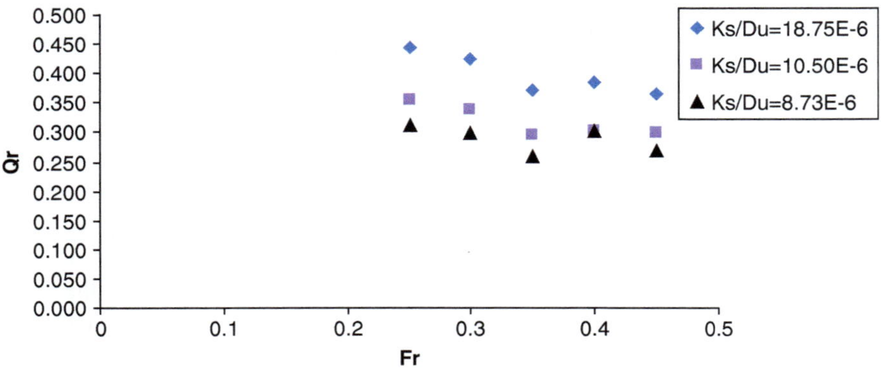

Fig. 4 The relation between the Froude number of the upstream and discharge ratio

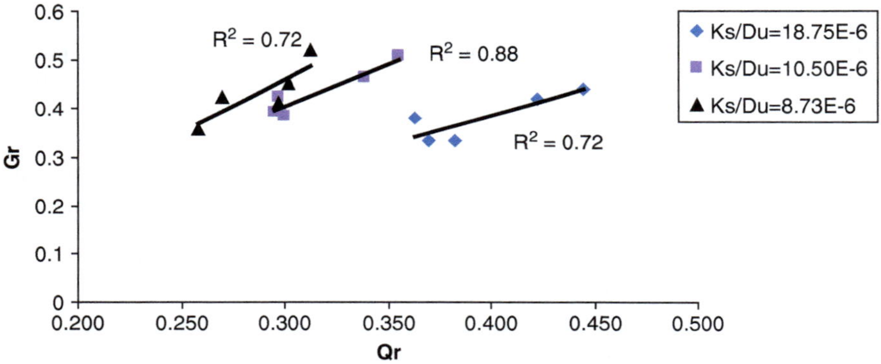

Fig. 5 The relation between Q_r and G_r in different roughness ratios

As said before, the sediment concentration in all the tests would be the same. Figure 5 shows the relation between Q_r and G_r.

It can be seen that at all three depths, the two ratios Q_r and G_r are in proportion to each other. It should be noted that the diversion flow ratios for each depth are chosen according to the flow free state, so for each depth they are positioned in a particular bound.

As one can see, the fitted slope line for a depth of 10 cm is less than for the other two depths. At this depth, due to the secondary current strength, when Q_r increases, the rate of increase in the sediments entry to the intake is less than at the other two depths. As stated by Raudkivi (1993), the secondary current strength and the infiltrating sediments decline along with the increase of the roughness ratio. The roughness ratio corresponding to a depth of 10 cm is high (k_s/D_u = 18.75E-6), followed by decrease of the infiltrating sediments. In Fig. 6, the results of this study

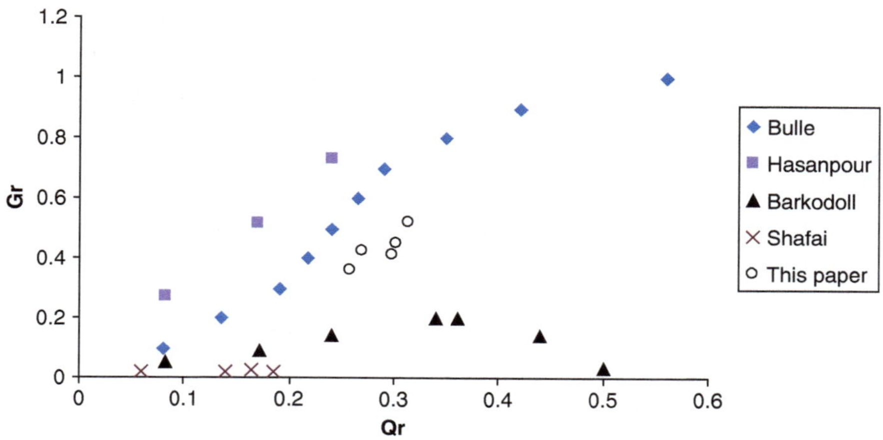

Fig. 6 Comparison between the results of this study and other

are compared with those of Barkdoll et al. (1999), Shafai Bajestan and Nazari (1999), Bulle (quoting from Schoklitsch 1937), and Hasanpour 2006). Of course, it is not so likely that the tests in different lab conditions be similar; however, by considering the process of changes, one can predict the effect of the wall slope as well as the suspended load. The main reason of the difference in the increase pattern of the infiltrating sediments along the increase of diversion flow ratio is the performance of the experiments with different Froude numbers under diverse geometric conditions of the main and branch channels.

In Hasanpour's (2006) work, an almost linear relation between the diversion flow ratio and the suspended load can be seen. There exists a turning point in Barkdoll's (1999) results from where on, any increase in the diversion flow ratio leads to a decrease in the infiltrating load ratio. The researches of Shafai and Nazari were done in a 90° bend intake with 60° intake angle, which causes much less sediment entry compared to other studies. Bulle made his experiments in 30° intake installed on a rectangular channel. Since in this study the tests are also performed with the same angle at a 25-cm depth with the most suspended load ratio, our results are in harmony with Bulle's.

A comparison makes it clear that with a constant infiltrating sediment ratio, the diversion flow ratio in this research is greater than that of Bulle, so G_r/Q_r ratio is less than in Bulle's research, which shows that slopping the channel wall has a positive role in the decrease of the suspended load infiltration. The infiltrating sediment ratio for depths of 10 and 20 cm is less than that for 25-cm depth; hence, the ratio G_r/Q_r at these two depths is surely smaller than Bulle's. The relation between G_r/Q_r and the Froude number at the upstream is presented in Fig. 7. It can be noticed that as the Froude number grows, in all the three depths we have a decline in the rate of infiltrating sediments to the intake, but in the bounded interval of Froude number 0.35–0.40 this rate reaches its minimum, and after these bounds it shows a relative

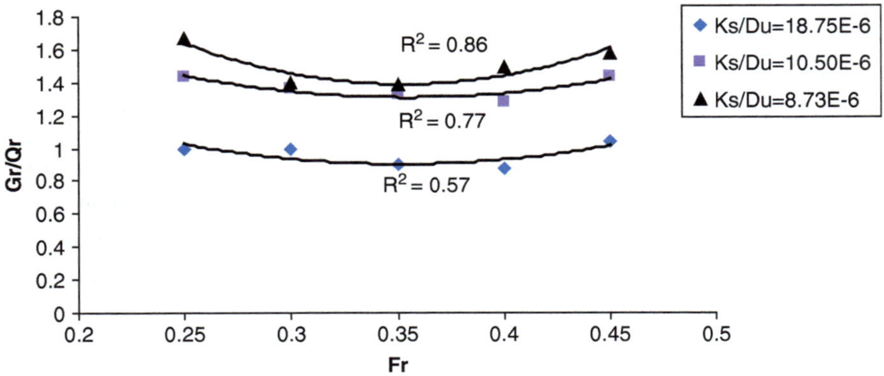

Fig. 7 The effect of Froude number of the upstream on the entering sediment ratio

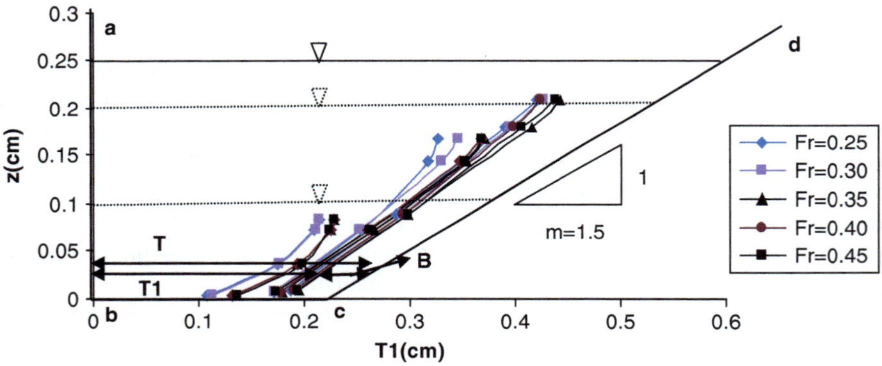

Fig. 8 The dividing stream surface (DSS) for different Froude numbers

increase. The results for the two depths, 20 and 25 cm, are similar, but the value of G_r/Q_r with a 10-cm depth shows a difference, the reason of which can be discerned in Fig. 8. This figure also shows the Dividing Stream Surface (DSS) for three depths: 10, 20, and 25 cm. It is seen that the stream tube dimensions in 20- and 25-cm depths are close to each other, and less than for 10 cm, owing to the fact that the diversion flow ratios at 20 and 25 cm are almost the same, but at a depth of 10 cm it is high (Table 1). At this depth ($k_s/D_u = 18.75\text{E-}6$), the transverse velocity distribution is such that G_r decreases too, making a considerable disparity between the value of G_r/Q_r ratio at the 10 cm and that for the other two depths.

According to Fig. 8, it is recommended that in the irrigation channels, after the determination of the diversion flow using the gates regulating the water surface, the depth of water should be so adjusted that the approaching Froude number falls into the interval 0.35–0.40, supplying a minimum G_r/Q_r.

5 Conclusions

In this research, with the help of some experiments, the entering suspended load to the 30° intake installed on the trapezoidal channel is studied. Using the obtained data, it is determined that the entering sediment ratio in a Froude number between 0.35 and 0.45 (in the upstream of the main channel) is minimal. Therefore, it is recommended that with the purpose of decrease in the suspended loads, the water depth in the irrigation channels to be so adjusted that the Froude number falls into this interval. Also it is proved that in a high roughness ratio, when the dividing flow ratio increases, the rate of increasing the sediments entry decreases. Generally, we conclude that the Froude number at the upstream of the main channel affects the diversion flow ratio. The diversion flow ratio, in turn, has its impression on the secondary current strength, and more importantly, on the stream tube dimensions, both of great importance in the rate of the suspended load entry to the intake.

Acknowledgement The authors are grateful to the Water Resources Management Company of Iran, Department of Applied Researches, for their financial support to this study.

References

Abassi AA, Ghodsian M, Habibi M, Neishabouri AAS (2002) Experimental investigation on sediment control in lateral intake using sill. In: Proceeding of the 13th IAHR-APD congress, vol 1, Singapore, pp 230–233

Barkdoll BD, Ettema R, Odgaard AJ (1999) Sediment control at lateral diversions: limits and enhancements to vane use. J Hydraul Eng, ASCE 125(8):826–870

Grace JL, Priest MS (1958) Division of flow in open channel junctions. Bulletin no. 31. Engineering experimental Station, Alabama Polytechnic Institute, Auburn

Hager WH (1984) An approximate treatment of flow in branches and bends. Proc Inst Mech Eng 198C(4):63–69

Hager WH (1992) Discussion of 'Dividing flow in open channels' by A. S. Ramamurthy, D. M. Tran, and L. B. Carballada. J Hydraul Eng 118(4):634–637

Hasanpour F (2006) On the behavior of the lateral intakes in the presence of composite submerged vanes and sill. Ph.D. thesis, College of Agriculture, Tarbiat Modares University, Tehran, Iran

Huang J, Weber LJ, Lai YG (2002) Three-dimensional numerical study of flows in open-channel junctions. J Hydraul Eng 128(3):268–280

Karami Moghadam M, Shafai Bajestan M, Sedghi H (2010) An experimental and numerical investigation at a 30 degree water intake in main channel with trapezoidal and rectangular section. J Sci Technol Agric Nat Resour (Isfahan University Technology, under publishing)

Law SW, Reynolds AJ (1966) Dividing flow in an open channel. J Hydraul Div 92(2):4730–4736

Murota A (1958) On the flow characteristics of a channel with a distributory. Technology Reports of the Osaka University, Japan, 6(198)

Neary VS, Odgaard AJ (1993) Three-dimensional flow structure at open-channel diversions. J Hydraul Eng, ASCE 119(11):1223–1230

Neary VS, Sotiropoulos F, Odgaard AJ (1999) Threedimensional numerical model of lateral-intake inflows. J Hydraul Eng 125(2):126–140

Novak P, Moffat A, Nalluri C (1990) Hydraulic structures. Pitman, London, p 546

Ramamurthy AS, Junying Qu, Diep Vo (2007) Numerical and experimental study of dividing open-channel flows. J Hydraul Eng, ASCE 133(10):1135–1144

Raudkivi AJ (1993) Sedimentation, exclusion and removal of sediment from diverted water – IAHR hydraulic structures design manual, No. 6. A.A. Balkema, Rotterdam

Razvan E (1989) River intake and diversion dams. Elsevier Science, New York

Schoklitsch A (1937) Hydraulic structures, vol 2 (trans: Shulits S). American Society of mechanical Engineers, New York, pp 722–751

Shafai Bajestan M, Nazari S (1999) The impression of the diversion angle of intake on the entering sediments to the lateral intakes at the vertical bond of river. J Agric Chamran Univ 22(1):47–66

Tanaka K (1957) The improvement of the inlet of the Power Canal. Transactions of the Seventh General Meeting of I.A.H.R., 1, 17

Taylor E (1944) Flow characteristics at rectangular open channel junctions. Trans ASCE 109:893–912

Thomson M (1949) Theoretical hydrodynamics. McMillan & Co, London

Weber LJ, Schumate ED, Mawer N (2001) Experiments on flow at a 90° open-channel junction. J Hydraul Eng 127(5):340–350

Yang F, Chen H, Guo J (2009) Study on "diversion angle effect" of lateral intake flow. 13th IAHR Congress, Vancouver, Canada

Estimation of River Banks Influence on Tachoida Shape at the Meridian

Zygmunt Meyer

1 Introduction

The concept of the so-called modified tachoida was presented in the author's previous paper (Meyer 2009a). The concept focuses on the problem of avoiding simplifications that were taken by Prandtl (1956) in formulating the so-called logarithmic tachoida. The proposed modified logarithmic tachoida makes it possible to define hydrodynamic equilibrium state for the flowing stream by introducing eddy viscosity coefficient of water at the bottom layer. The method also allows for the approximation of the model parameters, based upon data coming from measurements. The verification made for a set of data taken from the literature (Stone and Hotchkiss 2007) indicates that the tachoida measured at the meridian of the river differs from the one proposed by the author in the previous papers (Meyer 2009b). The reason is the assumption of a two-dimensional (2D), vertically plane model, taken for consideration. This suggests that three-dimensional (3D) model should be used to include the influence of river banks on the tachoida measured at the meridian of the river.

In the chapter, the mathematical description of the 3D approach is given and an example of the verification based upon experimental data is presented.

2 Mathematical Description of Phenomenon

Mathematical description of the water flow phenomenon when no tangential stresses are applied at the free water surface (no wind, no ice) was achieved under the following assumptions (Meyer 2009a):

$$\tau_{xz}(z) = \tau_{\mathrm{b}}\left(1 - \frac{z}{H}\right) \tag{1}$$

Z. Meyer
Department of Geotechnics, West Pomeranian University of Technology, Szczecin, Poland
e-mail: meyer@zut.edu.pl

P. Rowinski (ed.), *Experimental Methods in Hydraulic Research*, 215
Geoplanet: Earth and Planetary Sciences, DOI 10.1007/978-3-642-17475-9_14,

$$K(z) = K_0\left(1 + \frac{z}{z_0}\right) \tag{2}$$

$$\upsilon_x(0) = 0 \tag{3}$$

where H is the depth of the water, $K(z)$ is the eddy viscosity coefficient of water, K_0 is the eddy viscosity coefficient at the bottom, z, x is the basic set of coordinates (x directed horizontally, z vertically upward), z_0 is the linear measure of bed roughness, $\tau_{zx}(z)$ is the component of turbulent shear stress, and τ_b is the shear stress at the bottom level. These simplifications were discussed in the previous papers (Meyer 2009a, b). Under such assumptions, the solution has the following form:

$$\upsilon_x(z) = \frac{u_*}{k}\left[\frac{(1+U)\ln\left(1+U\frac{z}{H}\right)}{U} - \frac{z}{H}\right] \tag{4}$$

and

$$\upsilon_0 = \frac{1}{H}\int_0^H \upsilon_x(z)\mathrm{d}z = \frac{u_*}{k}\left[\frac{(1+U)^2\ln(1+U) - U(U+1)}{U^2} - \frac{1}{2}\right] \tag{5}$$

where

$$U = \frac{H}{z_0}, \quad u_* = \sqrt{gHI} \tag{6}$$

and similarly as in (Prandtl 1956):

$$z_0 = \frac{K_0}{k \cdot u_*} \tag{7}$$

where I is the slope of the free water surface, g is the acceleration due to gravity, k is the von Karmann constant, and u_* is the shear velocity. It can be noticed that for $x \gg 1$ (such cases often occur in practice):

$$\upsilon_x(z) = \frac{u_*}{k}\left[\ln U + \ln\frac{z}{H} - \frac{z}{H}\right] \tag{8}$$

and

$$\upsilon_0 = \frac{u_*}{k}\ln\left(\frac{U}{e^{3/2}}\right) \tag{9}$$

After dividing side by side, the basic relationship will be obtained in the following form:

$$\upsilon_x(z) = \upsilon_0 \left[1 + \frac{\frac{3}{2} - \frac{z}{H} + \ln\frac{z}{H}}{\ln\left(\frac{U}{e^{3/2}}\right)} \right] \tag{10}$$

The following symbols were used in the above equations: e = the base of Naperian logarithm, U = parameter of the model, $\upsilon_x(z)$ = water velocity, υ_0 = depth averaged water velocity. It is to be stressed that the von Karmann constant is still a matter of research (Nowh 1989).

Equation (10) gives good results until $z/H > 0.02$. Moreover, it can be noticed that for $z/H = 0.30701$, there is $\upsilon_x = \upsilon_0$. An example of the curve according to (10) is presented in Fig. 1.

Formula (10) is especially suitable for experimental data analyses, because if there is a set of field measurement data $\{\upsilon_i, z_i/H\}$, it is possible to calculate constant values of U and υ_0 in this formula, using the least square method.

$$\upsilon_0 = \frac{\sum\left(Y_i^2\right) \cdot \sum\left(\upsilon_i^2\right) - \left[\sum\left(\upsilon_i Y_i\right)\right]^2}{\sum\left(\upsilon_i\right) \cdot \sum\left(Y_i^2\right) - \sum\left(Y_i\right) \cdot \sum\left(\upsilon_i Y_i\right)} \tag{11}$$

$$\ln\left(\frac{U}{e^{3/2}}\right) = \frac{\sum\left(Y_i^2\right) \cdot \sum\left(\upsilon_i^2\right) - \left[\sum\left(\upsilon_i Y_i\right)\right]^2}{\sum\left(Y_i\right) \cdot \sum\left(\upsilon^2\right) - \sum\left(\upsilon_i\right) \cdot \sum\left(\upsilon Y_i\right)} \tag{12}$$

where

$$Y(z) = \frac{3}{2} - \frac{z}{H} + \ln\frac{z}{H}, \quad Y_i = Y(z_i) \tag{13}$$

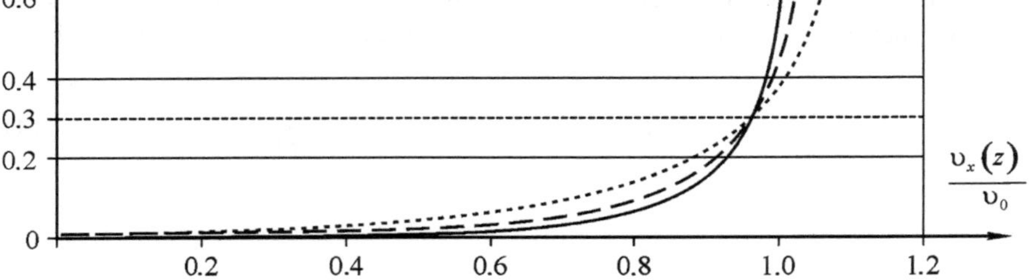

Fig. 1 Plot of curves $\upsilon_x(z, U)$ for different values of U

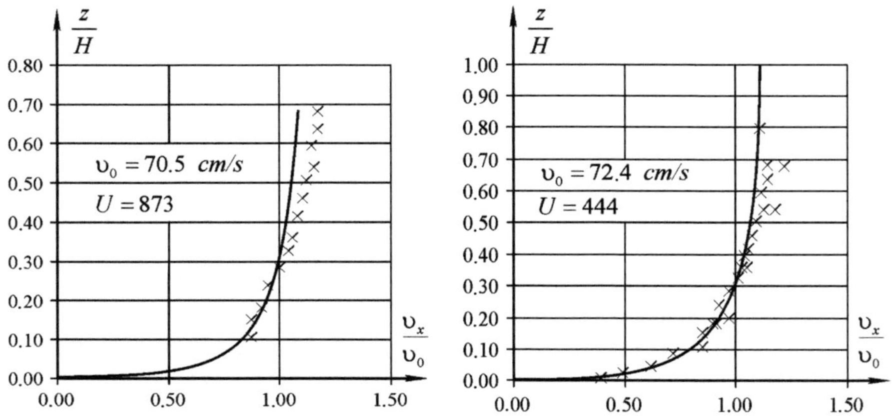

Fig. 2 Results of approximation using (10)–(12) for the modified tachoida (Data from Stone and Hotchkiss 2007)

The examples of results of such calculations are presented in Fig. 2. Data for calculation were taken from Stone and Hotchkiss (2007). The research performed by Stone and Hotchkiss (2007) was made at a broad natural river (width of ca. 100 m, and flow of ca. 300 m^3/s) whose cross-section shape was close to rectangular.

In Fig. 2, it can be seen that measured and calculated velocities differ from each other. This concerns especially the upper layer. The measured velocities at the water surface are bigger than the calculated ones. This comes from the influence of river banks and will be analyzed in the later part of this chapter.

3 Influence of River Banks upon Tachoida

The influence of river banks is a factor that affects the shape of modified tachoida. Taking the river banks into consideration requires a two-dimensional model, $\upsilon_x = \upsilon_x(z, y)$, to be introduced. This problem can be described in the following way: The two-dimensional problem needs a second component, the turbulent shear stress, to be included. So in addition to $\tau_{xz,}$ there is a new component, τ_{xy}. It is assumed that the components of shear stress can be described by the relationships:

$$\tau_{xz} = \rho \cdot K_z \cdot \frac{\partial \upsilon_x}{\partial z} = C_z \cdot f_1(y) \cdot \left(1 - \frac{z}{H}\right) \tag{14}$$

and

$$\tau_{xy} = \rho \cdot K_y \cdot \frac{\partial \upsilon_x}{\partial y} = C_y \cdot f_2(z) \cdot \left(1 - \frac{y}{B}\right) \tag{15}$$

where

$$C_z, C_y = \text{const} \tag{16}$$

and furthermore:

$$f_2(z) = \int \frac{\left(1 - \frac{z}{H}\right)}{K_z(z)} \cdot \mathrm{d}z \tag{17}$$

$$f_1(y) = \int \frac{\left(1 - \frac{y}{B}\right)}{K_y(y)} \cdot \mathrm{d}y \tag{18}$$

Moreover, as it has already been assumed,

$$K_z(z) = K_{0z}\left(1 + \frac{z}{z_{01}}\right) \tag{19}$$

$$K_y(y) = K_{0y}\left(1 + \frac{y}{z_{02}}\right) \tag{20}$$

and

$$z_{01} = \frac{K_{0z}}{k_z \cdot u_{*z}}; \quad z_{02} = \frac{K_{0y}}{k_y \cdot u_{*y}}$$

In the above equations, subindexes z and y indicate that the symbols are taken along certain axes. The scheme of the flow elements is given in Fig. 3. In the above equations, K_{0z} denotes the eddy viscosity coefficient at the bottom and K_{0y} denotes the eddy viscosity coefficient at the river bank. And furthermore, z_{01} is the linear measure of bed roughness, and z_{02} the linear measure at the river bank roughness; k_z and k_y denote von Karmann coefficients in vertical and horizontal directions, respectively.

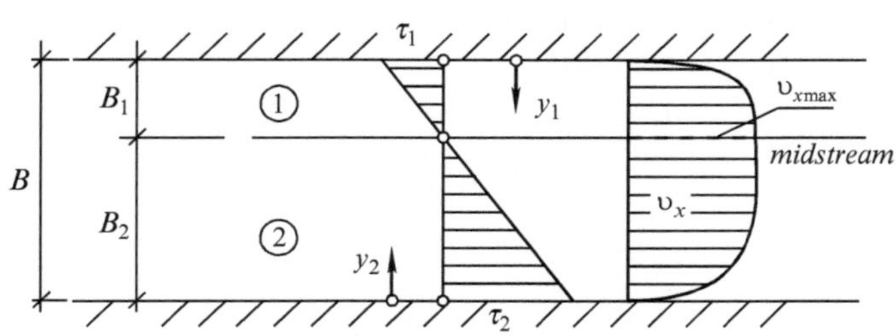

Fig. 3 Scheme of changes of flow elements in the horizontal plane

Moreover, it results from both the geometric relationships and earlier analysis of velocity changes in the horizontal plane (see Fig. 3) that:

$$\frac{\tau_1}{\tau_2} = \frac{B_1}{B_2} \tag{21}$$

It is convenient to assume for further calculation (to simplify notation) that the y axis is oriented from the point at the river bank toward the meridian. Then the velocity description for areas (1) and (2) is obtained in the following form:

For area (1)

$$\upsilon_x(y,z) = \upsilon_{01} \cdot \left[1 + \frac{\frac{3}{2} - \frac{y_1}{B_1} + \ln\frac{y_1}{B_1}}{\ln\left(\frac{U_1}{e^{3/2}}\right)}\right] \cdot \left[1 + \frac{\frac{3}{2} - \frac{z}{H} + \ln\frac{z}{H}}{\ln\left(\frac{U_3}{e^{3/2}}\right)}\right] \tag{22}$$

For area (2)

$$\upsilon_x(y,z) = \upsilon_{02} \cdot \left[1 + \frac{\frac{3}{2} - \frac{y_2}{B_2} + \ln\frac{y_2}{B_2}}{\ln\left(\frac{U_2}{e^{3/2}}\right)}\right] \cdot \left[1 + \frac{\frac{3}{2} - \frac{z}{H} + \ln\frac{z}{H}}{\ln\left(\frac{U_3}{e^{3/2}}\right)}\right] \tag{23}$$

and where υ_{01} and υ_{02} denote the averaged velocities in areas (1) and (2), respectively.

At the meridian, there must be $\upsilon_x(y_1,z) = \upsilon_x(y_2,z) = \upsilon_{\max}(z)$ for $y_1 = B_1$ or $y_2 = B_2$. Hence, there is an additional equation for calculating the relationship between υ_{01} and υ_{02}. The obtained relation reads:

$$\upsilon_{01} \cdot \left[1 + \frac{1}{2 \cdot \ln\left(\frac{U_1}{e^{3/2}}\right)}\right] = \upsilon_{02} \cdot \left[1 + \frac{1}{2 \cdot \ln\left(\frac{U_2}{e^{3/2}}\right)}\right] \tag{24}$$

Furthermore,

$$Q = \upsilon_0 \cdot B \cdot H = \upsilon_{01} \cdot B_1 \cdot H + \upsilon_{02} \cdot B_2 \cdot H \tag{25}$$

In the further analysis, it is convenient to put:

$$\varepsilon_i = \frac{1}{2 \cdot \ln\left(\frac{U_i}{e^{3/2}}\right)}, \quad i = 1,2,3 \tag{26}$$

which gives:

$$\upsilon_{02} = \upsilon_{01} \cdot \frac{1 + \varepsilon_1}{1 + \varepsilon_2} \tag{27}$$

and afterward:

$$\upsilon_0 = \frac{Q}{H\left[B_1 + B_2 \cdot \frac{1+\varepsilon_1}{1+\varepsilon_2}\right]} \tag{28}$$

The aim of this chapter is to define the river banks influence on tachoida at the meridian. We have:

$$\upsilon_x(y_1, z) = \upsilon_{01} \cdot \left[1 + \left(\frac{3}{2} - \frac{y_1}{B_1} + \ln\frac{y_1}{B_1}\right) \cdot 2 \cdot \varepsilon_1\right] \cdot \left[1 + \left(\frac{3}{2} - \frac{z}{H} + \ln\frac{z}{H}\right) \cdot 2 \cdot \varepsilon_3\right] \tag{29}$$

and putting $y_1 = B_1$ we get:

$$\upsilon_x(z, B_1) = \frac{Q}{BH} \cdot \frac{(1+\varepsilon_1)\cdot(1+\varepsilon_2)}{\frac{B_1}{B}\cdot(1+\varepsilon_2) + \frac{B_2}{B}\cdot(1+\varepsilon_1)} \cdot \left[1 + \left(\frac{3}{2} - \frac{z}{H} + \ln\frac{z}{H}\right) 2 \cdot \varepsilon_3\right] \tag{30}$$

Similar relationships will be obtained when area (2) is considered and $y_2 = B_2$ is assumed. The terms that appear in formula (30) have the following meaning:

$$\frac{Q}{BH} = \upsilon_0 \tag{31}$$

is the average velocity in the riverbed cross section. In the further analysis, we implement a coefficient k in the following form:

$$k = \frac{(1+\varepsilon_1)\cdot(1+\varepsilon_2)}{\frac{B_1}{B}(1+\varepsilon_2) + \frac{B_2}{B}(1+\varepsilon_1)} \tag{32}$$

In this way, it is possible to obtain $k = k(z)$, if n_1 and n_2, the Manning roughness coefficients of the river banks, vary with depth and the cross section is not rectangular:

$$\frac{\upsilon_x(z)}{\upsilon_0} = k(z) \cdot \left[1 + \frac{3}{2} - \frac{z}{H} + \ln\frac{z}{H}\right] \tag{33}$$

If $n_1 \neq n_2$, values of B_1 and B_2 must be first calculated in order to determine the location of the river meridian. This division will be found on the basis of the relationship:

$$\frac{B_1}{B_2} = \left[\frac{n_2 \cdot \ln\left(\frac{U_2}{e^{3/2}}\right)}{n_1 \cdot \ln\left(\frac{U_1}{e^{3/2}}\right)}\right]^{0,75} \quad \text{and} \quad B_1 + B_2 = B \tag{34}$$

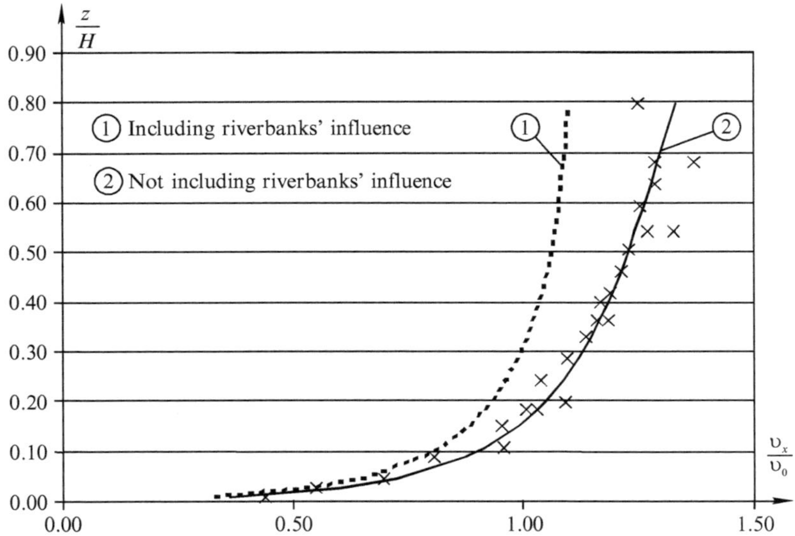

Fig. 4 Influence of river banks upon the diagram of modified tachoida (Data taken from Stone and Hotchkiss 2007)

We can also use the relationship between Manning roughness coefficient and the parameter U derived in the earlier paper (Meyer 2009a). So we have:

$$\begin{aligned} \ln U_2 &= \frac{3}{2}\left[-\ln n_2 + \sqrt{\ln^2 n_2 + 2 \cdot \ln n_2}\right] \\ \ln U_1 &= \frac{3}{2}\left[-\ln n_1 + \sqrt{\ln^2 n_1 + 2 \cdot \ln n_1}\right] \end{aligned} \tag{35}$$

The resultant approximation is shown in Fig. 4. It can be seen that there exists strong influence of river banks on the tachoida shape at the river meridian. And the problem cannot be treated generally as two dimensional; a three-dimensional approach must be applied. The results shown in Fig. 4 indicate that the river banks influence makes the flow more concentrated and so at the meridian we have higher water velocities, especially at the upper layer.

The author tried to use data presented in literature, that is, Helmiö (2001), to confirm the river banks influence. These data seem to confirm the model presented here.

4 Conclusions

1. The chapter presents the method for estimation of the river banks influence on the shape of tachoida at the river meridian.
2. It is a commonly accepted opinion that if the ratio of river's breadth to depth is big enough ($B/H > 30$), then a 2D model (i.e., in the z, x coordinates) can be

applied for the analysis. From the presented analysis it follows that even for the very broad river, there is an influence of river banks. Practically, the influence does not disappear. The author is conscious of the fact that the conclusions are drawn upon the turbulence model applied in the horizontal plane. So the analysis can be treated as an approach. But the results of experiments presented in the literature agree with the proposed model.

3. We also analyzed the river bank friction's influence upon the average flow velocity in the river cross section. In comparison to the velocity that was calculated by Chezy's formula, this value is smaller by ca. 6% according to the flow conditions. Every time, this value can be calculated from the formulae mentioned in this chapter.
4. Further analysis is needed to explain the role of movable bed in the creation of the eddy viscosity coefficient at the bottom and its influence on the river depth.

References

Helmiö T (2001) Friction measurements of ice cover: Theory and practice in river päntäneenjoki. In: Proceedings of the 2nd IAHR symposium on river, coastal and estuaries morphodynamics, Obihiro, Japan, 10–14 Sept 2001, pp 179–187

Meyer Z (2009a) Modified logarithmic tachoida applied to sediment transport in river. Acta Geophys 57(3):743–759

Meyer Z (2009b) An analysis of the mechanism of flow in ice-covered rivers. Acta Geophys 58(2):337–355

Nowh M (1989) The von Karmann coefficient in sediment laden flow. J Hydraul Res 27(4):477–499

Prandtl L (1956) Dynamika przepływów (Dynamics of flows; flows with free surface). Polish Scientific Publishers, Warsaw (in Polish)

Stone MC, Hotchkiss RH (2007) Evaluating velocity measurement techniques in shallow streams. J Hydraul Res 45(6):752–762

Statistical Analysis of Topography of Isvika Bay, Murchisonfjorden, Svalbard

Mateusz Moskalik and Robert J. Bialik

1 Introduction

It is common knowledge that the topography of Earth, other planets, and even moons and asteroids are formed by a variety of surface-forming processes acting at various spatial and temporal scales (Nikora and Goring 2005). The most important processes formed the topography of the bottom of bays in the region of Svalbard are tectonics and processes associated with glaciers, such as movement, erosion, sedimentation, and forming of glacial rivers. An analysis of these processes is vital not only for many aspects of geomorphology and sedimentology but also for coastal engineering and even fluvial hydraulic and hydrology, including the protection of coastal zones and the development and management of aquatic habitats.

In this chapter, we focus on the Isvika Bay bed topography by detailed statistical analysis based on the acoustic measurements carried out during Polish Polar Expedition (PPE) in 2009 (Salonen et al. 2010). In the first part of this chapter, the study area and measurement techniques are briefly described. In the next part, the results of statistical analysis are presented and investigated. For the purpose of this study, we followed the Nikora and Goring's (2004) approach proposed for the Mars topography. In addition, the comparison between the statistical analysis of Mars topography carried out by Nikora and Goring (2004, 2005, 2006) representing the global scale and topography of Isvika Bay representing the local scale is also shown. Such a study may help in verifying a methodology for identification of morphological forms and analyzing a leading mechanism in topography development in more complex regions such as Brepollen, also located at Svalbard archipelago, but in contrast to Isvika Bay, formed by six glaciers. This region will be analyzed in the near future, and because of its complexity we suspect very interesting results. We also believe that the proposed analysis, which is in fact a quite simple method based only on the statistical moments, may provide enough information about processes forming planetary surface in local as well as global scale.

M. Moskalik and R.J. Bialik (✉)
Institute of Geophysics, Polish Academy of Sciences, Ks. Janusza 64, Warsaw 01-452, Poland
e-mail: mmosk@igf.edu.pl; rbialik@igf.edu.pl

P. Rowinski (ed.), *Experimental Methods in Hydraulic Research*, Geoplanet: Earth and Planetary Sciences, DOI 10.1007/978-3-642-17475-9_15,

2 Study Area and Measurement Techniques

The Isvika Bay (80°N 18°40′E) is located inside Murchisonfjorden in the Vestfonna of Nordaustlandet, which is the northeastern island of the Svalbard archipelago (see Fig. 1). The analyzed bay was formed by ice-flow patterns three times in the past and now is under the influence of glacial rivers and marine sedimentation. The data used in this study were collected during the expeditions of PPE and the IPY Kinnvika project in summer 2009. Measurements were done with multibeam sonar "Sea Beam 1180" mounted on board of a research vessel Horyzont II. The vessel position and the sound velocity in water were taken into account during the analysis. Preliminary data analysis consisted of calculating the depth on the 1-m grid size. Ordinary Kriging interpolation method was used to supplement the lacking data. Processed data were used for further statistical investigation.

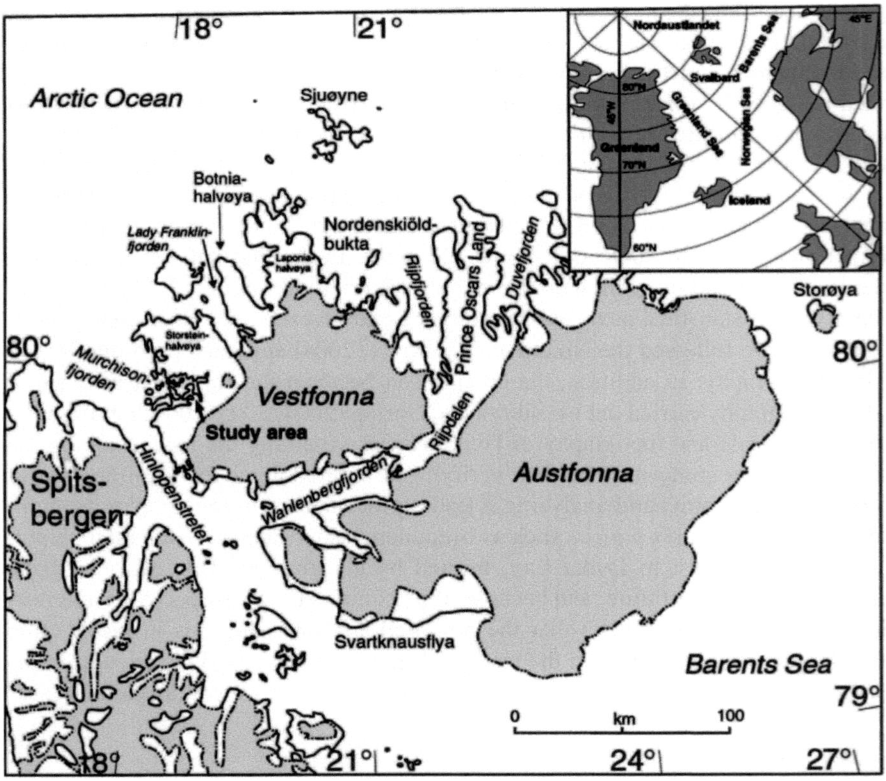

Fig. 1 Index map indicating the study area on the southeastern side of Murchisonfjorden, Vestfonna, Svalbard (Kaakinen et al. 2009)

3 Results

The statistical analysis was done based on the acoustic measurements for the $10 \times 10\ m^2$ cell scale. The maps of the mean elevation, variance (σ^2), skewness (Sk), and kurtosis (Ku) are shown in Fig. 2. The first parameter presented in Fig. 2a shows two distinct parts of the Isvika Bay, which are separated by an underwater ridge. The southwestern part is three times bigger and deeper than the northeastern one, which suggests that the processes forming the bed of smaller bay are connected with river sedimentation rather than with glacier erosion.

However, more interesting information comes from the variance of elevation, which is presented in Fig. 2b. This parameter is connected with power spectrum and may depend on the cell scale and grid size (Nikora and Goring 2004). Because this value gives important information about diversification of elevation, we therefore suggest that it may be associated with gradient of slopes. In order to analyze this relationship, the map of variance in logarithmic scale (Fig. 3a) was compared to the map of the slope inclination (Fig. 3b). In contrast to Fig. 2b, which is presented in the real scale, the adoption of the natural logarithmic scale (Fig. 3a) allows a clearer

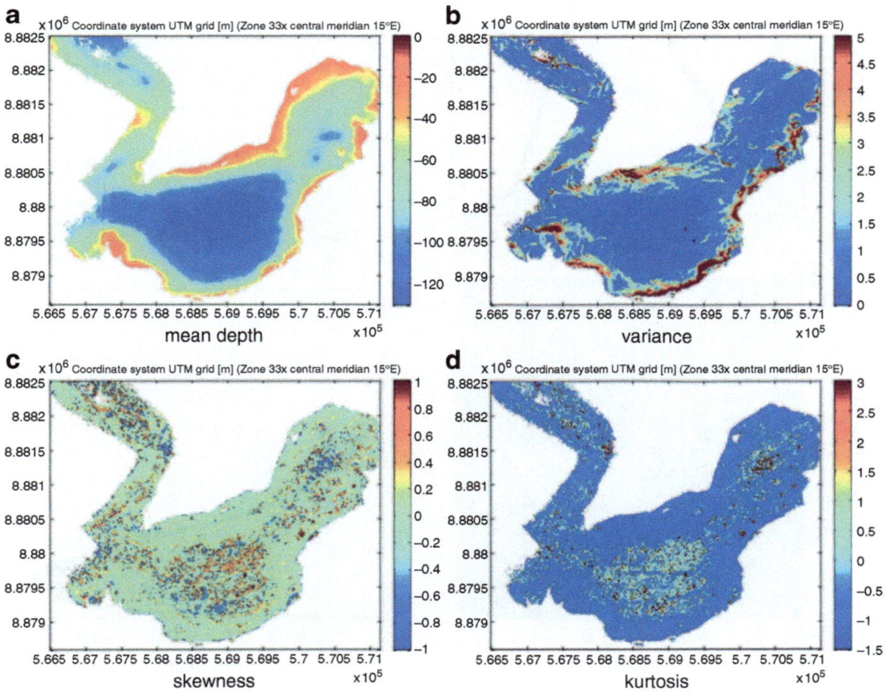

Fig. 2 Maps of the cell-scale statistical moments of surface elevation: (**a**) mean, (**b**) variance, (**c**) skewness, and (**d**) kurtosis

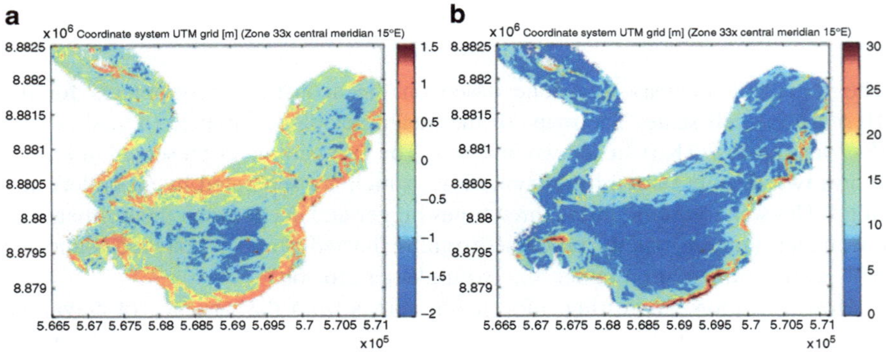

Fig. 3 Map of the variance in logarithmic scale (**a**) and map of the inclination slope (**b**)

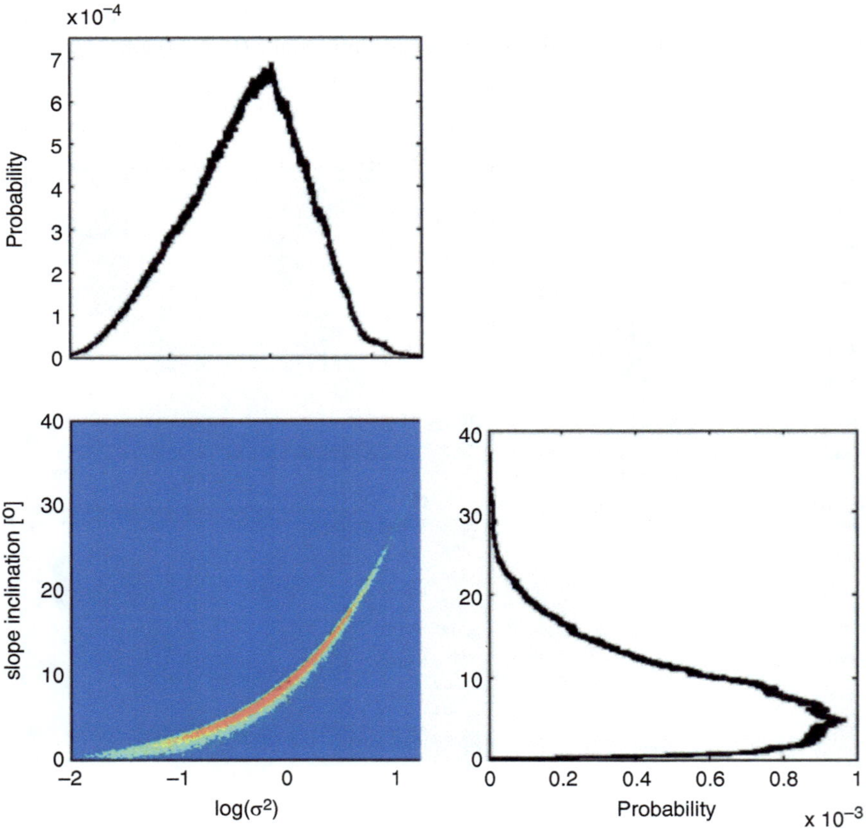

Fig. 4 The joint and marginal distributions of the variance in logarithmic scale and slope inclination

presentation of the regions in which the variance is of higher values than 3, which is directly connected with the slopes inclination (Fig. 3).

Figure 4 shows the probability distribution of the logarithmic value of variance and slope inclination together with their marginal distribution. It is easy to notice that there is a relationship between these two coefficients. We can also see in Fig. 4 that the surface of Isvika has slopes of inclination from 0° to 20° with the maximum value of PDF for 7.3°, which is related with variance from 0.05 to 10 with the maximum value of PDF for 1.

Figure 2c, d shows spatial distribution of skewness and kurtosis of surface elevation, respectively. It is interesting that the skewness coefficient is equal to 0 and kurtosis is lower than 0 only in the parts close to the slopes, which is directly connected with values of variance, which has the maximum values in these regions. It is also easy to see in Fig. 2d that the kurtosis may be connected with the erosion

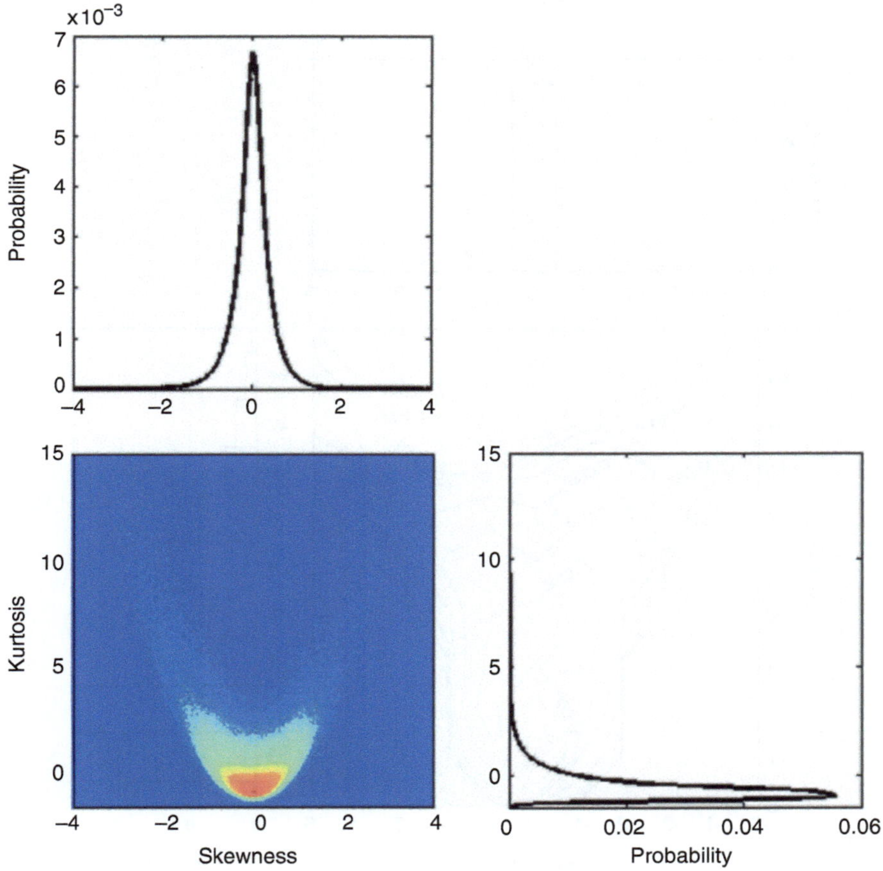

Fig. 5 The joint and marginal distributions of the cell-scale skewness and kurtosis coefficients for topography of Isvika Bay

forms such as couloirs, which may be recognized in this figure in the regions presented close to the slopes.

More interesting information is provided by the probability distribution $P(\mathrm{Sk}_z, \mathrm{Ku}_z)$ of the skewness and kurtosis coefficients of the Isvika surface elevation "z" together with their marginal distribution:

$$P(\mathrm{Sk}_z) = \int\limits_{\mathrm{DKu}} P(\mathrm{Sk}_z, \mathrm{Ku}_z)\mathrm{dKu}_z \text{ and } P(\mathrm{Ku}_z) = \int\limits_{\mathrm{DSk}} P(\mathrm{Sk}_z, \mathrm{Ku}_z)\mathrm{dSk}_z$$

which are presented in Fig. 5, where D denotes integrating domain. Similar plots have been already published in a number of studies associated with analysis of topography of the planets (Nikora and Goring 2004 (see Fig. 6) or Aharonson et al. 2001); however, for the best of our knowledge, it is the first time that such figure is presented in the relation to topography of exemplary polar region representing the

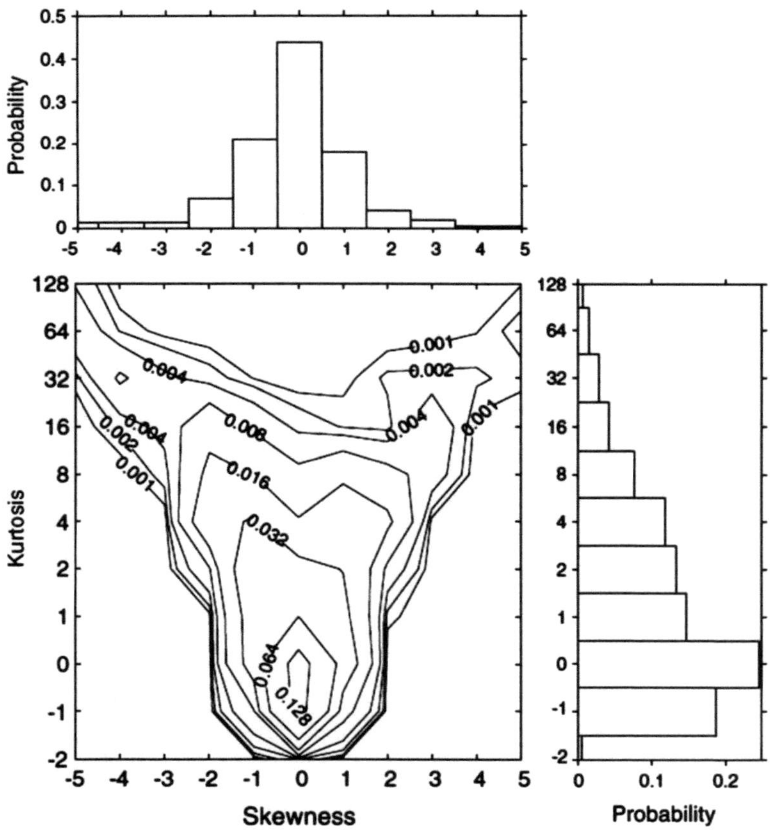

Fig. 6 The joint and marginal distribution of the cell-scale skewness and kurtosis coefficients for Mars topography (Nikora and Goring 2004)

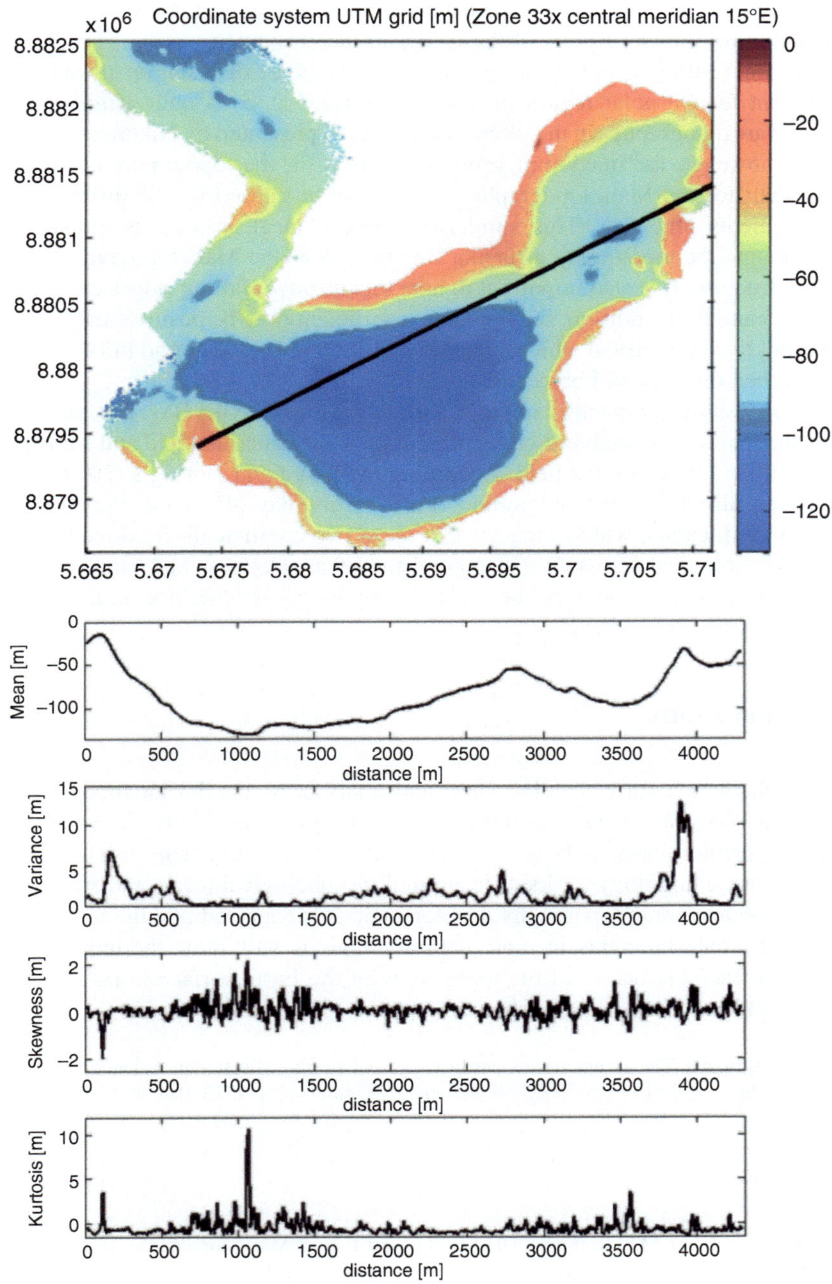

Fig. 7 Typical exemplary surface profile

local scale. The shape of the joint distribution suggests that the Gaussian model may be suitable for the description of the Isvika surface elevation.

It is interesting to note that the shape of the joint distribution of statistical coefficient for the polar region (in the local scale, Fig. 5) is quite similar to that one of Mars topography (in the global scale, Fig. 6) presented by Nikora and Goring (2004); however, the maximum value of kurtosis for the topography of Isvika is lower than for the Mars topography, which may be caused by the differences in measurements precision. This similarity between local as well as global scale suggests that the topography of Isvika Bay as well as the Mars topography have a fractal structure. It is also important to note that also two falling ridges can be seen in Figs. 5 and 6. In addition, the skewness coefficient may be positive and negative and is nearly symmetrical which suggests that if $Sk < 0$ we can find hills and when $Sk > 0$ then there should be deeps.

Figure 7 shows a typical example of surface profile of Isvika Bay. It is easy to see that some part of the analyzed bay represent the near-Gaussian field and some parts represent the non-Gaussian field. According to Nikora and Goring's (2004) analysis, it can also be seen that morphological forms may affect the skewness and kurtosis coefficients, which suggests that these two coefficients are directly linked with morphological forms. We also confirmed that skewness and kurtosis coefficients may be used to analyze the surface morphology for the local-scale topography represented by Isvika Bay.

4 Conclusions

In this chapter, a fully detailed statistical analysis of the Isvika topography is presented. The bay surface is thoroughly investigated and it is shown that the proposed method based only on the statistical moments may help in building and verifying a methodology for identification of morphological forms and analyzing a leading mechanism in topography development. It is expected that the future depth studies and measurements in more complex regions will show the unknown but suspected fractal structure of the topography of the Earth surface in polar regions which will be studied in the forthcoming paper.

Acknowledgments The research was partly supported by the Ministry of Science and Higher Education: Special Project No. IPY/279/2006 and by Grant No. N N525 350038

References

Aharonson O, Zuber MT, Rothman DH (2001) Statistics of Mars' topography from the Mars Orbiter Laser Altimeter: slopes, correlations, and physical models. J Geophys Res 106 (E10):23723–23735. doi:0148-0227/01/2000JE001403

Kaakinen A, Salonen V-P, Kubischta F, Eskola OK, Oinonen M (2009) Weichselian glacial stage in Murchisonfjorden, Nordaustlandet, Svalbard, Boreas. Int J Q Res 38(4):718–729. doi:10.1111/j.1502-3885.2009.00092.x

Nikora V, Goring D (2004) Mars topography: bulk statistics and spectral scaling. Chaos Solitons Franctals 19:427–439. doi:10.1016/s0960-0779(03)00054-7

Nikora V, Goring D (2005) Martian topography: scaling, craters, and high-order statistics. J Math Geol 37(4):337–355. doi:10.1007/s11004-005-5952-4

Nikora V, Goring D (2006) Spectral scaling in Mars topography: effect of craters. Acta Geophys 54(1):102–112. doi:10.2478/211600-006-0009-8

Salonen V-P, Kubischta F, Ojala AEK, Moskalik M (2010) Late Quaternary glaciomarine sedimentation in the Arctic bay Isvika, Nordaustalndet, Svalbard, ISC 2010 Mendoza

Assessing Validity of the Dead Zone Model to Characterize Transport of Contaminants in the River Wkra

Magdalena M. Mrokowska and Marzena Osuch

1 Introduction

Application of a model should always be accompanied with the awareness of its limitations in representing the reality. Besides, it is not only the model alone that is responsible for the quality of results, but also data used in a calibration process. In fact, both the errors incorporated in a model and those incorporated in data affect the model performance. Hence, parameter estimation should be performed cautiously.

When it comes to environmental models, it is often the case that many sets of parameters giving equally good predictions exist, which is called equifinality (Tychonoff and Arsenin 1977; Beven 2008). Then deterministic predictions without estimation of associated uncertainty could lead to incorrect results. Hence, taking into account uncertainty of results is crucial in a calibration process (Blasone et al. 2008; Ratto et al. 2001). In the chapter, an example of deterministic model calibration combined with assessment of solutions uncertainty is presented.

In this study the one-dimensional dead zone model has been applied. The dead zone model is an extension of the Fickian model, as it describes not only contaminants transport in a main channel, but also incorporates exchange of contaminants between the main channel and storage zones (Bencala and Walters 1983).

The objective of this chapter is to evaluate the validity of the one-dimensional dead zone model to characterize transport of pollutants basing on a tracer test performed on the River Wkra. In the chapter, we show that a deterministic solution to the dead zone model is not unique, as many solutions with acceptable fit to observations exist. As a result, a model output is associated with the uncertainty. To assess the uncertainty, and to identify its sources, the sensitivity analysis (SA) and uncertainty analysis are applied.

M.M. Mrokowska (✉) and M. Osuch
Department of Hydrology and Hydrodynamics, Institute of Geophysics Polish Academy of Sciences, Księcia Janusza 64, 01-452 Warsaw, Poland
e-mail: m.mrokowska@igf.edu.pl

P. Rowinski (ed.), *Experimental Methods in Hydraulic Research*, Geoplanet: Earth and Planetary Sciences, DOI 10.1007/978-3-642-17475-9_16,

2 Study Area and Tracer Test

The tracer test was conducted on the River Wkra, which is located in Central Poland, by the Institute of Geophysics, Polish Academy of Sciences, in 1997 (Owczarczyk et al. 1997). The study reach has a length of 5 km (between points A and B in Fig. 1). During the test, instantaneous release of a passive and conservative tracer – rhodamine B – was performed in cross-section P-0 (Fig. 1). Then, a tracer plume was sampled at five cross-sections below the injection point. Next, the time distribution of tracer concentrations was derived from measurements. The tracer test was performed in steady flow conditions for the following flow rates: $Q1 = 4.18\ \text{m}^3/\text{s}$, $Q2 = 3.97\ \text{m}^3/\text{s}$, $Q3 = 4.32\ \text{m}^3/\text{s}$, which resulted in 15 breakthrough curves (five curves per one flow rate).

3 Model Applied

Morphology of the reach under study indicates that pollutants transport could be affected by transient storage, which was described in previous studies of the reach (Rowiński et al. 2004). Thus, in the study the dead zone model, which incorporates exchange with storage zones (Bencala and Walters 1983), was applied.

The dead zone model applied in the study is represented by the set of equations (Bencala and Walters 1983; Runkel 1998, 2000):

$$\frac{\partial C}{\partial t} = -\frac{Q}{A}\frac{\partial C}{\partial x} + \frac{1}{A}\frac{\partial}{\partial x}\left(AK_x\frac{\partial C}{\partial x}\right) + \alpha(C_S - C) \tag{1}$$

$$\frac{\partial C_S}{\partial t} = \alpha\frac{A}{A_S}(C_S - C) \tag{2}$$

where A is the main channel cross-sectional area (m^2), A_S the storage zone cross-sectional area (m^2), C the main channel solute concentration (kg/m^3), C_S the storage

Fig. 1 The reach under study

zone solute concentration (kg/m^3), K_x the longitudinal dispersion coefficient (m^2/s), Q the volumetric flow rate (m^3/s), t the time (s), x the distance (m), and α is the storage zone exchange coefficient (1/s).

The first equation represents mass balance in a main channel, while the other one describes exchange of solute between a main channel and a storage zone.

According to the model, mass exchange between a main channel and storage zones occurs due to difference of concentrations between a main channel and storage zones, and it is proportional to the difference (Bencala and Walters 1983).

4 Methods

4.1 Deterministic Calibration of the Dead Zone Model

First of all, calibration was performed to specify parameter values that give the best fit of the simulated concentrations to the observed ones. Calibration was performed for flow rates $Q2$ and $Q3$ for four sub-reaches separately (Fig. 1: sub-reach one between P-1 and P-2 and so on). A breakthrough curve from an upstream cross-section of a sub-reach posed to be a boundary condition, and the output curve was computed for a downstream cross-section. As a result, eight calibrated breakthrough curves were obtained for further analyses. Results of tracer test for $Q1$ were used during verification, which is beyond the scope of this chapter.

During calibration, for each sub-reach the least square objective function was applied in the following form:

$$F = \sum_{t=1}^{t=n} \left(C_{\text{o,t}} - C_{\text{s,t}}\right)^2 \tag{3}$$

where n is the number of observed values, $C_{\text{o,t}}$ the observed concentration at a time step t, and $C_{\text{s,t}}$ is the simulated concentration at a time step t.

In the study, the Differential Evolution (DE) method (Price 2005), which is a global optimization technique, was applied.

Four parameters were estimated in the dead zone model: A – channel cross-sectional area (m^2), K_x – longitudinal dispersion coefficient (m^2/s), A_S – storage zone cross-sectional area (m^2), and α – storage zone exchange coefficient (1/s).

Ranges of parameter values for the DE method were chosen as follows: (0,100) for a channel cross-sectional area (A), a longitudinal dispersion coefficient (K_x), and a transient storage zone cross-sectional area (A_S) and (0,2) for a storage zone exchange coefficient (α). A starting point was set randomly. Optimal values of parameters are presented in Fig. 2. Parameter sets obtained by the DE method result in breakthrough curves that are in a good agreement with observations (Fig. 3).

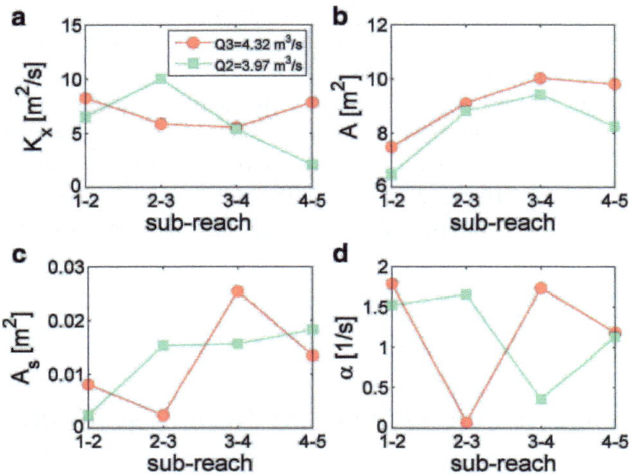

Fig. 2 Optimal values of the dead zone model parameters derived by means of differential evolution [DE] method for flow rates $Q2$ and $Q3$

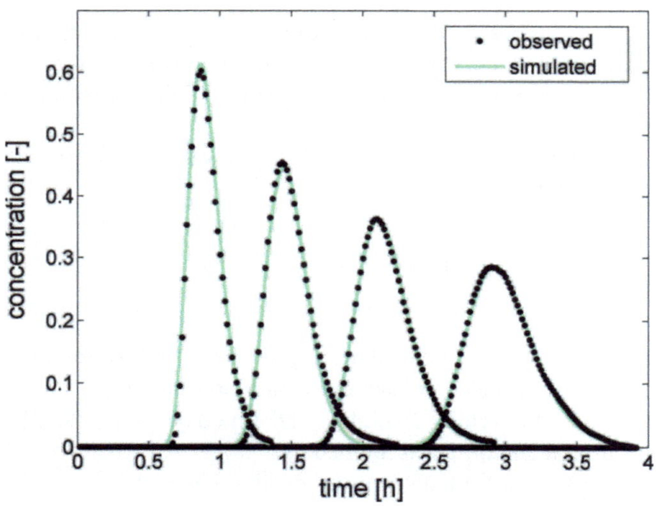

Fig. 3 Observed and simulated (differential evolution [DE] method) concentration–time distributions for flow rate $Q3$

Results obtained by DE method were evaluated by applying a coefficient of determination R^2. The coefficient of determination R^2 was derived for eight simulated breakthrough curves. The obtained values are between 0.991 and 0.999, which means that obtained parameter sets give acceptable results ($R^2 = 1$ for a perfect fit).

4.2 Monte Carlo Simulations: Evaluation of Response Surfaces

The Monte Carlo (MC) simulations were applied to prepare response surfaces in order to assess whether the unique optimal parameter values exist. First, 100,000 parameter sets composed of four parameters (K_x, A, A_s, and α) were created. Each parameter was drawn from a uniform distribution over the following intervals: K_x [1; 17], A [5; 12], A_s [0.01; 0.5], α [0.03; 3].

For each parameter set a simulation was computed, and a value of the least square objective function (the same as during calibration) was determined. Next, the response surfaces were created. An example of the response surface planes is shown in Fig. 4. Analysis of the response surfaces shows that it is possible to identify minimum values of objective function in the case of two parameters: K_x and A. On the other hand, unique parameter values do not exist for the objective function when it comes to A_s and α.

The dead zone model seems to be a case of the equifinality, since many sets of parameters exist that give equally good predictions. In this case, unique choice of parameters is not possible either by means of applied optimization method or by Monte Carlo simulations. The Monte Carlo simulations show that the choice of optimal parameter values is very uncertain.

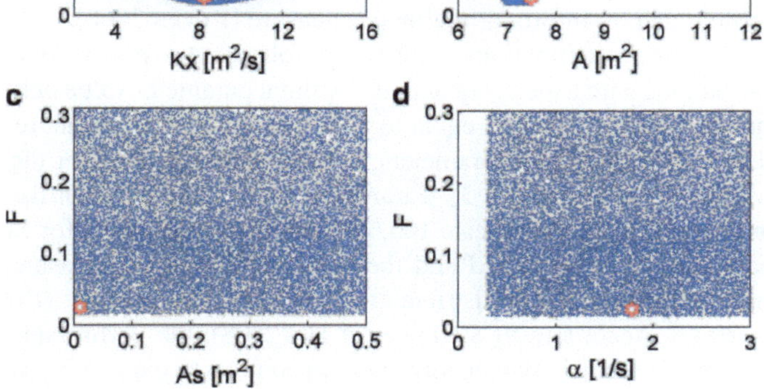

Fig. 4 Comparison between results of deterministic calibration and results of Monte Carlo (MC) simulations for sub-reach 1–2, flow rate $Q3$, on the basis of least square objective function F

4.3 Uncertainty Analysis

Taking into consideration that the dead zone model is deterministic and that equivalence of many parameter sets occurs, the uncertainty associated with the model predictions was estimated. Main sources of uncertainty are measurement errors, errors associated with model structure and numerical solution (Aronica et al. 1998; Ratto et al. 2001; Beven 2008; Blasone et al. 2008; Smith et al. 2008).

It is often the case that decomposition of a model error is difficult, especially when it comes to nonlinear hydrological models, and thus it is not possible to specify the statistical error model (Beven 2006; Smith et al. 2008). Because of the lack of knowledge about error model, the Bayesian Theorem does not apply and "pseudo-Bayesian" approach (Mantovan and Todini 2006) is often used (Beven 2006; Smith et al. 2008), which is called the Generalized Likelihood Uncertainty Estimation (GLUE). Comparing to Bayesian approach, in the GLUE method a performance criterion (informal likelihood function), which does not include an error model, is used instead of formal likelihood function derived from a stochastic error model (Smith et al. 2008). An informal likelihood measure evaluates acceptability of the parameter set according to the degree of simulated values fit to observed data and does not include statistical error model (Smith et al. 2008). It is required that a likelihood measure equals zero for unrealistic models and increases monotonically with the model fit to the data (Romanowicz and Beven 2006; Beven 2008).

Some scientists argue that GLUE methodology is not formally Bayesian and thus not statistically correct (Montanari 2005; Moradkhani et al. 2005; Mantovan and Todini 2006). However, Stedinger et al. (2008) showed that the GLUE methodology gives results consistent with "widely accepted and statistically valid analyses" when a statistically valid likelihood function is used.

According to the GLUE concept, it is possible to estimate a posterior distribution of parameters given a prior parameters distribution based on the knowledge of a modeled system and an informal likelihood measure (Beven 2008).

In the study, prior distributions were set as follows: for parameters K_x and A, normal distributions with a mean equal to the optimal parameter value derived by DE method and a standard derivation equal to 20% and 10% of the mean, respectively, for K_x and A; for the other two parameters, uniform distributions over the following intervals: A_s [0.01; 0.5], α [0.03; 3]. Parameters were sampled from the described distributions by MC method to create 100,000 parameter sets. Next, for each parameter set simulation was computed and the informal likelihood measure, which is proportional to the Gaussian distribution function, was applied (Romanowicz and Beven 2006). According to Stedinger et al. (2008), the performance measure implemented in this study is statistically valid, in contrast to many other widely used performance measures. Next, 95% confidence intervals were evaluated, which are illustrated in Fig. 5.

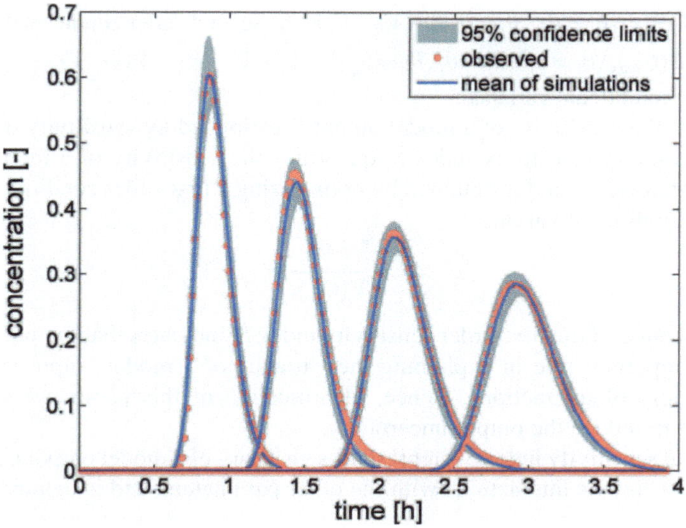

Fig. 5 Comparison of observed (*dots*) and simulated (*solid line*) concentrations of tracer at cross-sections 2, 3, 4, and 5 with 95% confidence bounds shown as *shaded areas*

4.4 Sensitivity Analysis

The sensitivity analysis (SA) aims at establishing the effect of uncertainty in model parameters on the uncertainty of model output. Hence, the SA is very useful in studying a model structure and in identifying sources of uncertainty (Ratto et al. 2001; Saltelli et al. 2004). The main objectives of the SA are: identification of irrelevant parameters to simplify a model, identification of parameters that affect the output in order to prioritize them in a calibration process; understand a model structure, i.e., identify interactions between parameters (Saltelli et al. 2004, 2008).

In the study, a global sensitivity analysis (GSA) method based on the Sobol' decomposition was applied. It is one of variance-based methods (ANOVA – ANalysis Of VAriance) (Archer et al. 1997; Sobol 2001).

In the chapter, the GSA is presented for the maximum concentration $C_{\max}$ as the model output and $\theta_i = [K_x, A, A_s, \alpha]$ as the model parameters. In this case, decomposition of an output unconditional variance $V(C_{\max})$ takes the following form:

$$V(C_{\max}) = V_{K_x} + V_A + V_{A_s} + V_\alpha + V_{K_xA} + V_{K_xA_s} + V_{K_x\alpha} + V_{AA_s} + V_{A\alpha} + V_{A_s\alpha} + V_{K_xAA_s} + V_{K_xA\alpha} + V_{K_xA_s\alpha} + V_{AA_s\alpha} + V_{K_xAA_s\alpha}$$

where $V_i = V[E(C_{\max}\backslash\theta_i = \theta_i^*)]$ is the first-order conditional variance (for fixed value of θ_i the mean of maximum concentration is calculated when other parameters vary. Next, the variance of results over all values of parameter θ_i is taken).

$V_{i,j} = V\left[E(C_{\max}\backslash\theta_i = \theta_i^*, \theta_j = \theta_j^*)\right] - V_i - V_j$ is the second-order conditional variance. $V_{i,j,k} = V\left[E(C_{\max}\backslash\theta_i = \theta_i^*, \theta_j = \theta_j^*, \theta_k = \theta_k^*)\right] - V_i - V_j - V_k - V_{i,j} - V_{i,k} - V_{j,k}$ is the third-order conditional variance.

In GSA, the sensitivity of a model output is estimated by sensitivity indices.

The first-order sensitivity index S_{i} quantifies the sensitivity of a model output $C_{\max}$ to parameter θ_i and is obtained by normalizing a first-order conditional by the output unconditional variance:

$$S_{\mathrm{i}} = \frac{V_i}{V(C_{\max})}$$

A high value of the first-order sensitivity index S_{i} indicates that the parameter θ_i plays an important role in explaining the variance of a model output singularly, independently of interactions. Hence, identification of this parameter should be prioritized to reduce the output uncertainty.

The total sensitivity index S_{Ti} represents sensitivity of a model output $C_{\max}$ to the parameter θ_i and its interactions with the other parameters, and is defined as:

$$S_{\mathrm{Ti}} = \frac{V(C_{\max}) - V\left[E(C_{\max}\backslash\theta_{-i} = \theta_{-i}^*)\right]}{V(C_{\max})}$$

where θ_{-i} are all parameters except of θ_i.

A high value of the total sensitivity index S_{Ti} corresponds to parameter θ_i affecting a model output both singularly and with interactions, which means that the parameter is relevant. A low value of the total effect S_{Ti} indicates that the parameter is unessential and its value could be fixed to simplify a model, and consequently, to reduce the uncertainty.

In the study, the Sobol' method, which uses Monte Carlo simulations, was applied to estimate sensitivity indices. In order to prepare Sobol' samples, Monte Carlo simulations of 100,000 parameter sets were performed. Parameters were drawn from the same distributions as described in Sect. 4.3.

Results of the SA are presented in Tables 1 and 2. According to the main effect S_{i}, parameters A and K_x are significant in explaining the output variation. Consequently, the uncertainty of results depends mostly on these two parameters. In other words, during model calibration the main effort should be put on evaluating A and K_x to reduce the uncertainty.

On the other hand, parameters characterizing transient storage, α and A_{s}, have no impact on the results, as both sensitivity indices are near to zero. It means that transient storage process is irrelevant in the studied case.

4.5 Simplification of the Dead Zone Model to the Fickian Model

As was shown in the sensitivity analysis, two model parameters, A and K_x, are sufficient to model contaminants transport in the studied reach, and α and A_{s} have

Table 1 First-order sensitivity index S_i of the dead zone model parameters

Sub-reach	Flow rate	K_x (m^2/s)	A (m^2)	A_S (m^2)	α(1/s)
1–2	$Q1$	0.296	0.702	0.015	0.000
	$Q2$	0.317	0.677	0.014	0.000
	$Q3$	0.318	0.682	0.012	0.000
2–3	$Q1$	0.315	0.680	0.008	0.000
	$Q2$	0.303	0.701	0.009	0.000
	$Q3$	0.315	0.684	0.007	0.000
3–4	$Q1$	0.310	0.690	0.007	0.000
	$Q2$	0.316	0.677	0.007	0.000
	$Q3$	0.318	0.680	0.006	0.000
4–5	$Q1$	0.314	0.681	0.009	0.000
	$Q2$	0.325	0.656	0.010	0.000
	$Q3$	0.313	0.684	0.006	0.000

Table 2 Total sensitivity index S_{Ti} of the dead zone model parameters

Sub-reach	Flow rate	K_x (m^2/s)	A (m^2)	A_S (m^2)	α(1/s)
1–2	$Q1$	0.298	0.701	0.015	0.000
	$Q2$	0.320	0.674	0.016	0.000
	$Q3$	0.320	0.682	0.012	0.000
2–3	$Q1$	0.320	0.682	0.008	0.000
	$Q2$	0.304	0.699	0.009	0.000
	$Q3$	0.318	0.684	0.008	0.000
3–4	$Q1$	0.313	0.690	0.007	0.000
	$Q2$	0.324	0.680	0.007	0.000
	$Q3$	0.323	0.681	0.007	0.000
4–5	$Q1$	0.320	0.683	0.009	0.000
	$Q2$	0.342	0.666	0.010	0.000
	$Q3$	0.319	0.685	0.007	0.000

no impact on the model output. Hence, the dead zone model could be simplified by neglecting transient storage parameters. Omitting transient storage parameters results in reducing the dead zone model to the Fickian model, which takes the form:

$$\frac{\partial C}{\partial t} = -\frac{Q}{A}\frac{\partial C}{\partial x} + \frac{1}{A}\frac{\partial}{\partial x}\left(AK_x\frac{\partial C}{\partial x}\right) \tag{4}$$

In Fig. 6 a comparison of the dead zone model results and the Fickian model results is presented. The comparison indicates that results of both models are in very good agreement with each other.

Additionally, the uncertainty analysis was performed for the Fickian model. The 95% confidence limits are the same as for the dead zone model (Fig. 5), which verifies the statement that transient storage parameters do not affect the output uncertainty, and the uncertainty is dependent on parameters A and K_x.

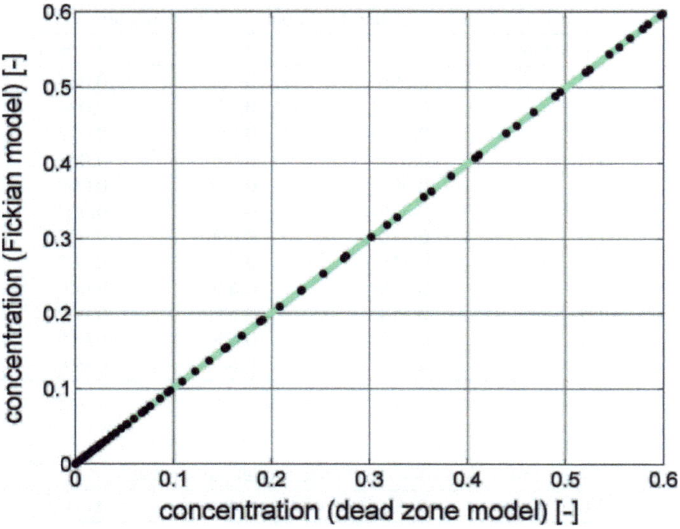

Fig. 6 Comparison of the dead zone model and the Fickian model results for sub-reach 1–2, flow rate $Q3$

In brief, the Fickian model is sufficient to model contaminants transport in the studied case (Mrokowska 2010).

5 Conclusions

To conclude, the results of the study indicate that calibration of reliable values of the dead zone model parameters is not possible for the studied case.

First of all, the Monte Carlo simulations combined with the performance measure (the least square objective function) showed that identification of a unique parameter set is not possible. It means that the choice of optimal set of parameters is uncertain.

To assess the model output uncertainty, the sensitivity analysis and uncertainty analysis were applied. According to the SA results, the model output is most sensitive to parameters A and K_x. Consequently, the uncertainty of results depends on these two parameters. More significantly, results of the SA showed that transient storage parameters are irrelevant in explaining the model output. Hence, the dead zone model could be simplified to the Fickian model by neglecting transient storage parameters (α and A_s).

In the studied case, the Fickian model proved to be sufficient to model contaminants transport.

References

Archer G, Saltelli A, Sobol IM (1997) Sensitivity measures, ANOVA-like techniques and the use of bootstrap. J Stat Comput Simul 58:99–120

Aronica G, Hankin B, Beven K (1998) Uncertainty and equifinality in calibrating distributed roughness coefficients in a flood propagation model with limited data. Adv Water Resour 22(4):349–365

Bencala KE, Walters RA (1983) Simulation of solute transport in a mountain pool and riffle stream: a transient storage model. Water Resour Res 19(3):718–724

Beven KJ (2006) A manifesto for the equifinality thesis. J Hydrol 320:18–36

Beven KJ (2008) Environmental modelling: an uncertain future? Routledge, London

Blasone R-S, Vrugt JA, Madsen H, Rosbjerg D, Robinson BA, Zyvoloski GA (2008) Generalized likelihood uncertainty estimation (GLUE) using adaptive Markov Chain Monte Carlo sampling. Adv Water Resour 31:630–648

Mantovan P, Todini E (2006) Hydrological forecasting uncertainty assessment: Incoherence of the GLUE methotology. J Hydrol 330:368–381

Moradkhani H, Hsu K-L, Gupta H, Sorooshian S (2005) Uncertainty assessment of hydrologic model states and parameters: sequential data assimilation using the particle filter. Water Resour Res 41(5):1–17

Montanari A (2005) Large sample behaviors of the generalized likelihood uncertainty estimation (GLUE) in assessing the uncertainty of rainfall-runoff simulations. Water Resour Res 41:W 08406. doi:10.1029/2004WR003826

Mrokowska MM (2010) Modelling of passive contaminants spreading in Wkra river based on a tracer test. Master thesis, Warsaw University of Life Sciences

Owczarczyk A, Wierzchnicki R, Palige J (1997) Wykonanie pomiaru znacznikowego dyspersji podłużnej na wybranym odcinku rzeki Wkry. Report IChTJ C/2/97 Institute of Nuclear Chemistry and Technology

Price KH, Storn RM, Lampinen JA (2005) Differential Evolution A Practical Approach to Global Optimization. Springer, Berlin Heidelberg

Ratto M, Tarantola S, Saltelli A (2001) Sensitivity analysis in model calibration: GSA-GLUE approach. Comput Phys Commun 136:212–224

Romanowicz R, Beven K (2006) Comments on generalized likelihood uncertainty estimation. Reliab Eng Syst Saf 91:1215–1321

Rowiński PM, Dysarz T, Napiórkowski JJ (2004) Estimation of longitudinal dispersion and storage zone parameters. In: Greco M, Caravetta A, Della Morte R (eds) River flow. Taylor & Francis, London

Runkel RL (1998) One-dimensional transport with inflow an storage (OTIS): a solute transport model for streams and rivers: U.S. Geological Survey water-Resources Investigation Report. 98-4018. http://co.water.usgs.gov/otis/documentation/primary/. Accessed 7 Sept 2009

Runkel RL (2000) Using OTIS to model solute transport in streams and rivers: U.S. U.S. Department of the Interior U. S. Geological Survey. Geological Survey Fact Sheet FS-138-99. http://co.water.usgs.gov/otis/documentation/primary/. Accessed 7 Sept 2009

Saltelli A, Tarantola A, Campolongo F, Ratto M (2004) Sensitivity analysis in practice – a guide to assessing scientific models. Wiley, Chichester

Saltelli A, Ratto AM, Andres T, Campolongo F, Cariboni J, Gatelli D, Saisana M (2008) Global sensitivity analysis. The primer. Wiley, Chichester

Smith P, Beven KJ, Tawn J (2008) Informal likelihood measures in model assessment: theoretic development and investigation. Adv Water Resour 31:1087–1100

Sobol IM (2001) Global sensitivity indices for nonlinear mathematical models and their Monte Carlo estimates. Math Comput Simul 55:271–280

Stedinger JR, Vogel RM, Lee SU, Batchelder R (2008) Appraisal of generalized uncertainty estimation (GLUE) method. Water Resources Research 44. doi: 10.1029/2008WR006822

Tychonoff AN, Arsenin VY (1977) Solution of ill-posed problems. Winston & Sons, Washington. ISBN 0-470-99124-0

Application of a Videometric Measurement System to Investigate Spatial Dike Breach

Lukas Schmocker

1 Introduction

1.1 Dike Erosion

The main causes of dike breaches are overtopping, internal erosion (piping) or loss of stability, yet most earth dikes fail due to overtopping as they are not designed to resist overflow and their resistance to failure caused by surface erosion is limited (Singh 1996). Dike overtopping starts normally at the lowest crest elevation or at a local dike discontinuity, e.g., at a bridge abutment. If water runs over the downstream dike slope, excessive surface shear stresses initiate the erosion process. Erosion starts if the induced shear stress exceeds the critical shear stress of the dike material. Once the initial breach is formed, the erosion process may be either plane or spatial. The latter case normally occurs in nature and includes both vertical and lateral erosion.

1.2 Research in the Past

Plane dike erosion has been investigated by, e.g., Powledge et al. (1989a, b), Chinnarasri et al. (2003), Dupont et al. (2007), Wu and Wang (2008), or Schmocker and Hager (2009). This dike erosion process is comparatively simple, as no three-dimensional (3D) flow patterns occur. Both the temporal water and sediment surfaces during the erosion process can be recorded across a glass sidewall of the hydraulic channel. Most past researches simply recorded the breach process using a standard CCD camera. However, given the spatial breach topography, an observation of the breach profiles across a channel sidewall is impossible, given the 3D erosion patterns.

L. Schmocker
Laboratory of Hydraulics, Hydrology and Glaciology VAW, Swiss Federal Institute of Technology ETH, CH-8092 Zurich, Switzerland
e-mail: schmocker@vaw.baug.ethz.ch

P. Rowinski (ed.), *Experimental Methods in Hydraulic Research*,
Geoplanet: Earth and Planetary Sciences, DOI 10.1007/978-3-642-17475-9_17,

Intrusive measurements are not applicable as they disturb the erosion process. Other methods were used to determine spatial breach profiles. Coleman et al. (2002) drained the reservoir at several stages of the erosion process and recorded the dried three-dimensional surface topography using a surface profile measurement system. They describe the breach profiles in detail and present a breach-discharge equation. However, the effort to determine these erosion profiles is comparatively high.

Pickert et al. (2004) used a Fringe Projection to continuously measure 3D breach profiles. An alternating black/transparent stripe pattern was produced on the dike surface using a slide projector and recorded with a video camera through the channel sidewall. The breach process could only be observed in the initial phase, however, as the observation through the sidewall failed for large breach widths. Rozov (2003) measured the breach process using washout indicators installed inside the model dam. The indicators work on the principle of breaking a circuit if it is reached by the moving sediment. A detailed analysis of the breach profiles was not provided.

A fast and reliable method to determine the 3D breach profile during a dike breach under laboratory conditions is still missing. To fill in this gap, an effort was made to apply a new non-intrusive videometric system, developed by AICON® 3D Systems Ltd[1] (Braunschweig, Germany). Preliminary experiences on the features of this system are presented below.

2 Test Setup

2.1 *Hydraulic Model*

The experiments were conducted in a glass-sided model channel at VAW (Fig. 1a). Its discharge capacity is $Q = 130$ L/s, it is 1.0-m wide and 9-m long, with a working section of roughly 5 m, and its bottom is horizontal. The channel is suited for the 3D tests described below. Test dikes of up to 0.50-m height can be erected, the tailwater is remote-controlled with a flap gate at the channel end and the flow features can be optically recorded either through the large channel windows or from the top. A horizontally graded, washed river gravel of median diameter $d_{50} = 1.14$ mm was used as sediment bed. The dike consisted of the same material and was erected above the sediment bed (Fig. 1b). For preliminary tests, the dimensions of the dike were: dike height $w = 0.20$ m, dike width $b = 1.0$ m, dike crest length $L_K = 0.10$ m with both the up- and downstream dike slopes $S_o = 1:3$ (V:H).

A triangular pilot channel was cut in the embankment crest along the channel sidewall, to initiate the breach process, with the channel wall simulating the breach centerline. This allowed recording the longitudinal breach profiles along the breach centerline and the corresponding flow depths. The reservoir filling was comparatively

[1]www.aicon.de

Fig. 1 Side view of (a) basic hydraulic model channel; (b) dike

fast and discharge was adjusted to ensure a more or less constant reservoir level during the breach process.

2.2 Videometric Measurement System

The stereoscopic-videometric AICON measurement system was used to determine the breach profiles. It was originally developed for the Federal Waterways Engineering and Research Institute (Bundesanstalt für Wasserbau, BAW, Karlsruhe, Germany) to continuously measure bed elevations and flow depths in a physical model of Elbe River (Godding et al. 2003). For the present research, the system was further developed and adapted by AICON 3D Systems Ltd.

The so-called AICON-method refers to *videometry* involving a rectangular grid projected onto a sediment surface using a high-power slide projector. The projector itself is mounted above the hydraulic channel. The sediment surface above and the bed forms below the water surface reflect a non-rectangular distorted 'grid' structure. The grid distance on the sediment surface was about 30 mm in the present test setup. The grid is recorded using three synchronised CCD cameras located above the channel. A fourth camera records the flow depth at the channel sidewall across the sidewall. The cameras have a resolution of about two million pixels and a recording frequency of 30 Hz. The digital images are directly transferred to the measuring computer. The test setup is shown in Fig. 2.

Before each measurement, the relative camera orientation as well as the camera parameters (inner orientation) are determined using a calibration panel. It consists of a number of measuring marks whose 3D-coordinates are known. Each measuring mark has a distinctive code number. The calibration panel is recorded several times at different positions with all four cameras and all required parameters calculated using the software AICON ProSurf® (Godding 1993) to define the 3D-position of

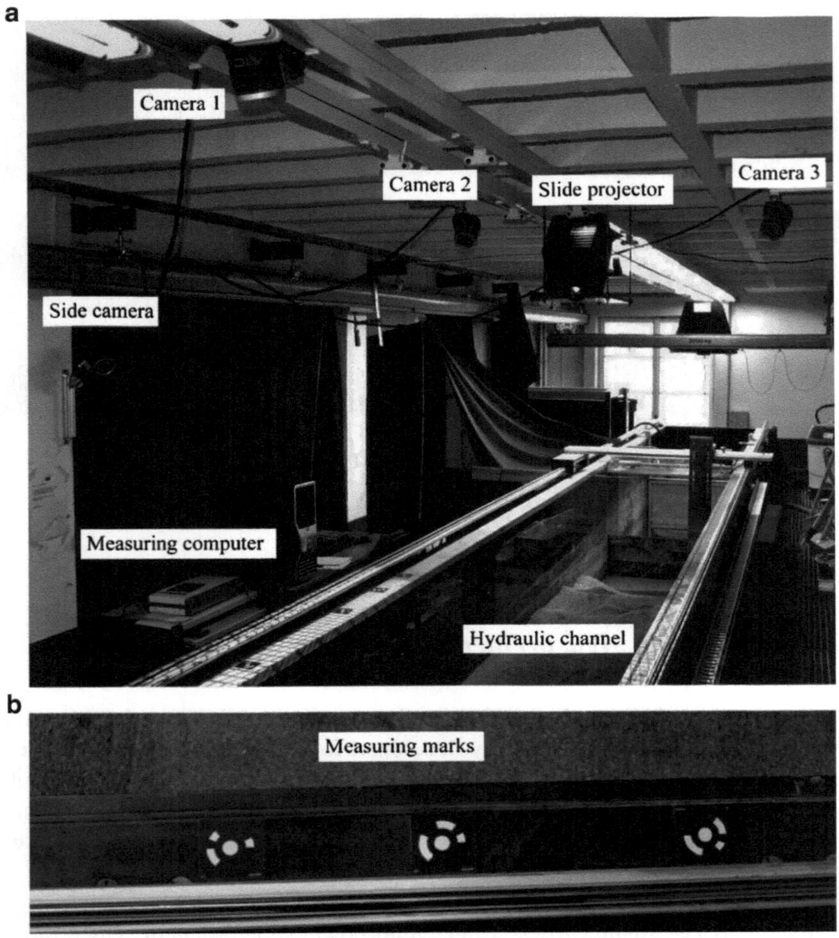

Fig. 2 The VAW test setup (**a**), and the measuring marks (**b**)

each camera. During a test, the projected grid is measured with the cameras and 3D-coordinates are calculated for each grid point.

To transform all calculated coordinates in an overall, superior coordinate system, the hydraulic channel itself is equipped with coded measuring marks (Fig. 2b). The coordinates of these marks are determined prior to the test. The coded marks on the hydraulic channel are recorded along with the projected grid and are assigned to 3D-coordinates, to allow for a 3D-Helmert-transformation of the measured grid points into the channel system.

During a test, the channel is only lit by the slide projector. The entire breach process, e.g., the projected grid, is recorded with the cameras. Figure 3 shows all four camera images before reservoir filling. These images are then processed using AICON ProSurf software. For each time step, the 3D-coordinates of the breach

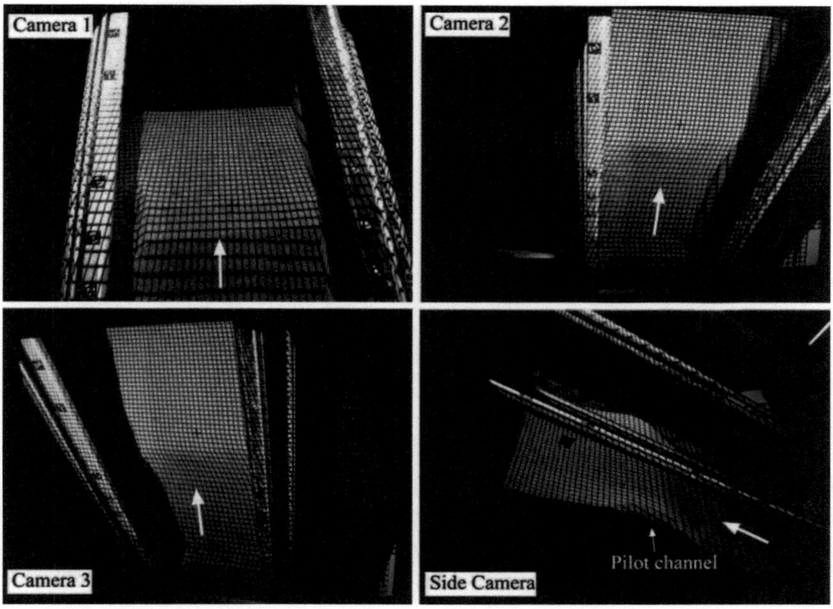

Fig. 3 Camera images before reservoir filling with projected grid on dike surface and measuring marks on channel side wall. The principal flow direction is indicated by →

profile are determined as follows: (1) Water elevation is manually determined from the side camera image to determine the distortion of the grid projection on the sediment surface below the water; (2) starting grid point is manually located in two of the three camera images; (3) based on the starting point, the corresponding grid points are automatically searched and determined in all three camera images; (4) using the camera calibration and channel orientation data, each grid point is assigned to 3D-coordinates.

3 Observations

The following refers to general observations during a breach process and especially the adaptability of the videometric recording system. The derived 3D-coordinates were further processed to maintain contour plots of the dike breach surface. Figure 4 shows breach profiles at four time steps. The origin of the coordinate system (x, y, z) was located at the downstream dike toe of the channel side wall. The contour plots show the downstream dike face up to the dike crest. The upstream dike face was not recorded in the present test setup. The breach process initiated at time $t = 0$ s with discharge entering the pilot channel, which is only partially visible as the grid spacing was too large in the present test. The breach develops then both vertically and laterally

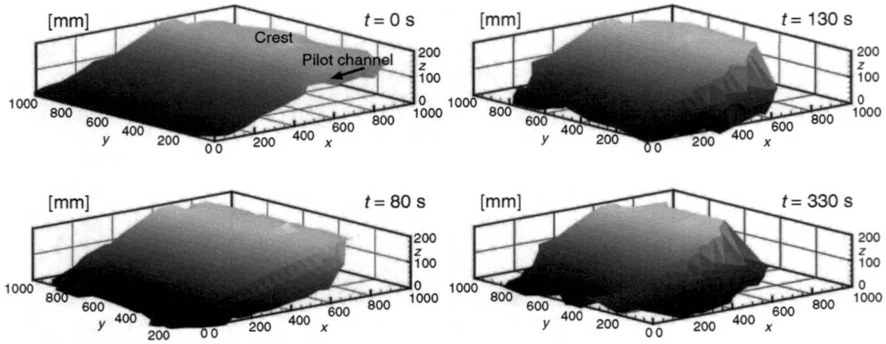

Fig. 4 Three-dimensional (3D) breach profiles at various times *t*

with time, in accordance to Coleman et al. (2002). During the entire test, the projected grid was recorded with a frequency of 1 Hz and afterwards evaluated using the AICON software. The quality and integrity of the breach profile depend essentially on the detected grid points of an image. There are several reasons that may disturb the visibility of the grid points and impede an automatic computer computation:

- Sediment colour changes as it gets wet. The contrast between the sediment and the projected grid varies therefore during the test.
- The projected grid is partly distorted due to inclined dike surfaces and the projector position.
- Light of projector is reflected on water surface, especially at large times under high outflow discharge. These reflections disturb the visibility of the sediment surface.
- Surface waves and suspended sediment may complicate the grid point detection due to reduced visibility (Fig. 5a).
- Breach side slope may be steep and grid detection therefore fails as the vertical displacement of the grid points is too large.

However, the first evaluations are surprisingly good, given the various factors disturbing the proper grid recording. A non-intrusive method is therefore available in principle to determine a sediment surface topography during a spatial dike breach. To increase grid visibility and to improve the automatic computation, the test setup is currently optimised. Further steps include a test series to account for the recording accuracy and a detailed analysis of the breach formation.

4 Conclusions

The videometric measurement system of AICON® 3D Systems Ltd. was used to determine the three-dimensional breach profiles during a spatial dike breach. This system allows for recording a breach sediment surface through the water surface

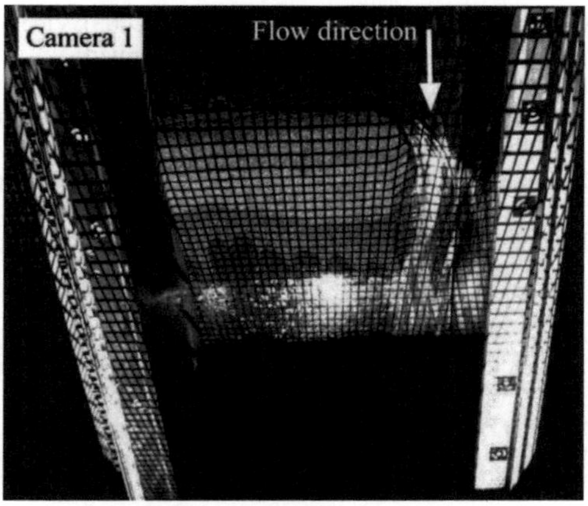

Fig. 5 Ongoing dike breach with projected grid and both surface waves and light reflections

using a rectangular grid projection and a set of four cameras. Preliminary tests confirmed the general adaptability of the AICON system to the dike breach problem. The three-dimensional breach profiles were directly computed from the camera images. Preliminary tests provide motivating results and a feasible, non-intrusive method seems therefore to be available for determination of a sediment surface topography during a spatial dike breach.

Several issues regarding the test setup have to be improved for a complete application of the technique, however. Surface waves and light reflections disturb the projected grid and complicate the optical grid visibility. The accuracy depends essentially on the water quality, on the reflection pattern by the water and on the sediment light adsorption characteristics. Black or white *artificial* sediment would be an optimum because of its excellent contrast qualities in addition to the small amount of impurities. An adaption of test setup and a detailed analysis of the derived data are currently in progress.

Acknowledgment The writer was supported by the Swiss National Science Foundation, Grant No. 200020-116680.

References

Chinnarasri C, Tingsanchali T, Weesakul S, Wongwises S (2003) Flow patterns and damage of dike overtopping. Intl J Sediment Res 18(4):301–309

Coleman SE, Andrews DP, Webby MG (2002) Overtopping breaching of noncohesive homogeneous embankments. J Hydraul Eng 128(9):829–838

Dupont E, Dewals BJ, Archambeau P, Erpicum S, Pirotton M (2007) Experimental and numerical study of the breaching of an embankment dam. In: Proceedings of 32nd IAHR congress, Venice, IAHR, Madrid, 1(178), pp 1–10

Godding R (1993) Ein photogrammetrisches Verfahren zur Überprüfung und Kalibrierung digitaler Bildaufnahmesysteme. Z Photogramm Fernerkund 2:82–90 (in German)

Godding R, Hentschel B, Kauppert K (2003) Videometrie im wasserbaulichen Versuchswesen. Wasserwirtschaft 93(4):36–40 (in German)
Pickert G, Jirka GH, Bieberstein A, Brauns J (2004) Soil/water interaction during the breaching process of overtopped embankments. In: Proceedings of 2nd International Conference on River Flow, Naples, Italy, pp 903–910
Powledge GR, Ralston DC, Miller P, Chen YH, Cloppner PE, Temple DM (1989a) Mechanics of overflow erosion on embankments I: research activities. J Hydraul Eng 115(8):1040–1055
Powledge GR, Ralston DC, Miller P, Chen YH, Cloppner PE, Temple DM (1989b) Mechanics of overflow erosion on embankments II: hydraulic and design considerations. J Hydraul Eng 115(8):1056–1075
Rozov AL (2003) Modeling of washout of dams. J Hydraul Res 41(6):565–577
Schmocker L, Hager WH (2009) Modelling dike breaching due to overtopping. J Hydraul Res 47(5):585–597
Singh VP (1996) Dam breach modelling technology. Kluwer, Dordrecht
Wu W, Wang SSY (2008) One-dimensional explicit finite-volume model for sediment transport with transient flows over movable beds. J Hydraul Res 46(1):87–98

Spatial Variability, Mean Drag Forces, and Drag Coefficients in an Array of Rigid Cylinders

T. Schoneboom, J. Aberle, and A. Dittrich

1 Introduction

Floodplain vegetation is an integral part of riparian ecosystems covering a wide range of conditions from highly flexible low grass to bushes and trees with rigid stems. However, vegetation also increases flow resistance, changes backwater profiles, and modifies sediment transport and deposition (e.g., Yen 2002). Thus, to meet both ecological and flood protection needs, it is crucial to understand the interplay of biological, physical, and chemical processes and how vegetation type, density, and pattern influence flow resistance and the turbulent flow field. In fact, the area of conflict between ecology and water resources has initiated an abundance of studies toward the hydrodynamics of aquatic and riparian ecosystems (e.g., Thorp et al. 2008; Nikora 2010 and references therein).

Various studies have shown that the resistance behavior of flexible vegetation elements differs substantially from the resistance behavior of rigid elements (e.g., Oplatka 1998; Järvelä 2002, 2004; Armanini et al. 2005; Wilson et al. 2008; Aberle et al. 2010). However, until today, there exists no universal approach for the determination of C_{D} for natural vegetation and species-specific resistance laws and parameters are still under development (e.g., Järvelä 2004). Therefore, most of the existing computational approaches for the determination of flow resistance due to vegetation are based on the assumption that riparian vegetation can be modeled by rigid cylinders (e.g., Baptist et al. 2007; Huthoff et al. 2007). This is reflected by the fact that most of the existing approaches require only plant density (expressed by the ratio of stem diameter d and averaged longitudinal and transverse plant spacing a_x and a_y, respectively) and the drag coefficient C_{D} as input parameters for the description of vegetation characteristics (e.g., Fischer-Antze et al. 2001; Beffa and Connell 2001). Simulating vegetation by rigid cylinders has the advantage that the drag coefficient C_{D} can be estimated relatively straightforward

T. Schoneboom (✉), J. Aberle, and A. Dittrich
Leichtweiß-Institute for Hydraulic Engineering and Water Resources, Technische Universität Braunschweig, Germany
e-mail: T.Schoneboom@tu-bs.de

P. Rowinski (ed.), *Experimental Methods in Hydraulic Research*,
Geoplanet: Earth and Planetary Sciences, DOI 10.1007/978-3-642-17475-9_18,

from the well known relationship between C_D and the cylinder Reynolds-number Re_d for an isolated emergent cylinder in an infinite wide flow (e.g., Nakayama 1999; Schlichting and Gersten 2006).

Although this relationship is valid only for a single isolated cylinder, it has also been used to estimate the drag coefficient for cylinders within a multicylinder array (e.g., DVWK 1991; Rowinski and Kubrak 2002; Baptist et al. 2007; Huthoff et al. 2007). However, the flow in a multicylinder array is influenced by the decay and spread characteristics of wakes forming at upstream cylinders (e.g., Petryk 1969; Li and Shen 1973; Lindner 1982; Wilkerson 2007). Studies carried out by e.g., Petryk (1969), Li and Shen (1973), Lindner (1982), Nepf (1999) and Poggi et al. (2004) showed that, in general, lower drag coefficients can be expected for a cylinder within a multicylinder array than for a single isolated cylinder. On the other hand, Kothyari et al. (2009) reported, based on drag force measurements of a single cylinder within a multicylinder array, larger drag coefficients than the ones expected for an isolated cylinder while Dunn et al. (1996) and Stone and Shen (2002) reported similar C_D-values for cylinders in multicylinder arrays and for an isolated cylinder. These contradicting results can partly be explained by differences in the definition of hydraulic variables (e.g., Statzner et al. 2006), the experimental setup (array density and cylinder Reynolds number), and the fact that drag forces were not directly measured in some of the studies. Especially the latter fact is of major importance, as the adequate determination of the drag coefficient must be based on the approach velocity on an individual cylinder.

Therefore, the objective of the present study was to test the computational approach presented by Lindner (1982) for the determination of the drag coefficient of cylinders in multicylinder arrays by comparing measured and computed drag forces and drag coefficients, taking into account the wake flow structure.

2 Background

The drag force F_D exerted by a single isolated cylinder can be estimated from:

$$F_D = 0.5 \rho C_D A_p u_c{}^2 \tag{1}$$

where ρ = fluid density, C_D = drag coefficient, A_P = plant projected area, and u_c = characteristic mean flow velocity of the undisturbed flow. For emergent cylindrical rigid elements, A_p corresponds to the product of water depth h and diameter d.

The drag coefficient C_D depends on the separation characteristics of the cylinder boundary layer and is a function of the cylinder Reynolds number $Re_d = u_c d/v$, where v denotes the fluid kinematic velocity (e.g., Nakayama 1999). Within the so-called subcritical flow regime ($Re_d \leq 4 \times 10^5$), the flow around a cylinder is laminar (e.g., Schlichting and Gersten 2006). The subcritical flow regime is relevant for most practical applications, and for $Re_d < 800$, the drag coefficient can be approximated according to $C_D = 3.07 Re_d^{-0.168}$, for $800 \leq Re_d \leq 8{,}000$ $C_D \approx 1.0$, and for $8{,}000 \leq Re_d \leq 4 \times 10^5$ $C_D \approx 1.2$ (e.g., Rowinski and Kubrak 2002; Schlichting and Gersten 2006). For large cylinder, Reynolds numbers ($Re_d > 4 \times 10^5$) C_D suddenly decreases to $C_D \approx 0.3$ in the so-called supercritical flow regime as the location of the separation point of the flow around the cylinder changes (e.g., Nakayama 1999).

However, the flow structure in a multicylinder array differs substantially from the flow structure around a single isolated element as wake flow and sheltering effects caused by upstream cylinders dominate the flow pattern. Petryk (1969), as well as Lindner (1982), showed that, for a regular arrangement of cylinders, C_D is a function of the cylinder diameter d, the cylinder spacing a_x and a_y, the approach velocity of an individual cylinder u_{0i}, and the slope S.

Petryk (1969) used the wake superposition concept to quantify the flow structure within a cylinder array. The superposition concept is based on the assumptions that (1) the channel is straight and has rectangular cross-section; (2) the flow in the channel is uniform and the flow approaching the cylinders is two-dimensional and fully developed; (3) the velocity defects caused by upstream cylinders are small; (4) the turbulence theory of Reichardt (1941)) is applicable and can be used to describe the influence of multiple wake-flows; (5) the cylinder diameters are small compared to the surrounding flow field; and (6) each wake flow can develop independent of the wake flow caused by upstream cylinders (Lindner 1982).

The distribution of the velocity defect u_d in a cross-section at a downstream distance Δx of a cylinder can be determined from (Lindner 1982):

$$u_{d,\Delta x}(z) = u_{\Delta x,\max} e^{-0.69\left(\frac{z}{b_{1/2,\Delta x}}\right)^2} \tag{2}$$

where z = transverse coordinate, $u_{\Delta x,\max}$ = maximum velocity defect at the center of the wake (averaged over flow depth), and $b_{1/2}$ = one-half of the wake width in which $u_{d,\Delta x} = u_{\Delta x,\max}/2$ (Li and Shen 1973; Lindner 1982). According to Petryk (1969), the maximum velocity defect can be estimated according to:

$$u_{\Delta x,\max} = -0.9\left(\frac{\Delta x}{C_D d}\right)^{-0.7}\left(\frac{1}{1+\frac{2g\Delta x S}{u_0^2}}\right)^{1.5} u_0 \tag{3}$$

where S = bed slope, g = acceleration due to gravity, u_0 = approach velocity of the cylinder causing the wake, and C_D = drag coefficient of the cylinder causing the wake. The wake width $b_{1/2}$ can be calculated according to:

$$b_{1/2,\Delta x} = 0.24 \Delta x^{-0.7} (C_D d)^{0.41} \tag{4}$$

Starting with the first row of cylinders for which $u_{01} = u_0$, the approach velocity of the cylinders in the second row can be calculated using (2)–(4). The approach velocity on the cylinders in the second row is required to estimate the approach velocities for the following cylinders. Hence, the approach velocity on the nth cylinder can be estimated from (Lindner 1982):

$$u_{0,n} = u_0 + \sum_{i=1}^{n-1} u_{\Delta x,i,\max} \left(\mathrm{e}^{-0.69\left(\frac{z_n - z_i}{b_{1/2,\Delta x,i}}\right)^2} - \frac{1}{B} \int_B \mathrm{e}^{-0.69\left(\frac{z - z_i}{b_{1/2,\Delta x,i}}\right)^2} \mathrm{d}z \right) \tag{5}$$

where B = flume width and z_n, z_i = transverse coordinate of the nth and ith cylinder, respectively.

In an infinite array of cylinders, the superposition of the velocity defect according to (5) can, in theory, be carried out infinitely. In order to reduce the computational effort, Lindner (1982) defined a threshold criterion so that the velocity defect is only considered if it is larger than 3% of the undisturbed flow velocity. Applying this criterion, Lindner (1982) found that $u_{0,n}$ is roughly constant for $n = 20$ cylinders.

Using (2)–(5), the depth-averaged velocity can be predicted at any point in the channel. Thus, it becomes possible to calculate the drag exerted by individual cylinders from $F_{\mathrm{D}i} = 0.5\rho C_{\mathrm{D}} A_{\mathrm{p}} u_{0,i}^2$, where the corresponding drag coefficient can be determined dependent on the cylinder Reynolds number $Re_{\mathrm{d},i} = u_{0,i}d/\nu$. On the other hand, using the cross-sectional averaged velocity u_0, the average drag force of the ith cylinder follows from $F_{\mathrm{D}i} = 0.5\rho C_{\mathrm{D},i} A_{\mathrm{p}} u_0^2$. Equating both expressions yields the following formulation for the estimation of the drag coefficient $C_{\mathrm{D},i} = (u_{0,i}/u_0)^2 C_{\mathrm{D}}$ (e.g., Li and Shen 1973). Lindner (1982) extended this formulation taking into account the blockage ratio according to the approach by Richter (1973) and taking into account the effect of gravity waves:

$$C_{\mathrm{D},i} = \left(1 + 1.9\frac{d}{a_y} C_{\mathrm{D}}\right) \left(\frac{u_{0,i}}{u_0}\right)^2 C_{\mathrm{D}} + \Delta C_{\mathrm{D}} \tag{6}$$

where ΔC_{D} = additional drag caused by surface waves around the cylinder. The additional drag depends on the Froude number of the averaged flow field and the flow depth in front of and just behind the cylinder (e.g., Lindner 1982; Pasche and Rouvé 1985).

For cylinder arranged with a constant transverse distance a_y, Lindner (1982) suggested reducing the computational effort by replacing B with a_y in (5) and assuming that the cylinder lines act as imaginary walls. Lindner (1982) pointed out that this simplification is valid only for $a_y/d > 10$ as below this threshold the wakes of two neighboring upstream cylinders influence each other, i.e., the wake-flows cannot develop independently. For the computations with a parallel cylinder setup, Lindner (1982) assumed that the cylinders are located in the center of the interval $-a_y/2$ to $a_y/2$ while, for a staggered setup, he assumed that the cylinders are located at the imaginary side walls. The assumption that the cylinders lines act as imaginary walls was the reason that blockage effects according to the Richter (1973) approach (see (6)) were considered by Lindner (1982).

Based on test computations, Lindner (1982) found that $u_{0,n}/u_0$ increases with increasing a_x/d but that for $a_x/d > 40$ the influence of the upstream cylinders is negligible. Based on test computations, Lindner (1982) further concluded that $u_{0,n}/u_0$ is larger for a staggered setup than for a parallel setup and that $u_{0,n}/u_0$ decreases with increasing mean flow velocity u_0 and decreasing slope S. Similarly, for a constant volumetric element concentration, Lindner (1982) showed that the disturbance of the flow is larger when the array is composed of cylinders with a small diameter than the disturbance caused by an array of cylinders with a larger diameter. It is also worth mentioning that Lindner (1982) validated his approach based on a flume study in which only flow velocities were measured. Therefore, the intention of the present study is to test the applicability of the Lindner (1982) approach to estimate the drag coefficient based on experiments in which drag forces were measured directly.

3 Experimental Setup

Experiments were carried out in a 32-m long, 0.6-m wide, and 0.4-m deep tilting flume in the laboratory of the Leichtweiß-Institute for Hydraulic Engineering and Water Resources, Technische Universität Braunschweig, Germany. In the experiments, the discharge Q was controlled by a valve and measured by an inductive flow meter. Water depth in the flume was adjusted by a tailgate located in a distance of 25 m to the flume inlet. Ten piezometers installed along the flume allowed for water level measurements. The water level measurements were used to calculate water surface slope from linear regression and water depth h. The bed roughness consisted of a rubber mat with 3-mm high pyramidal-shaped roughness elements.

The experiments were carried out with both in-line (L) and staggered (S) cylinder arrays and a density of 25 cylinders/m^2 (spacing between rows and between cylinders in a row $a_x/a_y = 0.2/0.2$ m). The plastic cylinders were 24.5-cm long and had a diameter of $d = 1$ cm. In order to ensure fully developed flow conditions, the cylinders were placed over a total length of 18.5 m starting at a distance of 6 m from the flume inlet, i.e., approximately 45 cylinder rows were located upstream of the test section. Figure 1 shows the investigated patterns in the 1.5-m long test section which was located at a distance of 15.1 m to the flume inlet. For the staggered pattern, some cylinders had to be placed close to the flume wall. To ensure a constant density, the cylinder placed near the flume wall alternated from the right to left wall (see Fig. 1).

Drag forces exerted on the vegetation elements were measured simultaneously with up to ten drag force sensors (DFS) described in Schoneboom et al. (2008). The DFS were mounted in a box below the flume bottom in the test section. This setup

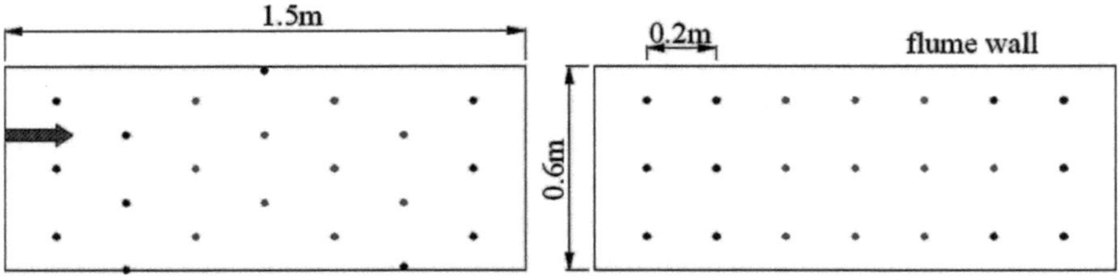

Fig. 1 Cylinder setup in the measurement section (*top view*). Left side staggered; right side in-line. Lateral and horizontal spacing 20 cm. Cylinders in *gray scale* were attached to drag force measurement systems

Table 1 Experimental boundary conditions, measured drag forces, coefficient of variation and measured and computed drag coefficients

Setup	u_0 (m/s)	h (m)	S (%)	Re_d (1)	$\langle F_D \rangle$ (N)	c_v (1)	C_{Di}[a] (1)	C_{Di}[b] (1)
In-line (L)	0.20	0.247	0.06	2,043	0.04	0.17	0.80	0.72
	0.30	0.251	0.13	3,001	0.09	0.10	0.86	0.72
	0.40	0.250	0.24	4,011	0.18	0.14	0.92	0.72
	0.50	0.251	0.36	5,002	0.27	0.06	0.87	0.72
	0.60	0.251	0.54	6,012	0.39	0.07	0.88	0.73
	0.69	0.252	0.72	6,944	0.51	0.06	0.87	0.73
Staggered (S)	0.20	0.246	0.07	2,032	0.05	0.12	0.98	1.12
	0.30	0.250	0.14	2,998	0.12	0.10	1.06	1.12
	0.40	0.252	0.26	3,973	0.21	0.06	1.08	1.12
	0.50	0.251	0.41	4,972	0.33	0.07	1.09	1.12
	0.59	0.253	0.62	5,938	0.47	0.09	1.09	1.12
	0.70	0.251	0.92	6,975	0.65	0.06	1.10	1.12

[a]Calculated on the basis of the measurements.
[b]Computed with Lindner (1982) approach.

ensured that the DFS did not disturb the flow and that they could be easily rearranged to match the corresponding cylinder arrangement. The cylinders attached to the DFS are highlighted in Fig. 1 (light gray elements).

The experiments were carried out with steady uniform and just submerged flow conditions with a constant water depth $h = 0.25$ m. In order to achieve uniform flow conditions, discharge and flume slope were adjusted so that the average water surface slope was identical to bed slope in the test section. In total, six different flow conditions were investigated for each arrangement. The experimental boundary conditions are summarized in Table 1.

The drag forces were measured with a sampling frequency of 200 Hz and a sampling interval of 60 s. Each measurement was repeated two times showing a high degree of reproducibility. A more detailed analysis of the time series showed that the mean value of the drag forces became, in general, stable after a sampling time of 30 s indicating that the sampling interval of 60 s was sufficient to obtain a stable mean value.

4 Results and Discussion

Figure 2 shows the measured time-averaged drag forces exerted by individual cylinders for the staggered (full symbols) and in-line (open symbols) arrangement as a function of the cylinder Reynolds number Re_d. The figure reveals a large variability of the measured drag forces for both arrangements with deviations from the spatial mean between −28% and +27% for the in-line and −23% and +19% for the staggered setup, respectively. The coefficient of variation c_v (or normalized random error, defined as the standard deviation of the drag force measurements divided by the spatially averaged drag force; Bendat and Piersol 2000) indicated that c_v was largest for the lowest velocities and became reasonably constant for the largest flow velocities (see Table 1). The analysis of the corresponding drag coefficients showed that C_{Di} deviated between −30% and +28% and −18% and +22% from the mean value for the in-line arrangement and staggered arrangement, respectively. The variability of the drag forces and C_{Di}-values showed that the spatially averaged drag force and drag coefficient C_D for a cylinder array cannot be estimated unambiguously from drag force measurements with a single cylinder. A similar result has been reported by Schoneboom et al. (2010) for drag forces of flexible artificial vegetation elements within vegetation arrays.

Figure 3 shows the spatially averaged drag forces $<F_D>$ and drag coefficients C_D as a function of the cylinder Reynolds number. Despite the observed spatial variability of drag forces in Fig. 2, the spatially averaged drag force $<F_D>$ followed a squared relationship with mean flow velocity for the in-line and staggered setup ($F_D \approx 1.14{u_0}^2$ and $F_D \approx 1.42{u_0}^2$, respectively, with $R^2 \approx 1$). Furthermore, Fig. 3 and the proportionality factors 1.14 and 1.42 showed that $<F_D>$ was consistently larger for the staggered than for the in-line pattern. This was also

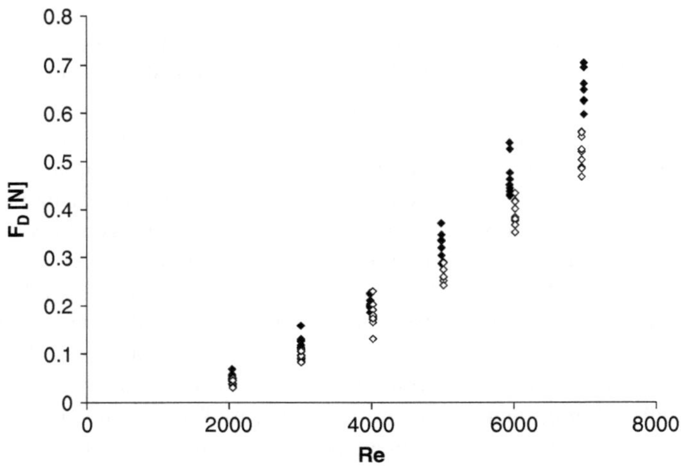

Fig. 2 Spatial variability of drag forces F_D for setup 20L (*open symbols*) and 20S (*full symbols*)

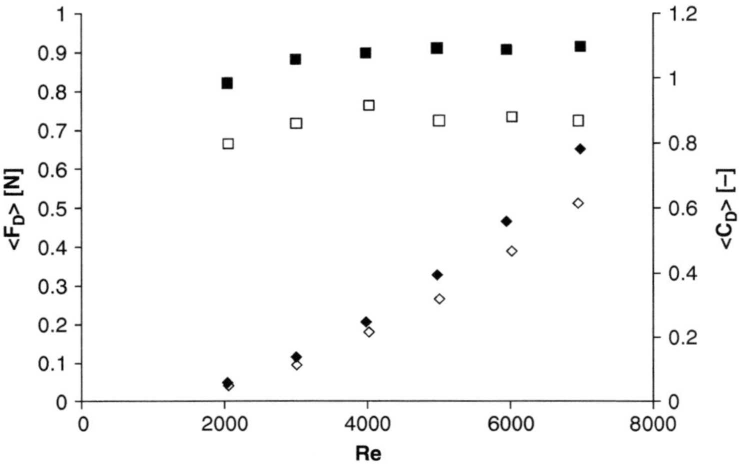

Fig. 3 Spatial averaged drag forces (*diamonds*) and drag coefficients (*squares*) dependent on stem-Reynolds number Re_d for the staggered (*filled symbols*) and in-line arrangement (*open symbols*)

reflected by the spatially averaged drag coefficients which were larger for the staggered than for the in-line arrangement (Fig. 3). A similar behavior has been reported by Li and Shen (1973) and Lindner (1982) for cylinder arrays and by Schoneboom et al. (2010) for flexible vegetation elements.

Figure 3 further indicated that the observed drag coefficients C_D were approximately constant for the two investigated arrangements. The mean C_D-values corresponded, for the range of cylinder Reynolds numbers varying between $2{,}000 \le Re_d \le 7{,}000$, to $C_D = 0.87$ and $C_D = 1.07$ for the in-line and staggered setup, respectively. The difference to the expected drag coefficient for a single isolated emergent cylinder in an undisturbed flow for the same cylinder Reynolds-numbers of $C_D = 1.0$ indicated the importance of the flow structure within cylinder arrays.

As aforementioned, the flow structure in a multicylinder array can be taken into account applying the Lindner (1982) approach. In the following application of this approach, the additional drag caused by surface waves around the cylinder ΔC_D was neglected as the experiments were carried out with just submerged flow conditions. Furthermore, according to the recommendation of Lindner (1982), the computations were based on the approach velocity on the 20th cylinder $u_{0,20}$ because the number of approximately 45 upstream cylinders exceeded this threshold. In fact, test computations showed that the approach velocity did not change significantly following the 20th cylinder confirming the conclusion of Lindner (1982).

The calculated drag coefficients are presented in Table 1 and showed that the drag coefficients for the in-line arrangement were underestimated by approximately 16% for the in-line setup and overestimated by 5% for the staggered setup. Given the large number of underlying assumptions, these deviations can be considered as rather small indicating the applicability of the Lindner (1982) approach. Furthermore, the maximum deviation of the measured drag coefficient of a single element from the calculated mean drag coefficient corresponded to 30%. These deviations were in a similar range to the deviations reported by Petryk (1969), although Petryk (1969) did not take into account blockage effects.

It is also worth mentioning that Lindner (1982) reported a decrease of $u_{0,n}/u_0$ with increasing u_0 and an increase of $u_{0,n}/u_0$ with increasing bed slope S. The hydraulic boundary conditions summarized in Table 1 show that in the present experiments, both u_0 and S were increased in order to achieve a constant water depth of $h = 0.25$ m. Hence, through the increase of u_0 lower ratios (and hence lower C_{Di}-values) would be expected while at the same time larger ratios $u_{0,n}/u_0$ (and hence larger C_{Di}-values) would be expected through the slope increase. Table 1 shows that calculated C_D-values were constant for the different hydraulic boundary conditions. Hence, it can be concluded that the two expected trends cancel each other out. This is also reflected by the constant range of drag coefficients C_D obtained from the measurements for each of the cylinder arrangements.

5 Summary and Conclusions

In this study, the spatial variability of drag forces within multicylinder arrays was investigated using experimental data from a laboratory study. Direct measurements of drag forces within the cylinder array showed that the spatial variability of time-averaged drag forces, deviating approximately $\pm 30\%$ from the spatial mean, is significant although the arrays were composed of identical elements. A similar result was obtained for the drag coefficients. The results further confirmed the findings reported in the studies of Petryk (1969), Li and Shen (1973), and Lindner (1982) that, for similar cylinder density and hydraulic boundary conditions, larger drag forces and drag coefficients are expected for a staggered setup than for an in-line setup.

These differences indicated that the wake flow structure plays an important role in multicylinder arrays. Therefore, the data were used to test the computational approach outlined by Lindner (1982) for the evaluation of the drag coefficients in a multicylinder array. This test revealed a good agreement between calculated and measured drag coefficients with deviations of the mean drag coefficient of approximately -16% for the in-line and $+5\%$ for the staggered setup.

The Lindner (1982) approach shows that drag forces and drag coefficients depend on the wake flow structure and hence on cylinder diameter d, cylinder spacing a_x and a_y, approach velocity of an individual cylinder u_{0i}, and slope S. Thus, estimating C_D values based on the single isolated cylinder analogy is only a crude approximation of the real drag coefficient.

It is worth mentioning that the presented experiments were carried out with limited hydraulic and geometrical boundary conditions. Hence, for a global and more detailed evaluation of the Lindner (1982) approach, further measurements would be required. However, the data used in this study were collected as a benchmark for comparisons with data from experiments with flexible vegetation elements (e.g., Schoneboom et al. 2010). Therefore, in our upcoming analyses, we intend to extend the Lindner (1982) approach to arrays of natural flexible vegetation elements taking into account plant-specific parameters as the results presented in this chapter clearly showed the applicability of the Lindner (1982) approach.

Acknowledgments This research was conducted under contract AB 137/3-1 from DFG (Deutsche Forschungsgemeinschaft).

References

Aberle J, Järvelä J, Schoneboom T, Dittrich A (2010) Flow resistance of rigid and flexible emergent vegetation revisited. 1st European IAHR congress, Edinburgh, UK, 4–6 May 2010 papers on memory stick

Armanini A, Righetti M, Grisenti P (2005) Direct measurement of vegetation resistance in prototype scale. J Hydraul Res 43(5):481–487

Baptist MJ, Babovic V, Rodriguez Uthurburu J, Keijzer M, Uittenbogaard RE, Mynett A, Verwey A (2007) On inducing equations for vegetation resistance. J Hydraul Res 45(4):435–450

Beffa C, Connell RJ (2001) Two-dimensional flood plain flow I: model description. J Hydrol Eng 6(5):397–405

Bendat JS, Piersol AG (2000) Random data: analysis and measurement procedures. Wiley, New York

Dunn C, Lopez F, Garcia M (1996) Mean flow and turbulence in a laboratory channel with simulated vegetation. Civil engineering studies hydraulic engineering studies no. 51, Hydrosystems Laboratory Urbana-Champaign

DVWK (1991) Hydraulische berechnung von fliessgewässern. Verlag Paul Parey, Hamburg/Berlin (in German)

Fischer-Antze T, Stoesser T, Bates P, Olsen NRB (2001) 3D numerical modelling of open-channel flow with submerged vegetation. J Hydraul Res 39(3):303–310

Huthoff F, Augustijn DCM, Hulscher SJMH (2007) Analytical solution of the depth-averaged flow velocity in case of submerged rigid cylindrical vegetation. Water Resour Res 43:W06413. doi:10.1029/2006WR005625

Järvelä J (2002) Flow resistance of flexible and stiff vegetation: a flume study with natural plants. J Hydrol 269:44–54

Järvelä J (2004) Determination of flow resistance caused by non-submerged woody vegetation. Int J River Basin Manag 2(1):61–70

Kothyari UC, Hayashi K, Hashimoto H (2009) Drag coefficient of unsubmerged rigid vegetation stems in open channel flows. J Hydraul Res 47(6):691–699

Li R-M, Shen HW (1973) Effect of tall vegetations on flow and sediment. J Hydraul Div 99(5):793–814

Lindner K (1982) Der Strömungswiderstand von Pflanzenbeständen. Mitt. Leichtweiß-Institut für Wasserbau, Technische Universität Braunschweig, No. 75 (in German)

Nakayama A (1999) Introduction to fluid mechanics. Butterworth-Heinemann, Oxford

Nepf HM (1999) Drag, turbulence, and diffusion in flow through emergent vegetation. Water Resour Res 35(2):479–489

Nikora V (2010) Hydrodamics of aquatic ecosystems: an interface between ecology, biomechanics and environmental fluid mechanics. River Res Appl 26:367–384

Oplatka M (1998) Stabilität von Weidenverbauungen an Flussufern Mitt. Versuchsanstalt für Wasserbau, Hydrologie und Glaziologie, No. 156, ETH Zürich, Switzerland (in German)

Pasche E, Rouve G (1985) Overbank flow with vegetatively roughened flood plains. J Hydraul Eng 111(9):1262–1278
Petryk S (1969) Drag on cylinders in open channel flow. Dissertation, Colorado State University, Fort Collins
Poggi D, Porporato A, Ridolfi L, Albertson JD, Katul GG (2004) The effect of vegetation density on canopy sub-layer turbulence. Boundary Layer Meteorol 111:565–587
Reichardt H (1941) Über eine neue Theorie der freien Turbulenz. ZAMM 21:257
Richter A (1973) Strömungskräfte auf starre Kreiszylinder zwischen parallelen Wänden. Dissertation, Karlsruhe, Univ., Fak. f. Bauingenieur- u.Vermessungswesen (in German)
Rowinski PM, Kubrak J (2002) A mixing-length model for predicting vertical velocity distribution in flows through emergent vegetation. Hydrol Sci J 47(6):893–904
Schoneboom T, Aberle J, Wilson CAME, Dittrich A (2008) Drag force measurements of vegetation elements. ICHE 2008, Nagoya, Japan, Papers on CD-ROM
Schoneboom T, Aberle J, Dittrich A (2010) Hydraulic resistance of vegetated flows: contribution of bed shear stress and vegetative drag to total hydraulic resistance. In: Proceedings of the international conference on fluvial hydraulics river flow 2010, Braunschweig
Schlichting H, Gersten K (2006) Grenzschicht-Theorie. 10., überarbeitete Auflage. Springer, Berlin Heidelberg
Statzner B, Lamoroux N, Nikora V, Sagnes P (2006) The debate about drag and reconfiguration of freshwater macrophytes: comparing results obtained by three recently discussed approaches. Freshw Biol 51:2173–2183
Stone BM, Shen HT (2002) Hydraulic resistance of flow in channels with cylindrical roughness. J Hydraul Eng 128(5):500–506
Thorp JH, Thoms MC, Delong MD (2008) The riverine ecosystem synthesis. Elsevier, Amsterdam
Wilkerson GV (2007) Flow through trapezoidal and rectangular channels with rigid cylinders. J Hydraul Eng 133(5):521–533
Wilson CAME, Hoyt J, Schnauder I (2008) Impact of foliage on the drag force of vegetation in aquatic flows. J Hydraul Eng 134(7):885–891
Yen BC (2002) Open channel flow resistance. J Hydraul Eng 128(1):20–39

Experimental Study on Gabion Stepped Spillway

Mahmood Shafai-Bejestan and Gh. Kazemi-Nasaban

1 Introduction

Rocks in their natural form are the most abundant and economical material in hydraulic engineering practice. Rocks have been used in dam construction, river engineering works, river intakes, etc. The size of rock in these practices depends on the hydraulic conditions, such as flow velocity, shear stress, hydrostatic, and hydrodynamic pressures. When these conditions are high, the required size of stone will be large and impractical. In such a case, the alternative is to tie the stones together by some means. Gabions are hexagonal mesh boxes filled with small-size stones. The advantages of gabions are (a) their stability, (b) low cost, (c) flexibility, and (d) porosity (Chinnarasri 2003). The gabions retain the advantages of rock fill in the flexibility and permeability, so that water pressure is minimized. The wire mesh container can be used for this purpose. Nowadays, the gabion is used for hydraulic engineering works such as revetments, channel linings, weirs, groins, and energy dissipation structures (Stephenson 1979a, b). One application of gabion is for constructing the stepped spillways. In this structure, the kinetic energy of flow is dissipated, as water flows downstream. The flow through the rock mixes with the flow over the crest, resulting in energy dissipation by jet impingement as well as due to friction loss through the rock fill. In addition, the energy is dissipated as flow cascades from one step to another. Three types of flow occur in gabion stepped spillway: (a) "napped flow," when flow cascades from one step to another; (b) "skimming flow," when the water fully flows through the steps; and (c) "pooled flow," when a step is provided at the end of each step (Sorenson 1985). Many empirical equations have been developed to distinguish three types of flow. Table 1 shows some of these formulas. In these equations, y_c is the critical depth, approximately equal to the flow depth at the crest or can be calculated as $y_c = \sqrt[3]{q^2/g}$. Symbols h and l are the height and length of each step, respectively. Figure 1 shows the definition of variables.

M. Shafai-Bejestan (✉) and Gh. Kazemi-Nasaban
College of Science and Water Engineering, Shahid-Chamran University, Ahwaz, Iran
e-mail: m-shafai@scu.ac.ir; Ghkazemnasaban@yahoo.com

P. Rowinski (ed.), *Experimental Methods in Hydraulic Research*,
Geoplanet: Earth and Planetary Sciences, DOI 10.1007/978-3-642-17475-9_19,

Table 1 Formulae for three types of flow

Type of flow	Formula	Author (s)
Nape flow	$\frac{y_c}{\hbar} = 0.092\left(\frac{\hbar}{\ell}\right)^{-1.276}$	Chanson (1994)
	$\frac{y_c}{\hbar} = 0.89 - 0.4\left(\frac{\hbar}{\ell}\right)$	Chanson (2001)
	$\frac{y_c}{\hbar} = 0.80 - (0.55)\left(\frac{\hbar}{\ell}\right)$	Chinnarasri (2003)
Skimming flow	$\frac{y_c}{\hbar} = 0.862\left(\frac{\hbar}{\ell}\right)^{-0.165}$	Yasuda and Ohtsu (1999)
Pooled flow	$\frac{y_c}{\hbar} = \left[0.55 - 0.16Ln\left(\frac{\hbar}{\ell}\right)\right]^6$	Aigner (2004)

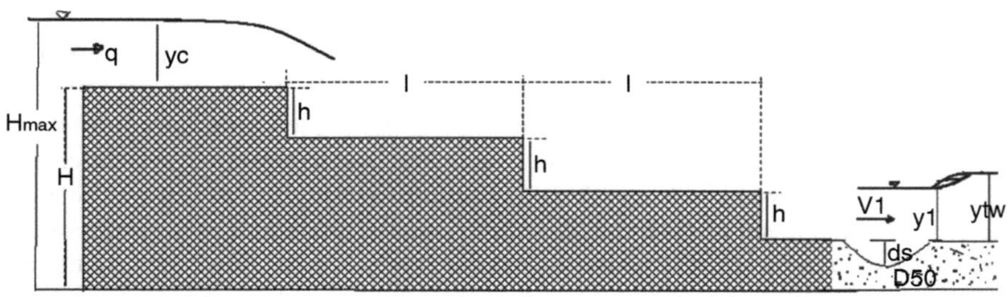

Fig. 1 Definition of variables used in this study

Many researchers have investigated the energy dissipation in stepped spillway, including the gabion stepped spillway. Kazemi-Nasaban (1996) and Peyras et al. (1992) have presented the following equations for the case of nape flow:

$$\frac{1}{1 - \frac{\Delta E}{H}} = a\left(\frac{q^2}{gH^2}\right)^b \tag{1}$$

in which ΔE is the amount of energy dissipation, H is the height of spillway, q is the unit discharge, a and b are coefficients which depend on the slope of spillway, as shown in Table 2:

For the case of pooled spillway, Aigner (2004) developed the following formula:

$$\frac{\Delta E}{Hs} = \frac{\frac{H}{y_c}}{\frac{H}{y_c} + 1.5} \tag{2}$$

in which H_{max} is the total upstream energy approximately equal to the sum of H and y_c.

Even though the cascading and impingement dissipate a lot of energy, the scour at the downstream end of the spillway remains to some extent a problem, especially

Table 2 Values of coefficients of a and b in (1)

Slope ($\hbar : \ell$)	a	b
1:1	0.238	−0.526
2:1	0.169	−0.654
3:1	0.208	−0.647
3.5:1	1.736	−0.279

when a stilling basin has not been provided at the downstream. The extent of such scour hole may result in the instability of the spillway or even its failure. Predicting the scour hole depth can help the engineers to design the spillway more safe. Review of literature reveals that no such study, to the knowledge of the authors, has been conducted. Therefore, it is the purpose of this study to conduct experimental tests and to develop a relationship for predicting the maximum scour hole depth, downstream of the gabion stepped spillways.

2 Theoretical Consideration

Consider a scour hole downstream of stepped spillway; for an equilibrium scour hole, the particles on the bed are at incipient motion. Forces exerted on the particles are the drag and lift forces and the submerged weight of particles. From the stability analysis of a particle at threshold condition one can obtain (Shafai-Bajestan et al. 1995):

$$\frac{d_s}{D_{50}} = f\left(\mathrm{SN}, \frac{y_1}{D_{50}}\right) \tag{3}$$

where d_s is the maximum scour hole depth, D_{50} is the median particle size of sediment, and SN is the stability number defined as follows:

$$\mathrm{SN} = \frac{V_1}{\sqrt{g(G_s - 1)D_{50}}} \tag{4}$$

In these equations, y_1 and V_1 are the flow depth and flow velocity just downstream of the spillway, G_s is the particle specific gravity and g is the acceleration of gravity.

3 Experimental Setup

Equation 3 is a general relation for predicting the scour hole depth. To determine the coefficients of this equation, a series of experimental tests have to be conducted. The experimental tests were conducted in a flume 50-cm wide, 8-m long and 1.5-m

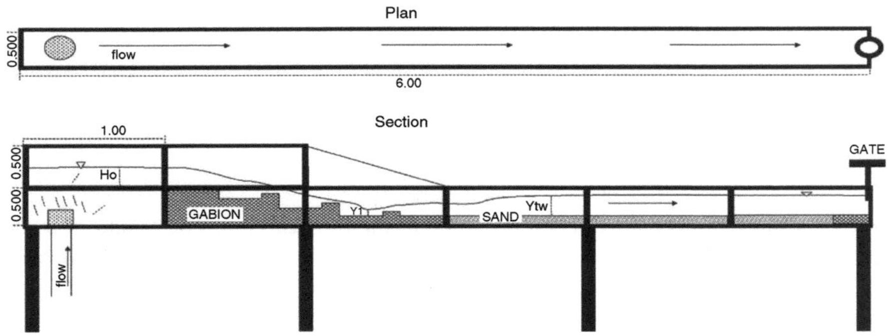

Fig. 2 Plan and section view of the experimental flume

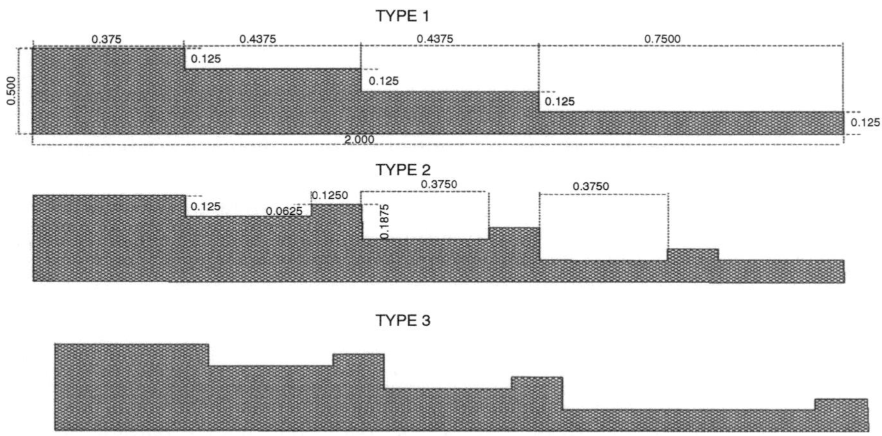

Fig. 3 Three types of stepped spillway tested in this study

high, in the hydraulic laboratory of the Shahid-Chamran University. Figure 2 shows the plan and section view of the flume. Three types of gabion stepped spillways (Fig. 3) were tested under different flow conditions. Two different sizes of bed materials, A and B, were placed inside the gabion mesh and three types of bed materials, C, D, and E, were placed downstream of the spillway. Figure 4 shows the size distribution of these materials.

After placing the desired gabion spillway and bed material, the required flow discharge was established by simultaneous adjusting of the inflow valve and the downstream gate. In all tests, the flow type was the nape or pooled flow. Upon the establishment of the desired flow and tail water depth, the flow characteristics and the scour depth were measured. During the test, when little or no removal of bed material from the scour hole was observed, usually 90 min from the start of the test, the pump was shut down. At the same time, the downstream gate gradually was

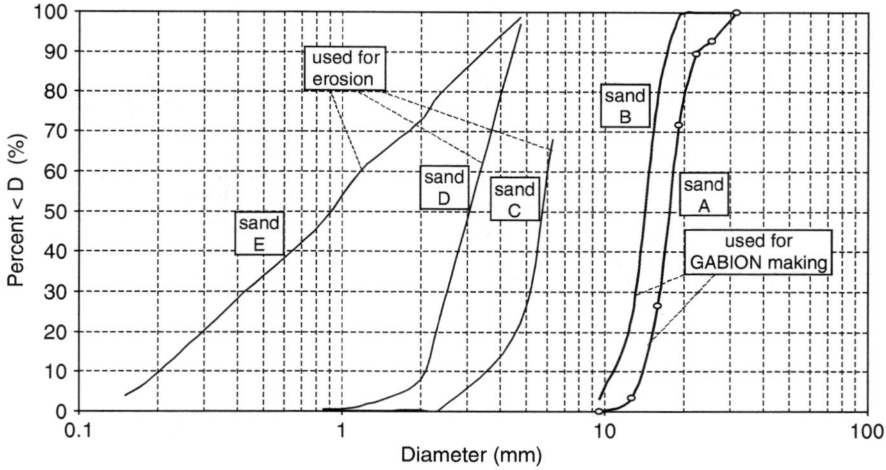

Fig. 4 Size distribution of rock and bed material

Table 3 Summary of the results

Type of spillway	Q (L/s)	y_1 (cm)	y_{tw} (cm)	V_1 (m/s)	D_{50} (mm)	SN	d_s (mm)
Type (1)	19.91	20	20	0.199	3.0	0.904	1
Simple stepped spillway	19.91	15	15	0.265	3.0	1.205	2
	19.91	13.5	11.5	0.295	3.0	1.339	5
	19.81	13.5	9.5	0.293	3.0	1.332	17.9
	41.35	19.0	24	0.435	3.0	1.975	3.6
	41.35	15.0	17	0.551	3.0	2.502	20.0
	57.95	16.5	21.75	0.702	3.0	3.188	20.4
	20	12.5	11.5	0.320	0.90	2.651	18
	41.35	15.0	17.0	0.551	0.90	4.568	21
	58.4	16.25	21.5	0.719	0.90	5.955	36
Type (2)	19.13	7	10.5	0.547	3	2.480	5.5
Pooled stepped spillway.	41.35	11.5	17	0.719	3	3.263	27
End sill in the middle	57.95	16	21	0.724	3	3.287	22.6
of downstream apron	19.91	8.5	11.5	0.468	0.9	3.881	25.7
	41.35	7.5	17	1.103	0.9	9.136	59.7
	57.95	10	21	1.159	0.9	9.603	44.0
Type (3)	19.91	15	11.7	0.265	0.9	2.199	18.2
Pooled stepped spillway	41.35	14.5	17	0.570	0.9	4.725	28
and sill in the end of downstream apron	58.4	14.0	21.5	0.834	0.9	6.912	34.5

closed to prevent sediment particles to be washed away and enter the scour hole. At the end of each test, the scour dimensions were recorded using a bed profiler. The results are shown in Table 3. Figure 5 shows the longitude profile of scour for three types of spillways for flow discharge of 20 L/s.

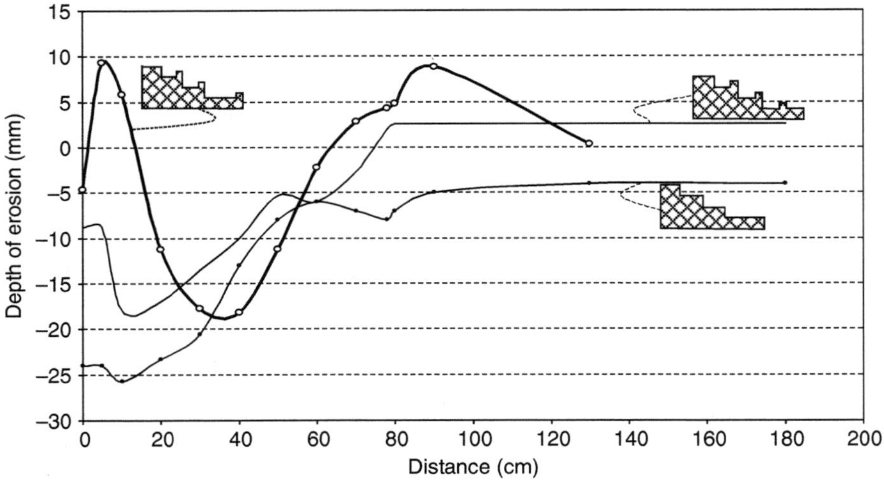

Fig. 5 Longitude profile of scour hole flow discharge equal to 20 L/s

4 Results and Discussion

To predict the scour dimensions downstream of the gabion stepped spillway, experimental program was conducted. Three types of spillways with a slope of 1(V):3.5(H), were tested under different flow conditions and two bed material sizes. The results obtained from these tests reveal that the scour hole dimensions in type I spillway, or simple spillway, is generally larger than the other two types of spillway. This is because the energy dissipation in pooled stepped spillway is greater than the simple stepped spillway. By analyzing data obtained from this study, three equations were developed by regressional method. These equations are as follows:

1. For simple gabion stepped spillway:

$$\frac{ds}{D_{50}} = 0.125\left(\frac{y_1}{D_{50}}\right)^{1.39}(\mathrm{SN})^{0.95} \tag{5}$$

$$r^2 = 0.97$$

2. For pooled stepped spillway (end sill at the middle of downstream apron):

$$\frac{ds}{D_{50}} = 0.43\left(\frac{y_1}{D_{50}}\right)^{0.98}(\mathrm{SN})^{0.92} \tag{6}$$

$$r^2 = 0.84$$

3. For pooled stepped spillway (end sill at the downstream end of apron):

$$\frac{ds}{D_{50}} = \left(\frac{y_1}{D_{50}}\right)^{0.94} (\mathrm{SN})^{0.61} \tag{7}$$

$$r^2 = 0.99$$

5 Design Procedure

The following procedure is presented to estimate the maximum scour hole depth:

1. Upon knowing the unit discharge and the height of spillway, determine the energy loss through stepped spillway by applying one of the equations presented in Table 2. For example, for the case of $\hbar/\ell = 3.5$, we get:

$$\frac{1}{1 - \Delta E/H} = 1.736\left(\frac{q^2}{gH^3}\right)^{-0.28} \tag{8}$$

2. Determine the amount of energy downstream of the spillway from (9):

$$E_1 = \left(\frac{3}{2}\right) y_c - \Delta E \tag{9}$$

   in which $y_c = (q^2/g)^{1/3}$
3. By simultaneously solving equations of energy (10) and continuity (11), one can obtain y_1 and V_1:

$$E_1 = \frac{V_1^2}{2g} + y_1 \tag{10}$$

$$q = V_1 y_1 \tag{11}$$

4. Determine the stability number from (4).
5. Determine the scour depth applying the proper equation (5–7, whichever applies).

6 Conclusion

In this study, three types of gabion stepped spillways under different flow conditions and bed materials were tested. Based on stability analysis of a particle at the point of incipient motion, a general formula was developed to predict the scour hole depth. With the help of regression analysis technique and the use of experimental data, three equations were developed for prediction of scour depth downstream of stepped spillways. From these equations, one can predict the scour depth. It was found that the scour depth downstream of simple gabion stepped spillway is greater than the scour depth for pooled stepped spillway. This is because the scour depth is directly proportional to the kinetic energy at the downstream end of the spillway. The kinetic energy at the end of simple stepped spillway is larger than the other because there is less energy dissipation through this structure as compared to other types of stepped spillways. A procedure for designing a gabion stepped spillway is presented.

References

Aigner (2004) Hydraulic design of pooled step cascades. In: Proceedings of the XXIX IAHR-congress, Beijing, China, 16–21 Sept 2001. http://rcswww.urz.tu resden. De/dainger/CASCADE.htm

Chanson H (1994) Comparison of energy dissipation between nape and skimming flow regimes on stepped chutes. J Hydraul Res 32(2):213–218

Chanson H (2001) Hydraulics design of stepped spillway and downstream energy dissipators. Dam Eng 11(4):205–242

Chinnarasri Ch (2003) Energy loss through stepped gabion spillways. In: Proceeding of 2nd regional conference on energy technology towards a clean environment, Thailand, pp 436–441

Kazemi-Nasaban Gh (1996) Flow characteristics through gabion stepped spillway. M.Sc thesis, Department of Hydraulic Structures, Shahid – Chamran University, Ahwaz, Iran (in Farsi)

Peyras L, Royet P, Degoutte G (1992) Flow and energy dissipation over stepped gabion weirs. J Hydraul Eng ASCE 118(2):707–717

Shafai-Bajestan M, Sayahi A, ML Albertson (1995) Scour hole depth downstream of the SAF stilling basin. In: Varma CVJ, Rao ARG (eds) Proceedings of 6th international symposium on river sedimentation, New Delhi, pp 635–639

Sorenson RM (1985) Stepped spillway hydraulics model investigation. J Hydraul Eng ASCE 111(12):1461–1472

Stephenson D (1979a) Gabion energy dissipaters. In: Proceeding of the 13th ICOLD congress, New Delhi, Q50, R.3, pp 33–43

Stephenson D (1979b) Rock fill in hydraulic engineering. Elsevier Scientific Publishing Co., Oxford, UK, pp 215

Yasuda Y, Ohtsu I (1999) Flow resistance of skimming flow in stepped channels. In: Proceeding of the 28th international association for hydraulic research (IAHR) congress, Graz, Austria

Bed Load Transport in a Physical Scale Model of Two Merging Mountain Streams

Sarah Simonett and Volker Weitbrecht

1 Introduction

Modeling bed load transport of a nonuniform grain size distribution in physical scale models is associated with various problems. The different behaviors of the grain size fractions mainly in terms of the initial motion (Shields 1936), the type of transport such as bed load or suspended load, the effect of cohesion and the development of bed forms (Simons and Richardson 1996; Bogárdi 1974), might involve that geometrically scaled model grains feature different characteristics to nature. The appropriate choice of the model grain size distribution is important for the accuracy of the test results. Pugh (2008), as well as Gill and Pugh (2009) propose a scaling, guaranteeing similitude of particle's settling velocity. Settling velocity is relevant in determining when a particle will remain at rest or how far it will travel as suspended load (Pugh 2008). In the present physical scale model, bed load transport is the main process. This paper proposes a methodology to simulate bed load transport in physical scale models minimizing scale effects. It is based on the correct simulation of initial motion of sediments where the critical shear velocity is the determining parameter. The methodology is composed of the following changes of the input data: first, conversion of the grain size distribution (Sect. 3.2) and secondly, adaptation of the sediment input (Sect. 3.3) including the reduction of the sediment input and the extension of the flood hydrograph. A similar work based on this approach has been conducted by Weitbrecht and Rüther (2009).

The herein presented results are motivated by the flood protection project Meiringen. Meiringen, a mountain village, is situated in the Canton of Bern, Switzerland, at about 600 m altitude. In August 2005 Meiringen got heavily flooded by the two lined mountain torrents, Alpbach and Milibach, converging in

S. Simonett (✉) and V. Weitbrecht
Laboratory of Hydraulics, Hydrology and Glaciology (VAW), Swiss Federal Institute of Technology, ETH Zurich, CH-8092 Zurich, Switzerland
e-mail: simonett@vaw.baug.ethz.ch; wv@ethz.ch

P. Rowinski (ed.), *Experimental Methods in Hydraulic Research*,
Geoplanet: Earth and Planetary Sciences, DOI 10.1007/978-3-642-17475-9_20,

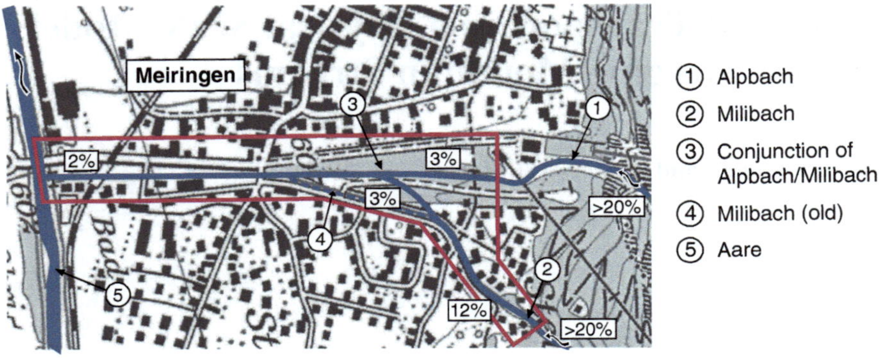

Fig. 1 Model perimeter of the physical scale model Meiringen. The numbers in percentage correspond to the slope

Table 1 Characteristics of Milibach and Alpbach above the conjunction

Type of rainfall	Flood event	Milibach		Alpbach above conjunction	
		Discharge (m^3/s)	Bed load (kg/s)	Discharge (m^3/s)	Bed load (kg/s)
Persistent precipitation	HQ_{100}	4	190	20	300
	EHQ	6	280	30	2,000
Thunderstorm	HQ_{100}	15	610	34	1,140
	EHQ	22	1,850	51	3,280

the middle of the village and subsequently flowing together into the river Aare (Fig. 1). It is assumed that the flood in 2005 corresponded to a return period of 100 years (Table 1). Persistent precipitation induced a mobilization of the unstable subsurface of the mountainous catchment area, leading to a high sediment input into the streams. While the sediment was transported in the steep, upper part of the catchment area, the transport capacity was too small in the flat and populated area of Meiringen. Sediments deposited in the channel and reduced the flow section, causing an overbank flow. Deposition initially took place mainly at the slope changes and bridges. The Milibach (Fig. 1) got blocked with boulders of diameters larger than 1 m. As Meiringen is located on the alluvial fan of the two channels, the overtopped water and sediment went directly down through the village. Sediment was deposited everywhere, even inside houses.

In order to protect Meiringen from future flooding, a flood mitigation project was proposed by local engineers and tested at the Laboratory of Hydraulics, Hydrology and Glaciology (VAW) in Zurich. The project includes sediment traps installed in the upper part of the catchment area to reduce the sediment input into the populated area. In spite of these measures, a lot of sediment will reach the channels in Meiringen during exceptional floods (Table 1). Therefore, the sediment transport

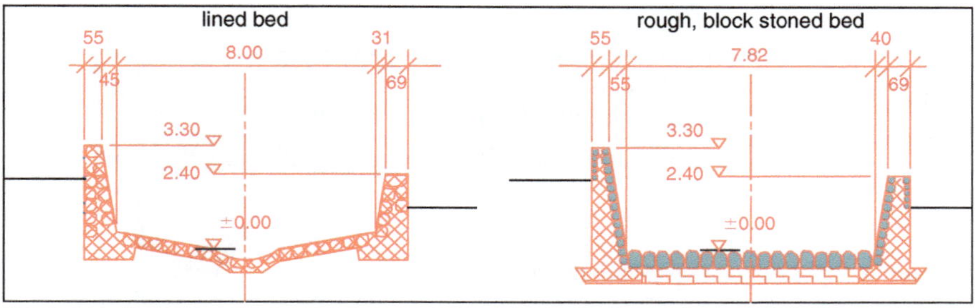

Fig. 2 Cross section of the lined bed and the rough, block stoned bed in the Alpbach

capacity of the channels in the populated area will be additionally enhanced by the following modifications:

- Increase of the flow section by increasing the channel width and banks and dismounting bridges with low freeboard
- Increase of the bed slope of Milibach by moving the confluence further upstream (Fig. 1)

At the same time, the existing lined channel bed of the Alpbach is considered to get changed to a rough, block stoned bed in order to gain an ecologically more valuable channel (Fig. 2). The drawback, however, is the smaller transport capacity in comparison to the lined channel.

As the catchment areas of Milibach and Alpbach are small, the discharges react immediately to rainfalls and two different types of flood events can be distinguished: long-lasting floods (30 h), as in the year 2005, arising from persisting but weak precipitation with a lot of mobilized material due to the long wetting, and short floods (3–5 h) arising from thunderstorms with a high discharge but little total sediment input. The discharge and sediment hydrographs of floods arising from persistent rainfalls are rather stationary, whereas floods of thunderstorms exhibit a single peak. Table 1 depicts the peak loads of the design floods HQ_{100} and the extreme floods EHQ examined in the model tests.

2 The Physical Model

The aim of the physical model tests was to prove and optimize the planned measures in terms of increased flood safety for the design flood HQ_{100} and extreme flood EHQ as well as in terms of sediment transport. In case of EHQ, defined as $1.5 \times HQ_{100}$, certain areas are accepted to be flooded. The evaluation of the rough and lined Alpbach channel in terms of flood safety and maintenance expenses for bed load excavation after a flood was of particular significance. Furthermore, the confluence (Fig. 1 (3)), the influence of the backwater of the river Aare (Fig. 1 (5)), and the construction of the sediment retention basin dam (Fig. 3 (7) and (8)) were

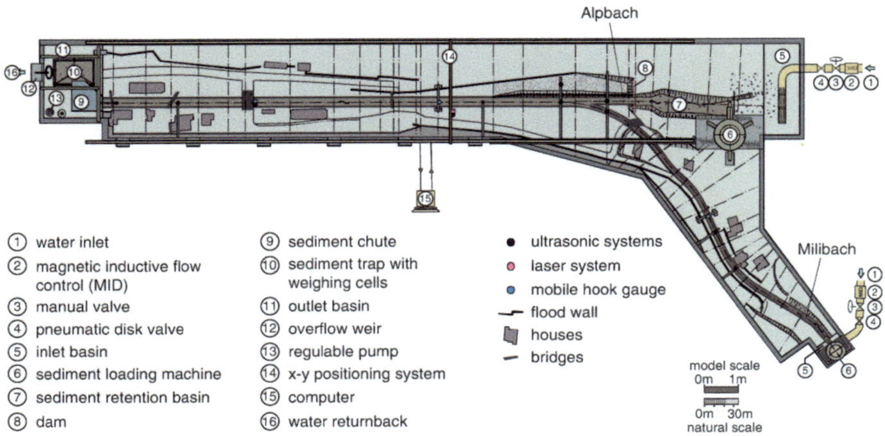

Fig. 3 Physical model of the torrent Alpbach and torrent Milibach

investigated. In this chapter, however, the focus is on the modeling of bed load, other aspects and results of the model are just briefly mentioned.

The physical model has been built with Froude similitude and a scale of 1:λ with $\lambda = 35$. The project perimeter includes the Alpbach from the end of the sediment retention basin and the Milibach from the beginning of the entrance to the populated area and is in nature about 35 m wide and 700 m long (Fig. 1). The model itself is about 10 m wide and 21 m long, including the two inlet basins for the Alpbach and Milibach (Fig. 3 (5)) as well as the outlet basin (Fig. 3 (11)).

A particular characteristic of the physical model of Meiringen is given by the distinct water and sediment hydrographs of the Milibach and Alpbach. In order to run these hydrographs, discharge and sediment input are PC-controlled by regulating the gate valve of the intake (Fig. 3 (4)) and the sediment loading machine (Fig. 3 (6)), respectively. The sediment is added to the channels in a short distance downstream of the water inflow.

The outlet basin is situated where the Alpbach meets the river Aare (Fig. 1). The water level in the outlet basin is controllable by a regulated pump (Fig. 3 (13)) and an overflow flap gate (Fig. 3 (12)) to guarantee a constant Aare water level while simulating a flood hydrograph. This allows investigating the effect of backwater into the Alpbach.

The transported sediment that reaches the outlet basin is forwarded by a chute (Fig. 3 (9)) into a submerged sediment trap (Fig. 3 (10)). This trap is connected to three weighing cells allowing continuous measuring of sediment output, analogue to Tamagni et al. (2008). The sediment input to the model is continuously measured by the loading machine and recorded. The deposition in the Alpbach is measured by scanning the channel after every experiment with a laser distance meter mounted on an automatic *x*–*y* traversing system (Fig. 3 (14)). Deposition in the Milibach and sediment which overtopped the banks and deposited beyond the channels are collected, dried, and weighted subsequent to each experiment. By means of these measurements, a bed load balance is established for each experiment. Water levels are measured continuously by ultrasonic systems at ten different cross sections along the channels.

3 Modeling Bed Load

3.1 Problems Arising from Conventional Scaling

Modeling sediment transport implies the conversion of the natural grain size distribution into model scale. Doing this, the anomaly of the critical dimensionless shear stress Θ_{cr} depending on the particle Reynolds number Re_{*d}, as depicted in the Shields' diagram (Fig. 4), has to be considered. $Re_{*d} = u_{*cr}d/\upsilon$ with υ being the kinematic viscosity and u_{*cr} being the critical dimensionless shear velocity:

$$u_{*cr} = \sqrt{\Theta_{cr} \cdot (\rho_s/\rho_w - 1) \cdot g \cdot d} \tag{1}$$

For $Re_{*d\,cr} > 200$, Θ_{cr} is considered as a constant value of 0.047. For $Re_{*d\,cr} = 2 - 200$, Θ_{cr} varies between 0.03 and 0.047 and for $Re_{*d\,cr} < 2$, the relation $\Theta_{cr} = 0.1/Re_{*d\,cr}$ is valid.

In gravel bed rivers with highly turbulent supercritical flow conditions, as Alpbach and Milibach, Re_{*d} is to a great extent higher than 200 for the majority of grain fractions. The conventional, geometrical conversion of the grain sizes into model scale by the multiplication with the scale 1:λ, however, leads to a reduction of Re_{*d} and might involve that some scaled grains show values of $Re_{*d} < 200$. As Θ_{cr} is hence smaller than 0.047 (Fig. 4), these grains get transported by a smaller

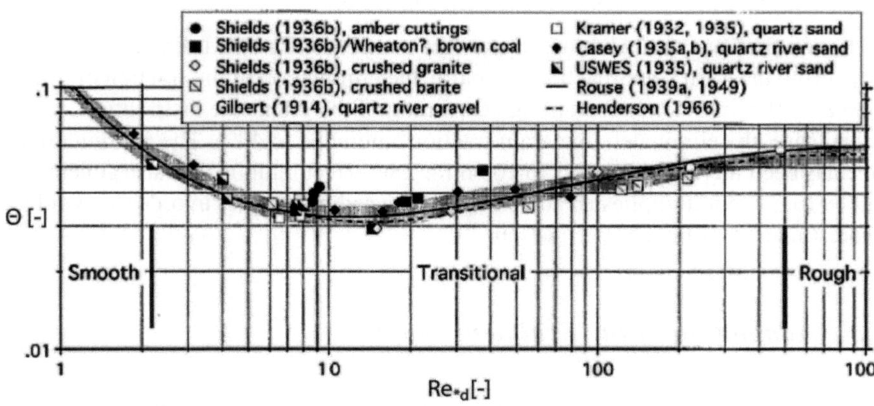

Fig. 4 Shield's diagram (Adapted from Buffington 1999)

discharge in comparison to nature and a falsification of the sediment transport occurs. Moreover, by scaling grain sizes, bed forms could be generated which in nature would not occur and thus form roughness and the total resistance would be increased in the physical model.

3.2 Conversion of the Grain Size Distribution into Model Scale

In order to avoid the above-mentioned scale effects and thus to achieve a similar transport rate at similar discharges, Zarn (1992) suggests an adjustment of the scaling. His approach assures that the reduction of the shear velocity from nature into model scale is consistent with the velocity scale $1/\sqrt{\lambda}$ according to Froude:

$$u_{*\text{model}} = \frac{u_{*\text{nature}}}{\sqrt{\lambda}} \tag{2}$$

The suggested conversion of the grain size distribution from nature to model scale takes place in three steps:

1. Conversion of the grain size distribution into model scale by the multiplication with the scale $1{:}\lambda$, according to $d_{\text{model}} = d_{\text{nature}}/\lambda$
2. Coarsening of the grain sizes with $Re_{*d\ \text{cr}} = 2$–200, belonging to $d = 0.22$–4 mm. According to (1) and (2), the following relation is valid:

$$d_{\text{model}} = \frac{1}{\lambda} \cdot \frac{\Theta_{\text{cr nature}}}{\Theta_{\text{cr model}}} \cdot d_{\text{nature}} \tag{3}$$

   with $\Theta_{\text{cr nature}}$ being 0.047 and $\Theta_{\text{cr model}}$ being between 0.03 and 0.047, depending on the grain size.
3. Elimination of the grain sizes with $Re_{*d\ \text{cr}} < 2$, belonging to $d < 0.22$ mm, as cohesive properties of the material might play a role.

Figure 5 shows the resulting coarser grain size distribution following the three steps from the Zarn's approach. Principally, the elimination of the fraction of grains with $d < 0.22$ mm leads to a strong coarsening of the grain size distribution. In the actual case, the eliminated fine grain fraction amounted to 40%. The medium grain size has been increased from 51 to 105 mm. This effect inhibits the false generation of bed forms, but it implies an adjustment of the sediment input in order to avoid an overestimation of the coarse fraction.

3.3 Adaptation of the Sediment Input

The coarser the sediment, the smaller the transport capacity. Therefore, the coarsening of the grain size distribution according to Zarn (1992) would result in a smaller

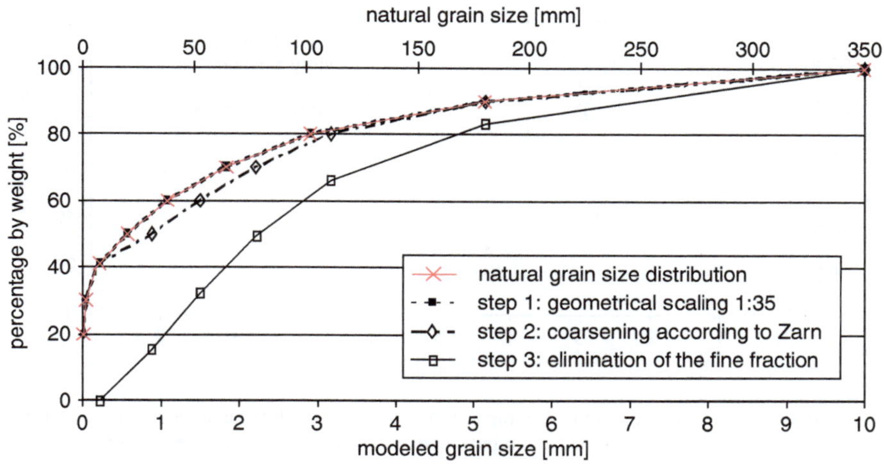

Fig. 5 Natural grain size distribution (upper *x*-axes) and converted grain size distribution in model scale (lower *x*-axes)

transport and a higher deposition, if the sediment input is not adapted. In order to guarantee correct transport processes, it is here proposed that the ratio between the sediment input at each time step of the load curve and the theoretical maximum transport capacity is kept identical in nature and the model (4). Consequently, the equilibrium slope at each time step equals in nature scale and in model scale.

$$\frac{\text{Bed load input}_{\text{nature}}}{\text{Max. transport capacity}_{\text{nature}}} = \frac{\text{Bed load input}_{\text{model}}}{\text{Max. transport capacity}_{\text{model}}} \tag{4}$$

The theoretical maximum transport capacity in the nature and the model has been calculated with the formula of Smart and Jäggi (1983) for steep channels with a gradient between 3% and 20%. Due to the coarser material, an approx. 50% smaller maximum transport capacity results in the model. Thus, the sediment input per time step, and hence the total bed load, is diminished by half in comparison to the original input. Figure 6 shows the original input for the natural material (a) and the reduced input for the coarsened material (b), both in natural scale. The discharge underlies no change.

As a consequence of the reduced sediment input in the model, both, the sediment volume per time step and the total volume that exceed the transport capacity (Fig. 6 (II)), are smaller than in nature (Fig. 6 (I)). This leads to a lower deposition height and water level. Overtopping would thus occur later in the model than in nature. To compensate for this effect, a compensation of the reduction of the sediment load is needed and the volume must be increased. For this purpose, we propose to extend the duration of the model discharge and sediment hydrograph by an extension factor, guaranteeing a correct Froude-scaled total bed load. The extension factor γ is calculated from the relation of the total bed load of original and reduced input of

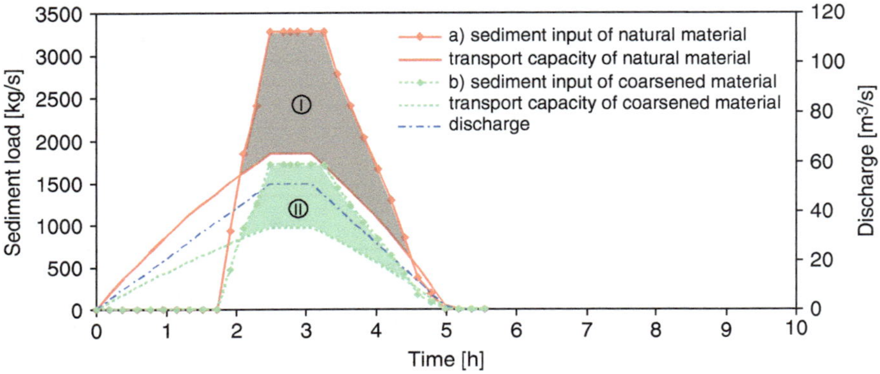

Fig. 6 Reduction of the sediment hydrograph of the flood event EHQ of a thunderstorm due to the coarsening of the grain sizes: (**a**) original input (—◆—), (**b**) reduced input (--◆--). *Colored areas*: (I) calculated deposition volume of the natural sediment, (II) calculated deposition volume of the coarsened sediment

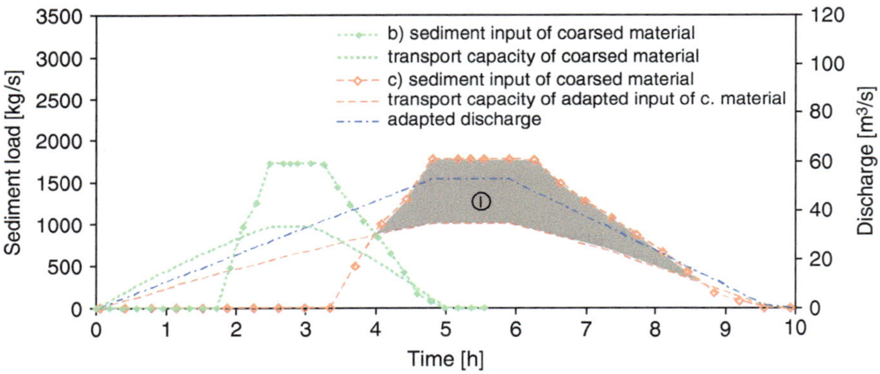

Fig. 7 Extension of the discharge and sediment hydrograph of the flood event EHQ of a thunderstorm by the extension factor $\gamma = 1.93$, (**b**) reduced input (--◆--), (**c**) adapted input (--◇--). (I) Calculated deposition volume of the coarsened material (*colored area*). Note: (**b**) and (I) of Figs. 6 and 7 are equal

Alpbach and Milibach (5). Figure 7 shows the reduced input (Fig. 7b), correspondent to (Fig. 6b), and the extended hydrograph (Fig. 7c). The total volume of the extended hydrograph (Fig. 7c) and the original hydrograph (Fig. 6a) is equal.

$$\gamma = \frac{\text{Total bed load of original input of Alpbach and Milibach}_{\text{natural material}}}{\text{Total bed load of reduced input of Alpbach and Milibach}_{\text{coarsened material}}} \tag{5}$$

Converting the model results to natural scale, the extension of the hydrograph must be considered, as the time is not only multiplied by the Froude factor $\sqrt{\lambda}$ but also by the inverse extension factor $1/\gamma$.

4 Results of the Model Tests

The comparison of the model tests of a reduced, not extended hydrograph (Fig. 7b) and an adapted, extended hydrograph (Fig. 7c) of a HQ_{100} long-lasting event with an Aare water level of HQ_{30} showed the same behavior of deposition and erosion both in time and space (Fig. 8). Deposition took only place in the section where the Alpbach was influenced by the backwater effect of the Aare. Upstream of it, the sediment got transported in the channel.

As the hydrograph of a HQ_{100} long-lasting event is rather stationary, the equilibrium slope of both model tests is reached at the same time, independent of the extension of the hydrograph. In comparison to unsteady hydrographs of thunderstorms, the extension by the extension factor γ (5) does not significantly influence the recorded water level, but influences the total bed load balance.

In Alpbach and Milibach, the design flood events HQ_{100} of both rainfall types (Table 1) do not lead to an overtopping. Local depositions take place in the channels but do not endanger flood safety. Concerning the extreme flood events EHQ, the maximum transport capacity is exceeded in both channels and deposition reduces the flow section such that water and sediments overtop the banks and overflow the area between the channels and the flood walls.

The rough bed (Fig. 2) behaves stable under all examined flow conditions. In relation to the lined channel, the water level and the deposition height are increased.

Flow is supercritical in both configurations. However, in the lined channel, the energy level is higher. Thus, the hydraulic jump at the backwater of the river Aare is stronger exhibited. In both bed configurations, the sediment which is transported through the channels deposits at the hydraulic jump and initiates a forward and backward migration of deposition. In comparison to the rough channel, stationary waves are generated at the deposition front in the lined channel, as depicted in Fig. 9. The stationary waves migrate with the deposition front up- or downstream, depending if deposition or erosion is taking place. Due to the interaction of the waves with the bed, antidunes are generated (Kennedy 1963). The differences

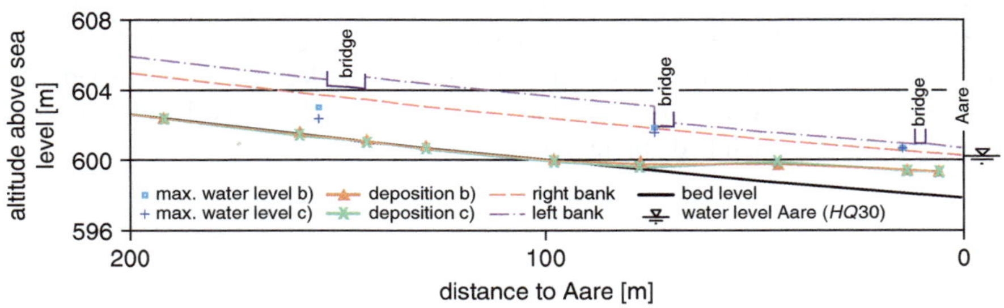

Fig. 8 Results of the model tests of a not extended (**b**) and an extended (**c**) hydrograph of a HQ_{100} long-lasting event with an Aare water level of HQ_{30}: bed level at the end of the event (deposition) and maximum measured water level during the flood. Depicted is the lowest section of the Alpbach, from 0 to 200 km

Fig. 9 Hydraulic jump at the interface between deposition and free bed in the lined Alpbach and standing waves propagating downstream of the hydraulic jump

between the rough and lined bed, however, are only observable as long as the channel beds are not completely covered by sediments. Covering of the channel bed leads to similar roughness in both bed configurations and accordingly to similar water and energy levels.

The model tests further showed that the planned confluence with an angle of approx. 20° (Fig. 1 (3)) works well for all examined combinations of discharge in the two rivers, generating only small shockwaves. The outlet dam (Fig. 3 (8)) of the sediment retention basin (Fig. 3 (7)) has been optimized by building a symmetric inflow trumpet which provides a continuous flow without breakaways at the dam. If the sediment retention basin is aggraded, a drop at the outlet dam is established. The energy dissipation at the drop generates a high stationary wave involving scour protection in the rough bed configuration.

5 Conclusion

The physical scale model Meiringen is a complex system. The two merging mountain torrents, with different discharge and sediment hydrographs, require correct and time-dependent regulation of the input parameters. Due to the many different research objectives regarding flood safety, the effect of backwater, and the function of the confluence and the dam, a lot of relevant processes take place at the same time at different locations in the model and require the constant observation of more than one person as well as installed monitoring cameras. Because of the relevance of the sediment processes for the research objectives, the correct simulation of the grain size distribution is important. From conventional, geometrical scaling by the multiplication with the scale factor $1/\sqrt{\lambda}$, model effects emerge and, unfortunately, there is no other proved practice for simulating bed load transport in physical models. The approach of Zarn (1992) – as applied in this chapter – is based on a correct simulation of initial motion of all grain fractions. It is composed of three steps (Fig. 5). By step 1, the natural grain size distribution is geometrically scaled. By step 2, the grains of $d = 0.22$–4 mm are coarsened to get Froude-scaled shear velocity, and by step 3, the grains of $d < 0.22$ mm are eliminated in order to avoid cohesive effects. Steps 2 and 3 coarse the grain size distribution. In the actual case of Meiringen, the eliminated fine fraction of step 3 amounted to 40% involving a strong coarsening on the grain size distribution. As the transport capacity of coarse material is smaller, a lower transport rate would occur in the model using the coarsened material and deposition would be overestimated compared to nature. To gain a correct simulation of the transport processes in the model, a reduction of the sediment input is consequently implied. The reduction is proposed in a way that the use of the theoretical maximum transport capacity is equal in model and nature scale (4). This leads to a sediment input that is not Froude-scaled (Fig. 6). If no further correction would be conducted, less sediment would be added in total and an underestimation of the deposition height and likewise of the bed level and the water level would occur. Therefore, the model hydrograph is extended in time by the extension factor γ (5). The extension guarantees correct bed load input per time step and total volume and thus, a simulation of deposition height, slope, and water level as well as a bed load balance which is concordant in the nature and model (Fig. 7).

Converting the model results to natural scale, the extension of the hydrograph must be considered by multiplying the model time both by the Froude factor $\sqrt{\lambda}$ and by the inverse extension factor $1/\gamma$.

Using the extension factor γ based on the sum of the bed load of the Alpbach and Milibach, two simplifications are made. It is first assumed that the relation of the input at each time step is equal to the relation of the total input and second that the Milibach and Alpbach exhibit the same extension factor. Due to these simplifications, the sediment input per time step is finally simulated with an accuracy of 5%. Taking into account the overall uncertainties of the model with respect to input data, e.g., expected hydrograph or sediment sizes, and other model inaccuracies, it can be concluded that the proposed methodology of the grain size conversion (Sect. 3.2) and the adaptation of the sediment input (Sect. 3.3), including the reduction of the input and the extension of the flood hydrograph, leads to reasonable results.

However, it must be considered that the proposed method is only valid for systems exhibiting turbulent flow in each zone of interest. The proposed method is not valid for systems composed of different kind of flow, e.g., flow with low Reynolds number in basins.

References

Bogárdi J (1974) Bestimmung der Grenzzustände der Geschiebebewegung. Wasserwirtschaft 7:205–212

Buffington JM (1999) The legend of A.F. Shields. J Hydraul Eng 125(4):376–387

Gill TW, Pugh CA (2009) Sediment transport similitude for scaled physical hydraulic modeling. In: 33rd IAHR congress: water engineering for a sustainable environment, Vancouver, British Columbia, Canada, pp 5444–5451

Kennedy JF (1963) The mechanics of dunes and antidunes in erodible beds. J Hydraul Res 12(2):521–546

Pugh CA (2008) Sediment transport scaling for physical models. Appendix C. Sedimentation engineering: processes, measurements, modeling, and practice. No. 110. ASCE, New York

Shields A (1936) Anwendung der Aehnlichkeitsmechanik und der Turbulenzforschung auf die Geschiebebewegung. Mitteilung 87. Versuchsanstalt für Wasserbau und Schiffbau, Berlin

Simons DB, Richardson EV (1996) Resistance to flow in alluvial channels. Professional Paper 422-J, U.S.Geological Survey, GPO, Washington

Smart GM, Jäggi MNR (1983) Sediment Transport in steilen Gerinnen. Mitteilung 64. Laboratory of Hydraulics, Hydrology and Glaciology, Swiss Federal Institute of Technology, Zurich

Tamagni S, Janisch T, Hager WH (2008) Bed stabilization with unstructured block ramps: physical model investigations. In: River flow conference 2008, Turkey, pp 2075–2081

Weitbrecht V, Rüther N (2009). Laboratory and numerical model study on sediment transfer processes in an expanding river reach. In: 33rd IAHR congress: water engineering for a sustainable environment, Vancouver, British Columbia, Canada, pp 5436 – 5443

Zarn B (1992) Lokale Gerinneaufweitung – eine Massnahme zur Sohlstabilisierung der Emme bei Utzendorf. Mitteilung 118. Laboratory of Hydraulics, Hydrology and Glaciology, Swiss Federal Institute of Technology, Zurich, pp 98–103

Cross-Section Changes in the Lower Part of a Mountain River After the Flood in Spring 2010, as Presented by Means of CCHE2D Program

Andrzej Strużyński, Maciej Wyrębek, Mateusz Strutyński, and Krzysztof Kulesza

1 Introduction

The turn of spring and summer 2010 in Poland was characterized by extremely heavy rainfall which led to two major floods affecting the territory of the whole country. Measurements of morphological changes in a mountain river Białka in the Tatra Mountains on 12–14 May 2010 turned out to be made immediately before the flood. Inventory measurements conducted afterwards, i.e., on 1–2 July 2010 illustrated the river activity during the flood. On the basis of the measurements and hydrological data obtained from the Institute of Meteorology and Water Management (IMGW), a two-dimensional CCHE2D computer model was launched, in which a simulation of the river bottom changes during the flood wave passing was conducted. The Białka River has a bedload composed of pebbles reaching even 0.5 m in diameter. The model of bedload transport was launched using Wu, Jia, and Wang equation (Wu et al. 2000).

2 Researched Object

The Białka River flows in the Tatra Mountains region in South Poland. It is formed from joining the Biela Voda (Slovakia) and the Rybi Potok (Poland). According to the research conducted in 1955–2004, regarding the biodiversity, in her lower reach the Białka River is the first class, i.e., the highest class river (Kulesza et al 2004).

A. Strużyński (✉), M. Wyrębek, and M. Strutyński
Department of Hydraulic Engineering and Geotechnics, Agriculture University in Kraków, 24/28 Mickiewicza Ave, 30-059, Kraków, Poland
e-mail: rmstruzy@cyf-kr.edu.pl; m.wyrebek@gmail.com; mstrutynski@ar.krakow.pl

K. Kulesza
Institute of Meteorology and Water Management, Board of Water Management Systems, Kraków, Poland
e-mail: k.kulesza@imgw.pl

P. Rowinski (ed.), *Experimental Methods in Hydraulic Research*, 287
Geoplanet: Earth and Planetary Sciences, DOI 10.1007/978-3-642-17475-9_21,

The headstreams of the river are located at an altitude of 1,420 m a.s.l. The area of the Białka catchment is 229.9 km^2 with a length of 41 km. It flows into the artificial reservoir Czorsztyn at an altitude of 529 m a.s.l. The mean annual flow is about 5 $m^3\ s^{-1}$, whereas spring floodwaters reach the flows of about 500 $m^3\ s^{-1}$. In the analyzed, mouth part, the Białka is a braided river with usually one to three arms. The presented material covers 2.5 km of the river from its mouth. A 1-m high concrete step is situated at a distance of 360 m from the river mouth. The width of the cross-sections of the measured riverbed at this mouth section fluctuates from 75 to 255 m. The river bottom slope at this section is 1.04% on the average, which causes the river to become inaccessible at above 60–70 cm flow depth. The bottom ordinates for individual arms in the same cross-section may differ by more than 10 cm. The bedload is mostly composed of granite grains with diameters reaching more than 50 cm.

3 Measuring Methods

The measurements before the floodwater (m-1) were conducted on 12–14 May 2010, and after floodwater (m-2) on 1–2 July 2010. Geodetic measurements of the bankfull channel and measurements of bed material granulation were carried out.

3.1 Surveying Measurements

The measurements of the riverbed in the m-1 measurement cycle were conducted using scattered points condensed in selected cross-sections, whereas during m-2 measurements, we focused on making the cross-sections. In this way, m-1 measurements in situ were carried out at 1,019 measurement points, whereas during m-2 measurements 470 pickets were marked. The data accumulated prior to floodwater occurrence made it possible to prepare 25 cross-sections used in the process of computer modeling (Fig. 1). The measurements conducted after the floodwater passage enabled a comparison of the bottom configuration changes. The analyses were performed in six cross-sections: the first was situated 10 m upstream from the river step (340 m upstream from the Białka mouth), the second was 111 m upstream from cross-section I, the sixth was 50 m upstream from the road bridge on the border of Dębno and Frydman villages (700 m upstream from the Białka mouth), the third was 300 m upstream from the bridge, the fourth was 315 m, and the fifth was 390 m upstream from the bridge.

3.2 Granulometric Measurements

Because of the character of the river hydraulics, a photographic method of measuring the dimension and shape of the grains was chosen to determine the bedload granulation. Application of the photographic method has been possible because the

Fig. 1 The view of the studied object and its location

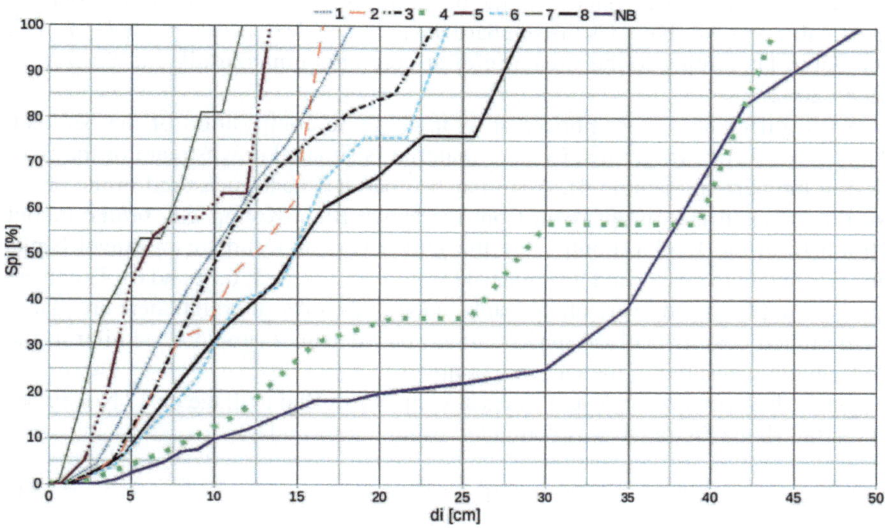

Fig. 2 Granulation curves in the Białka lower reach

Białka is a mountain river in which changes of bottom configuration depend on the movements of bed material. The photographs were made in eight locations covering the bottom, banks, and gravel bars. In total, the measurements of *a* and *b* dimensions of 2,422 grains were made. Beside a small amount of fine fractions, in the river bottom there are also gravels and stones, usually with diameters of 15–20 cm and bigger grains, whose *b* dimension (grain diameter) reaches over 50 cm. Granulometric curves of the bed material are shown in Fig. 2. The curves marked with numbers from 1 to 8 were made in April and May of 2010. The collection localities are described in Table 1. The "NB" curve was drawn on the basis of the measurement

Table 1 Compiled parameters characterizing the bedload in the Białka mouth area (d_m is the mean size of grains covering riverbed, d_i are characteristic diameters of bedload fractions, and δ is the standard deviation of sieve curve

Location	Sample number	d_m (cm)	d_{16} (cm)	d_{84} (cm)	d_{50} (cm)	δ (–)
Right riverbank	1	11	4.3	16	9.8	1.93
10 m upstream the river step	2	12.1	5.3	15.8	12.5	1.73
Shallow flow	3	12.9	5.6	20.5	10.3	1.91
Flow current downstream the step	4	30.2	11.6	42.1	28.5	1.92
Gravel bar	5	8.21	3.1	12.7	5.9	2.02
Flow current upstream	6	15.5	7.1	22.6	14.8	1.78
Fine material on the bank	7	6.4	1.7	10.7	5.1	2.51
Rough material on the bank	8	17.1	6.5	26.7	14.8	2.03
200 m upstream from the bridge	NB	35	14.6	43	37	1.72

carried out in Nowa Biala village on 18 November 2009. The material has been mixed. No location was found with the bed armoring described by δ of an approximate value of 1.3 (where δ is the standard deviation of sieve curve: $\delta = (d_{84}/d_{16})^{0.5}$; see Table 1) pointing to the existence of armored bottom (Bartnik and Strużyński 1999, 2002).

Some differences were noticed in granulometric composition of the bedload. The curves form two groups. Coarser material is situated downstream of the river step in the current (sample 4). A similar bedload has been measured in the "NB" locality situated 200 m upstream from the Nowa Biala-Krempachy bridge (about 10 km from the Białka mouth). On the gravel bars and midstream gravel banks one can find both coarse and fine material (samples: 1, 2, 3, 5, 6, 7, and 8). The finest grains (samples 5 and 7) are on the bank and gravel banks, although there are places where much coarser bedload may be found on the bank (sample 8), similar to the bed sample measured in the current (sample 6).

On the basis of sizes *a*, *b*, and *c* measured in Nowa Biala village on 18 November 2009, the shape factor SF was computed using the formula

$$\mathrm{SF} = \frac{c}{\sqrt{ab}} \tag{1}$$

where *a* is the biggest, *b* the average (sieved), and *c* is the smallest diameter.

The grain size ranged from 2 to 49 cm. The grain shape factor fluctuated from 0.17 to 0.93 reaching the average value of 0.59. The grains measured in the mouth reach in April and May 2,010 were characterized by the shape factor values from 0.4 to 0.7 with the value 0.6 describing the whole sample.

3.3 Hydrological Data

The course of the flood water in May 2010 registered by IMGW station in Krakow is illustrated by observations from the Trybsz gaging point (Fig. 3).

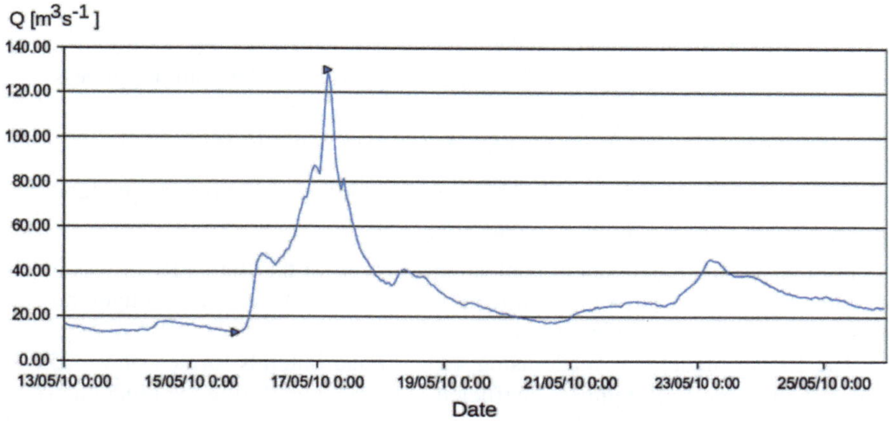

Fig. 3 Changes of flow observed at the Trybsz station during flood flow on the Białka River in May 2010

The station is located 10.6 km upstream from the Białka mouth. In this reach the catchment is narrow, which causes a slight increment of the immediate catchment, but the river station does not register the outflows from the Trybska River and its tributary – Czarnogórska River, with their mouth just below Trybsz village. In result, the station monitors a 25 km^2 smaller catchment. Taking into consideration the catchment increment causes that the maximum of expected flow might have reached the value of 147 $m^3\,s^{-1}$. According to the IMGW data, the flow occurred on 17 May 2010 at 5.00 a.m. According to Punzet equation (Carpathian formula) it is a flow of a 30% occurrence probability. In the hydrological sequence for the years 1995–2004 the minimal flow on Trybsz 2 station is 1.24, the maximum is 174.15, and the mean flow is 4.76 $m^3\,s^{-1}$ (Kulesza et al. 2004). The 174.15 $m^3\,s^{-1}$ flow occurred in the summer season.

4 Computer Modeling Using CCHE2D Program

CCHE2D is a two-dimensional computer model, making computations using finite element method (Zhang 2006). It makes it possible to compute water flows in open channels considering logarithmic velocity distribution. It enables computations of slow- and fast-changing flows.

Moreover, it is possible to conduct the computations of bedload and suspended load transport and changes of the bottom configuration (Wu 2001). The authors of the program calibrated the computations of bedload transport for coarse fractions using Wu, Wang, and Jia's equation (Wu et al. 2000)

$$\phi_{bk} = 0.0053\left[\left(\frac{n'}{n}\right)^{3/2}\frac{\tau_b}{\tau_{ck}} - 1\right]^{2.2} \tag{2}$$

where

ϕ_{bk} is the nondimensional bedload transport capacity

$n' = \frac{d_{50}^{1/6}}{20}$ is the Manning's coefficient corresponding to the grain roughness

n is the Manning's coefficient, τ_b is the bed shear stress

$\tau_{ck} = 0.03(\gamma_s - \gamma)\left(\frac{p_{hk}}{p_{ek}}\right)^{0.6}$ is the critical shear stress

$p_{hk} = \sum_{j-1}^{N} \frac{p_{bj}d_j}{(d_k+d_j)}$ and $p_{ek} = \sum_{j-1}^{N} \frac{p_{bj}d_k}{(d_k+d_j)}$ are the hiding and exposure probabilities for the kth size class of bed material, respectively.

For this reason, the model of bedload movement for the Białka River was based on this very formula (the other available were: SEDTRA, Wu et al., Engelund and Hansen's and Ackets and White's).

The Białka model was made using 40,050 quadrilaterals formed on $I = 50 \times J = 801$ lines mesh. Roughness coefficient n of the flow during the model activity was computed using Wu and Wang equation (1999) based on the diameter of bed material and relative roughness. In order to improve the precision of the model, the k-ε equation was applied for computing water flow, whereas modeling of bed material transport was conducted for three layers with a minimum thickness of 0.3 m, enabling fast changes of the bottom configuration. The primary granulation applied in the model was described by ten fractions measured on measurement point 3. The thickest bedload layer provided a source of uniformly grained material. Initial conditions of the bed material movement and the amount of transported material in the selected reference cross-section were computed using the TRANS program (Bartnik 1992) based on MPM-B formula, developed by Bartnik (Bartnik and Kopka 1991).

5 Presentation and Interpretation of Results

5.1 Cross-Sections

The floodwater which occurred in the spring 2010 in many cross-sections caused considerable changes in the bottom configuration measured in situ. Also the results of modeling supplement these characteristics, although they do not represent precisely the morphodynamic changes which occurred. The results were presented graphically in Fig. 4.

In the cross-section V an incision into the concave bank occurred to a depth of about 23 m. At the same time, a forming accumulation zone was observed on the left bank. Therefore, concentrated cross-section incised itself into the right bank.

A change of the river arms from cross-section V, in cross-section IV led to the appearance of a developed river channel on the riverbed axis and a well-developed left fork. No changes of the right bank part of the riverbed occurred.

The width of the riverbed between cross-sections IV and III has doubled and there is a flat gravel bank partially reinforced by vegetation in the middle. A division

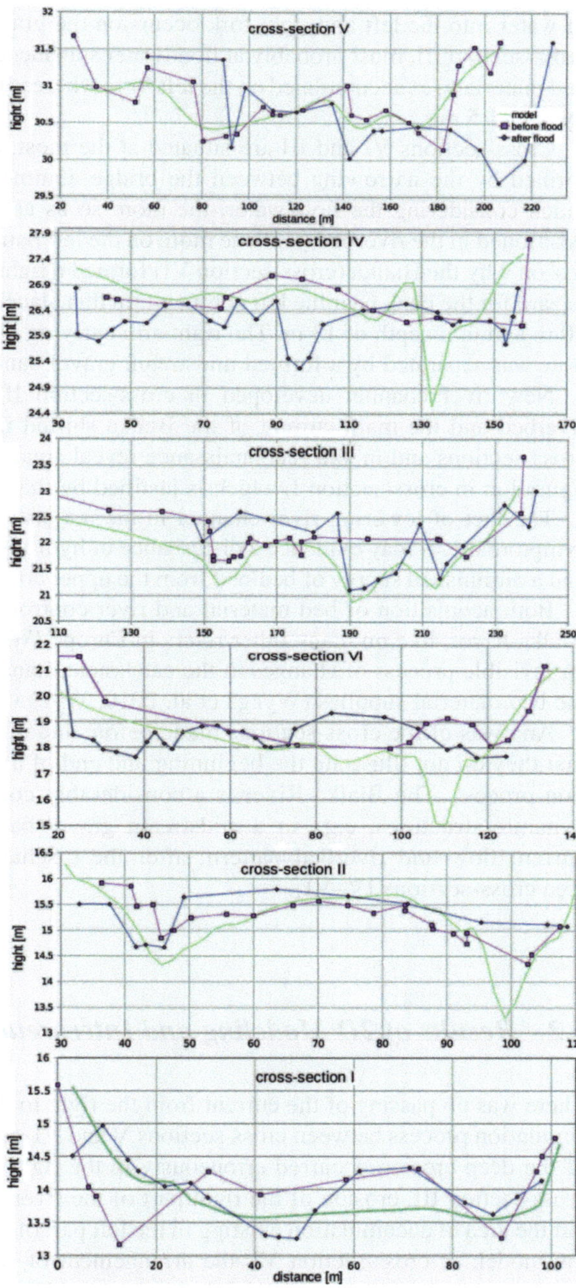

Fig. 4 Changes of selected cross-sections after floodwater passing in May 2010 measured in situ and determined as a result of computer modeling

of water into the left and right fork occurs on the gravel bank during the flood. In cross-section III, most probably at floodwater subsidence, a considerable amount of bed material was accumulated on the left fork, whereas the right branch was lowered by over 1.5 m.

Cross-sections VI and III are situated at the most sore point of the back water formed by the narrowing between the bridge abutments 75 m apart. This is not much considering the floodwater, the more so as also the two bridge pillars are positioned in the riverbed (and one more on the left bank). This might have been the reason why the Białka (cross-section VI) formed a tight meander currently posing a hazard for the bank near the left abutment. In this place, the bank was scoured along 40 m and to a depth of 13 m. The right arm carrying also a considerable portion of flow was separated by a formed midstream gravel bank about 40-m wide.

New river channel developed in cross-section II in the middle part of the riverbed and the main current of the Białka shifted into it. Measurements of the cross-sections and in situ reconnaissance reveal small changes of the bottom configuration in cross-section I, which is justified by the river step effect.

The fact of lowering river channel in the majority of studied cross-sections is symptomatic. It may evidence a disturbance of hydrodynamic balance in the Białka and a diminished supply of bedload from the upper cross-sections (Krzemień 2006).

Both acquisition of bed material and river control works are conducted on the Białka River, like on many other rivers in Europe (Wyżga et al. 2008). Moreover, an invisible process of changes in the catchment management occurs, diminishing the bed material supplies (Wyżga et al. 2010; Wyżga 2008).

Analysis of the cross-sections made before and after floodwater passing show that they do not illustrate the beginning and end of a single erosion or accumulation process. The Białka River is a considerably complicated system, in which dynamic structures, e.g., in a midstream gravel bank, form depending on the current flow and riverbed pattern, after the discharge and new arms develop (see cross-sections IV–VI).

5.2 Results of 2D Modeling and Interpretation

There was no passing of the current from the right to the left bank in the computer simulation process between cross-sections V and VI, and in result the modeled area of the deep erosion occurred erroneously in the right part of the cross-section. In cross-section III, erosion of the right part of the riverbed was computed correctly, but the area of accumulation existing in the left part of the riverbed did not appear in the model. In cross-section VI, the arrangement of streams of the modeled flow caused not only the scour of the right bank gravel bar but also an incision appeared there, not registered in nature. No phenomenon of midstream gravel bank forming or left bank scouring appeared either. In cross-sections I and II, the results of modeling are satisfactory or correct.

Computer simulation did not illustrate the occurrence of riverbed processes in horizontal pattern. Despite activating the module enabling modeling fast bed changes with unsteady flow, both simulated erosion and bed material accumulation processes were of vertical character and did not cause a disappearance or formation of new forks. No distinct processes of side erosion were observed.

The causes of modeling errors might be: the application of 2D model containing only the module generalizing vertical water velocities distribution as logarithmic one, limiting the number of nodes to over 40,000, application of the formula for computing bed material transport not calibrated for diameters exceeding 12.8 cm, and occurrence in nature of areas of increased flow roughness which caused oversize grains or wood debris. The presented model did not consider the existing transport of suspended load.

5.3 Bedload Transport

Results of bedload transport at the moment of peak floodwater passing obtained from CCHE2D program were compared in cross-section VI with the results of computations conducted using TRANS program. The modeled bedload transport in the individual nodes was added up giving the result of transport in the whole cross-section equal to 27.18 kg s^{-1}. Transport computed using TRANS program was converted into the 1 m of the transportation zone width yielding the result of 4.69 kg m^{-1} s^{-1}. The value converted into the model transportation zone width in both existing flow streams of 15 m gives the results of 70 kg s^{-1}. The transportation zones determined on the basis of the cross-section after floodwater together have a similar width (16 m) but are slightly differently situated. The model, despite the fact that the deepest part of the cross-section is on the right, quite correctly points to the occurrence of more intensive transport in the river channel axis, about 10 m from the left bank. Analysis of cross-section VI indicates that the left transportation zone should be situated 40 m to the right in the area of eroded bank. The critical depth for the Białka bedload was computed as 0.78 m.

6 Conclusions

An analysis of flows for the many-year period (1995–2004) demonstrates that the interpreted floodwater may appear twice in 15 years. This has been confirmed by the age of the bush vegetation covering gravel bars in the area of the Białka riverbed.

Floodwaters may appear in spring or in summer.

Bedload starts to move during floodwater and rapid changes occur in the horizontal and vertical pattern of the watercourse. Prior to the 2010 floodwater, the bed surface was not armored. One should expect an easy movement of the bedload because frequent weather breakdowns with heavy rains in the Tatras may prevent forming a stable cover.

Cross-sections V, IV, and VI are mostly prone to erosion. In cross-section II the process of material accumulation prevailed. During the floodwater in 2010 a massive amount of bedload was transported. It should be noticed that in the upper reaches the Białka incises itself into the bedrock. This is a process which drastically changes the flow regime in mountain rivers. For this reason, reparation measures should be undertaken in the Białka catchment.

In the lower reach, the Białka is a braided river. It forms a complicated multi-bed pattern which poses a difficulty for the process of modeling hydromorphological changes during flood with transported bedload.

In most cases, the CCHE2D model was not a correct representation of horizontal changes in the elevation of a bed composed of rough grains. Instead, the energy of the modeled flow caused linear erosion and accumulation processes. Possibly, making a model of a shorter river reach with a dense mesh would improve the correctness of illustration of bank erosion in rivers of this type.

The amount of transported bedload in the model is comparable with the TRANS method, calibrated for mountain rivers.

Acknowledgments The research reported in this article was performed within the project N N305 186537: The description of balanced mountain rivers and streams state on the basis of natural morphological parameters.

References

Bartnik W (1992) Hydraulika rzek górskich z dnem ruchomym. Początek ruchu rumowiska wleczonego (Hydraulics of streams and mountain rivers with movable bed. The incipient motion of bedload transportation), Rozprawa Hhabilitacyjna nr 171, Zesz. Nauk. AR w Krakowie, Kraków, pp 122 (in Polish)

Bartnik W Kopka (1991) Methods of measuring incipient motion of bed in mountain rivers. In: Drs. Shou-Shan Fan, Yung-Huang Kuo (eds.) Proceedings of the Fifth Federal Interagency Sedimentation Conference, Federal Energy Regulatory Commission, Las Vegas, Nevada, USA, pp 4-170–4-174

Bartnik W, Strużyński A (2002) Estimation of hydraulic parameters of armored layer forming in mountain rivers and streams, Advances on hydro-science and engineering, ICHE, and Warsaw University of Technology, on CD-ROM

Bartnik W, Strużyński A (1999) Determining hydrodynamic balance in mountain stream floods. In: Third International Symposium on Ecohydraulics, IAHR and Utah State University, Salt Lake City, published on CD-ROM

Krzemień K (2006) Badania struktury i dynamiki koryt rzek karpackich (Study of the structure and the dynamics of Carpathian rivers), Infrastruktura i Ekologia Terenów Wiejskich, nr 4/1/2006, s.131–142 (in Polish)

Kulesza et al (2004) Ustalenie warunków referencyjnych dla rzeki Bialka (Describing the reference conditions for Bialka River), IMGW in Kraków, manuscript (in Polish)

Wu W (2001) CCHE2D sediment transport model (version 2.1), technical report no. NCCHE-TR-2001-3. School of Engineering, The University of Mississippi, Oxford, pp 48

Wu W, Wang SSY (1999) Movable bed roughness in alluvial rivers. J Hydraul Eng ASCE 125 (12):1309–1312

Wu W, Wang SSY, Jia Y (2000) Nonuniform sediment transport in alluvial rivers. J Hydraul Res IAHR 38(6):427–434

Wyżga B (2008) A review on channel incision in the Polish Carpathian rivers during the 20th century. In: Habersack H, Piégay H, Rinaldi M (eds) Graver-bed rivers VI: from process understanding to river restoration. Elsevier, Amsterdam, pp 525–555, ISSN: 0928-2025

Wyżga B, Zawiejska J, Radecki-Pawlik A (2008) Description of incision of rivers and it's influence on the hydraulics of flood flows – examples from Catpathian rivers. Landform Anal 9:402–405

Wyżga B, Hajdukiewicz L, Radecki-Pawlik A, Zawiejska J (2010) Eksploatacja osadów z koryt rzek górskich – skutki środowiskowe i procedury oceny. Gospodarka Wodna, Warszawa, Sigma-Not 6:243-249

Zhang Y (2006) CCHE-GUI – Graphical users interface for NCCHE model user's manual – version 3.0, technical report no. NCCHE-TR-2006–02. School of Engineering, The University of Mississippi, Oxford, pp 158

Application of Hydrodynamics Model for a Case Study of the Kolbudy II Reservoir Embankment Hypothetical Failure

Michał Szydłowski

1 Introduction and Case Study Description

The Polish law requires an analysis of catastrophic flood events following hypothetical failures of dams and reservoir embankments. According to the regulations of the Polish Minister of Environment concerning technical conditions for hydrotechnical structures and their location, the catastrophic flood ranges must be determined for dams with total water head greater than 2 m or water capacity greater than 0.2 million cubic meters. The chapter presents an example of numerical simulation of flood wave propagation resulting from hypothetical embankment break of the Kolbudy II reservoir, being part of the Bielkowo hydro-power plant.

The Bielkowo power plant is an element of hydro-power system of the Radunia River (Majewski et al. 2005). The Radunia River is situated in the Pomeranian Voivodship (Poland) and is controlled by the Gdańsk Board of Water Management. The hydrotechnical structures on the river are owned and operated by ENERGA Elektrownie Straszyn Ltd.

The hydrotechnical system of the Bielkowo hydropower plant is presented in Fig. 1 and can be described, after Jarzębińska (2005), as follows. The system consists of control weir Kolbudy, diversion channel, two reservoirs (Kolbudy I and Kolbudy II), concrete power tunnel, surge tank, steel penstock, and hydropower plant operating on the head of 44.25 m. The total water head was formed due to 4-km long water diversion between two cross-sections of the Radunia River on the steep 8 km segment of the river. The weir Kolbudy forms the reservoir Kolbudy I and controls the flow between the river and diversion channel. The channel carries water to the Kobudy II reservoir with the maximal discharge of 20 m^3/s. A total length of the canal is 1,350 m, and its typical cross section is trapezoidal. At the outlet from the channel to the reservoir, a second small control weir is located. Reservoir Kolbudy II was formed by constructing four earth dams closing the natural lowland

M. Szydłowski
Civil and Environmental Engineering, Gdańsk University of Technology, Narutowicza 11/12, 80-233 Gdańsk, Poland
e-mail: mszyd@pg.gda.pl

P. Rowinski (ed.), *Experimental Methods in Hydraulic Research*,
Geoplanet: Earth and Planetary Sciences, DOI 10.1007/978-3-642-17475-9_22,

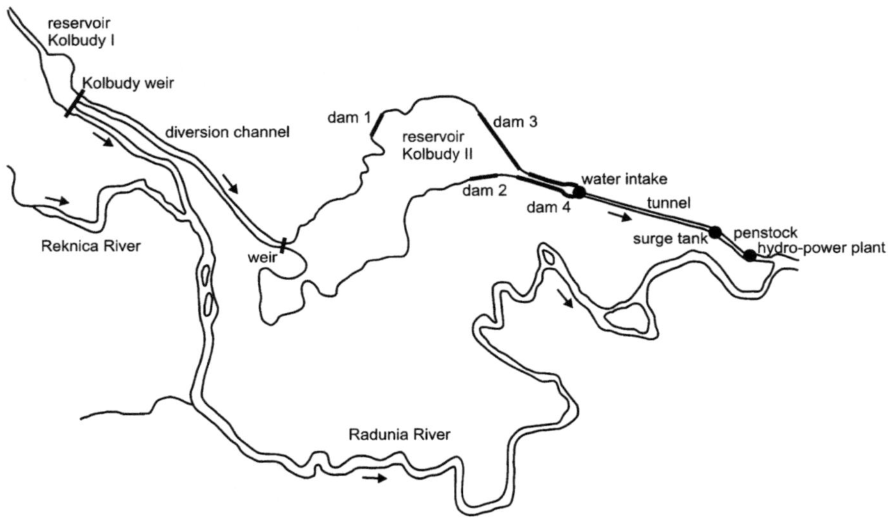

Fig. 1 Bielkowo hydropower plant hydrotechnical system

area near the Kolbudy town. The total length of the embankment is about 1,560 m. The dams are sealed with clay cores and covered using Ehlers-type hexagonal concrete plates. Locally, the dikes reach the height of 7 m. The crest elevation is 87.50 m a.s.l. and maximal water level in reservoir is equal to 86.3 m a.s.l. The area of the reservoir Kolbudy II is about 54 ha and the maximal depth is 9 m. There is no information about reservoir bathymetry. The total volume of the reservoir Kolbudy II is about 1.2×10^6 m^3.

The water from the Kolbudy II reservoir is supplied to the Bielkowo hydropower plant. The water intake is situated at the end of the reservoir, at the inflow section formed with the embankments. The water is conveyed in the concrete tunnel of a total length of 834 m and a diameter of 3.6 m. The tunnel is located underground. At the end of the tunnel the surge tank is situated, that protects the pipeline against the water hammer effect. Then the water is turned into the 166 m long steel penstock of 3 m diameter leading to the power plant. Bielkowo hydropower plant is equipped with three Francis turbines of 2.4 MW and discharge 7.2 m^3/s each. From the turbines, the water flows to the tailrace channel and finally to the Radunia River.

The failure of the dam 1 (Fig. 1) is considered in the chapter as an example of numerical simulation of a catastrophic flood.

2 Digital Terrain Model (DTM) and Dam-Break Outflow Hydrograph

The numerical simulation of the catastrophic flood is composed of several elements. At the first step, the digital models of elevation (DEM) and land development must be elaborated. The geodesic work was done by the Local Geodesy and Cartography

Fig. 2 Digital elevation model

Company (OPGK) in Gdańsk. The data needed for the description of the terrain elevation and coverage were elaborated using the aerial photographs forming the orthophotomap. Finally, the DEM with a resolution of 5 $\times$ 5 m was interpolated (Fig. 2). Then, the main land cover categories were identified on the aerial photographs and the Manning roughness coefficient distribution map was prepared. The following values of the roughness coefficient were adopted in the land cover model: free surface water bodies, like rivers, creeks, and lakes – 0.025; meadows – 0.035; arable lands – 0.05; bushes and forests – 0.11; roads – 0.012; built-up areas – 0.15.

In the next step of numerical simulation of catastrophic flood, the location of embankment failure must be identified or assumed and the water discharge hydrograph through the broken dike has to be estimated. In this test case, the location of the breach was assumed in the middle of the dike. Then the computation of water discharge from the reservoir was undertaken as a unique problem separated from the flood routing. In order to estimate the outflow hydrograph, the reservoir was considered as a flat pond. It was impossible to analyze the hydrodynamics of the Kolbudy II reservoir because of the lack of information about the bathymetry of the reservoir.

In order to predict the outflow from the reservoir it is necessary to describe the failure mechanism. This problem is very complex for the earth dams. In this work it was decided to omit the detailed definition of breach formation process. The dike break was assumed to be sudden, instantaneous, and total. The breach width and height were imposed constant ensuring the highest flood hazard. Following the United States Bureau of Reclamation (USBR), estimating the breach width up to five times longer than water depth above final breach elevation (Wahl 2004), the breach was assumed 15-m long with the bottom at the elevation corresponding to the total erosion of the embankment. The final breach bottom level was imposed equal to 83.0 m a.s.l. and the initial water level in reservoir was 86.3 m a.s.l. The hydrograph of the water outflow through the breach is presented in Fig. 3. It was estimated using known reservoir stage-area relation and applying the simple broad crest flow equation for calculation of discharge inside the breach.

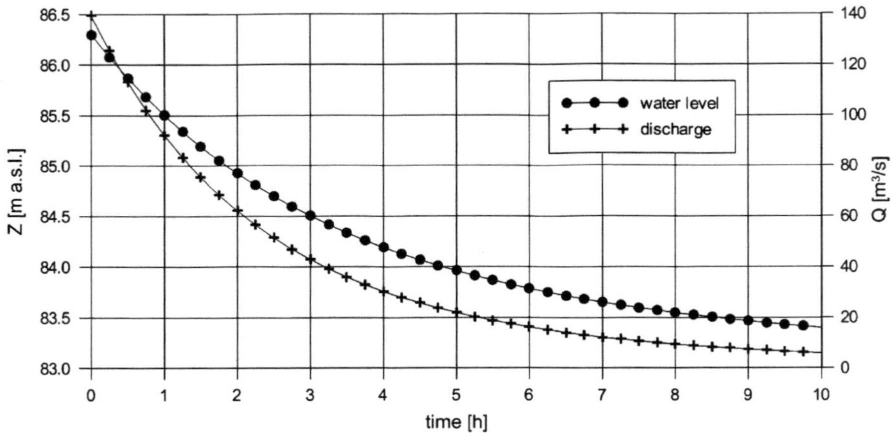

Fig. 3 Outflow hydrograph

3 Hydrodynamics Model and Solution Method

The mathematical model used in the Kolbudy II reservoir case study is a system of shallow water equations. The model can be presented in conservative form as (Abbott 1979):

$$\frac{\partial \mathbf{U}}{\partial t}+\frac{\partial \mathbf{E}}{\partial x}+\frac{\partial \mathbf{G}}{\partial y}+\mathbf{S}=0 \tag{1}$$

$$\mathbf{U}=\begin{pmatrix} h \\ uh \\ vh \end{pmatrix},\quad \mathbf{S}=\begin{pmatrix} 0 \\ -g\,h\left(S_{ox}-S_{fx}\right) \\ -g\,h\left(S_{oy}-S_{fy}\right) \end{pmatrix} \tag{2a,b}$$

$$\mathbf{E}=\begin{pmatrix} uh \\ u^2h+0.5g\,h^2 \\ uvh \end{pmatrix},\quad \mathbf{G}=\begin{pmatrix} vh \\ uvh \\ v^2h+0.5g\,h^2 \end{pmatrix} \tag{2c,d}$$

In the system of (1) and (2): t = time; (x, y) = horizontal coordinates; (u, v) = depth-averaged velocities in x and y directions; h = local depth; S_{ox}, S_{oy} = bed slopes in x and y directions; S_{fx}, S_{fy} = bottom friction terms in x and y directions defined by Manning formula; g = gravity. Equation 1 can be presented in another vector form as (LeVeque 2002):

$$\frac{\partial \mathbf{U}}{\partial t}+\nabla\mathbf{F}+\mathbf{S}=0 \tag{3}$$

where vector $\mathbf{F}$ is defined as $\mathbf{Fn} = \mathbf{E}n_x + \mathbf{G}n_y$ and $\mathbf{n} = (n_x, n_y)^{\mathrm{T}}$ is an unit vector.

In order to solve the mathematical model of free surface water flow (3), a numerical method of partial difference equations integration has to be implemented. To integrate the model in space, the finite volume method (LeVeque 2002) was applied. This method requires to calculate the fluxes of mass and momentum through the computational cells (volumes) interfaces. They were computed using the Roe scheme (Roe 1981). Detailed description of the method is available in the literature (Toro 1997); therefore, it is not presented in this chapter. The solution of (3) must be completed with time integration scheme. The two-step explicit scheme of finite difference method is used in solution algorithm. The computational code for numerical simulation of the flash floods was prepared at the Hydroengineering Department of Gdańsk University of Technology (Szydłowski 2007).

4 Flood Numerical Simulation

The geometry of potential flood region is presented in Fig. 4. It was covered with the unstructured triangular mesh composed of 12,463 elements. The mesh was locally refined along the reservoir embankments to better represent the relief of the floodplain and complex structure of the flow near the breach. In this region, the lengths of the element sides were equal to 5 m. They were increasing up to 50 m with the distance from the reservoir embankments. The additional mesh refinement was applied along an expected flood path.

The reservoir was excluded from the computational domain. The boundary of the computational domain along the reservoir embankment (except for the breach) was assumed closed. The other boundaries were treated as the open ones. Initially, at the beginning of the flood simulation, the surface of the floodplain was assumed dry. At the broken dike segment, the hydrograph (Fig. 3) was assumed as a boundary condition. The calculations were carried out with the time step $\Delta t = 0.1$ s and the total simulation time was equal to 1.5 h.

Fig. 4 Geometry of flow area and numerical mesh

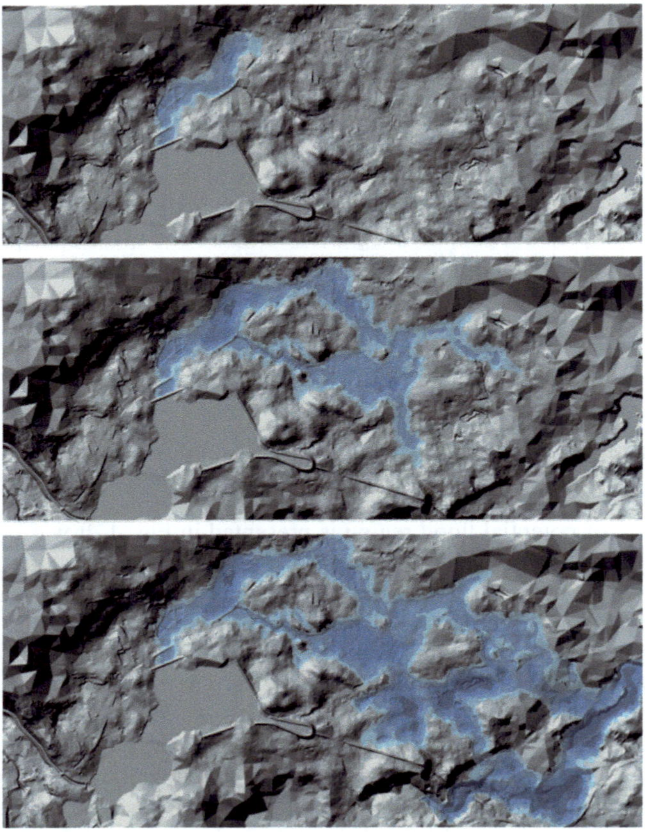

Fig. 5 The flood zone after $t = 0.25$ h, $t = 0.75$ h and $t = 1.5$ h

The results of flood simulation are presented in Fig. 5. The pictures present the flood range evolution and the water depth at the moments $t = 0.25$ h, $t = 0.75$ h and $t = 1.5$ h after the reservoir dike collapse.

As it can be seen, after embankment breaking the water body is released from the reservoir and the flood wave propagates in a horizontal plane. The flood expands adequately to the embankment breach location and the relief of the floodplain. The flood area is divided into four zones (shades of blue color), where depth is less than 1 m, in the range of 1–2 and 2–4 m, and greater than 4 m. Such depth zoning is suggested by the Polish National Water Management Authority and can be directly used for hazard zones identification.

After total simulation time $t = 1.5$ h the filling of the surface retention of flood area is observed. At this time, the simulated flood wave front is reaching the Radunia River valley and the Goszyńskie Lake being the reservoir of the Straszyn hydropower plant – the next one in the Radunia cascade line.

The computations presented in Fig. 5, as well as the other simulation results (flow velocities, flood wave front propagation speed, and time of water arrival) estimated for the several variants of the reservoir embankment breaking were used by the ENERGA Elektrownie Straszyn Ltd. (an owner and operator of the Bielkowo hydropower plant) to create the flood risk maps. The maps have been used by the Gdańsk Board of Water Management and the Disaster Recovery Center to manage the flood risk in the Gdańsk region.

5 Conclusion

Numerical simulation of the dam breaching and the catastrophic flood are important for reducing threats due to potential reservoir embankments failures. The results of hypothetical flood simulation after the Kolbudy II reservoir embankment break prove that the two-dimensional Saint-Venant model is a proper solution to analyze the spatial pattern of the risk of flooding. The model provides information about depth and duration of inundation as well as velocities and direction of water flow. These parameters combined with the information about inhabitants number and buildings construction and value can give an approximation of flood risk.

The Bielkowo hydropower plant is one of the eight power plants located on the Radunia River. Considering the cascade form of the Radunia hydropower system, the mathematical modeling of flood routing along whole Radunia River valley is necessary to manage the flood risk in this region. The need of the complex analysis of the flooding after collapse of one of the hydropower plants is a result of danger of failure of the next dam due to overflowing. The extreme flood following the failure of the first dam could damage some hydrostructures along the cascade increasing the flood hazard in the river valley. Such a situation should be simulated and the complex catastrophic flood risk maps for the Radunia catchment ought to be elaborated. At this moment the possibility of flood analysis for the Radunia catchment is limited due to incomplete DEM of the catchment area and poor reservoirs bathymetry data.

Acknowledgments This chapter presents some results obtained in the project "A mathematical model of the hydraulic effects of the Bielkowo hydropower derivation channel embankments breaks and Kolbudy II reservoir dams failures," which has been financed by the ENERGA Elektrownie Straszyn, ul. Hoffmanna 5, 83-010 Straszyn, Poland.

References

Abbott MB (1979) Computational hydraulics: elements of the theory of free-surface flows. Pitman, London, pp 324

Jarzębińska T (2005) Hydraulic power plants on the Radunia River. In: Open channels flows in view of water framework directive. Proceedings of XXV International School of Hydraulics, Institute of Hydro-Engineering of the Polish Academy of Sciences, Gdańsk

LeVeque RJ (2002) Finite volume method for hyperbolic problems. Cambridge University Press, New York, 558 pp
Majewski W, Jarzębińska T, Jasińska E, Wołoszyn E, (2005) Characteristics of the Radunia River and its catchment in view of WFD. In: Open channels flows in view of water framework directive. Proceedings of XXV International School of Hydraulics, Institute of Hydro-Engineering of Polish Academy of Sciences, Gdańsk
Roe PL (1981) Approximate Riemann solvers, parameters vectors and difference schemes. J Comput Phys 43:357–372
Szydłowski M (2007) Mathematical modeling of flood waves in urban areas, vol 86, Monographs of Gdańsk University of Technology. Gdańsk University of Technology, Gdańsk, 149 pp (in Polish)
Toro EF (1997) Riemann solvers and numerical methods for fluid dynamics. Springer, Berlin, 592 pp
Wahl TL (2004) Uncertainty of predictions of embankment dam breach parameters. J Hydraul Eng 130:389–397

Some Observations on the Similarity of Tracer Data from a Small River

S.G. Wallis and J.R. Manson

1 Introduction

The water industry's understanding of pollutant transport in rivers is aided by undertaking tracer experiments. These usually consist of the discrete release of tracer followed by the measurement of tracer concentration-time profiles at several locations downstream of the tracer release site. The commonest analysis of such data uses the method of moments to estimate the mean flow velocity (*U*) and the dispersion coefficient (*D*) for the hydraulic conditions pertaining at the time of the experiment. By undertaking experiments over a wide range of flow rates, the results can be used to predict *U* and *D* at any flow rate by a simple regression analysis. Such predictions of *U* and *D* are a necessary requirement for implementing numerical solutions of the advection–dispersion equation, which are often the cornerstone of water quality modelling software. For several decades it has been realised that the method of moments is unreliable when used with incomplete concentration-time profiles, and the method has become out of favour because it appears to require extremely good-quality tracer data to avoid its inherent weakness. However, the aim of this chapter is to show that such pessimism is misplaced. To achieve this, some insights are provided from a cumulative analysis applied to non-dimensional tracer concentration-time profiles derived from a recent series of tracer experiments. Furthermore, by exploiting the fact that the non-dimensional profiles have very similar shapes, a pooling of information from only a few high-quality tracer experiments enables reliable estimates of *U* and *D* to be obtained from poor-quality tracer experiments.

S.G. Wallis (✉)
School of the Built Environment, Heriot-Watt University, Edinburgh, UK
e-mail: s.g.wallis@hw.ac.uk

J.R. Manson
School of Natural Sciences and Mathematics, Richard Stockton College of New Jersey, Pomona, NJ, USA

P. Rowinski (ed.), *Experimental Methods in Hydraulic Research*,
Geoplanet: Earth and Planetary Sciences, DOI 10.1007/978-3-642-17475-9_23,

2 Tracer Experiments

Six tracer experiments were carried out during the autumn of 2009 in the Murray Burn, which runs through the Heriot-Watt University campus at Riccarton in Edinburgh. The same stretch of the river had been used during an earlier set of tracer experiments (see, e.g., Wallis 2005). A map of the study area is shown in Fig. 1. In the 2009 experiments, the tracer (Rhodamine WT) was injected at Site 1 and concentration-time profiles were measured at Sites 3 and 4, which are 183 m apart. For each experiment, water samples were collected from the centre of the stream at regular time intervals and taken back to the laboratory for analysis.

The water samples were analysed in a controlled environment laboratory using the following procedure. Firstly, the water samples and tracer standards (made by serial dilution from the same stock tracer solution used to prepare the injected tracer) were stored overnight in the laboratory to allow them all to be at the same temperature. Secondly, the following morning, a Turner Designs Field Fluorometer was calibrated using the standards, and thirdly, the water samples were analysed. To enable any machine drift or temperature fluctuations to be taken into account, some of the water samples and calibration standards were reanalysed at the end of the sample analysis and the laboratory temperature was recorded several times during the analysis. The calibration of the fluorometer was linear and it remained stable during the water sample analysis, but it varied between experiments. The laboratory temperature was stable at about 20°C for all the experiments. Initial data analysis consisted of removing the background signal and converting the resultant fluorescence values to tracer concentrations. The background signal for each concentration-time profile was estimated from the end of the trailing edge.

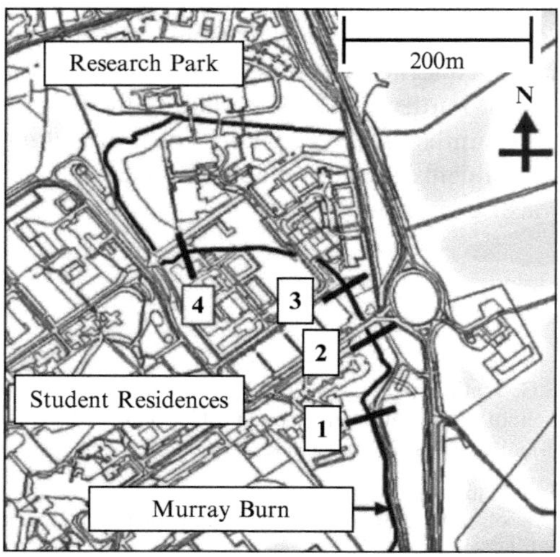

Fig. 1 Map of study area: the numbers show the sites from the earlier study

The sampling was designed to capture complete and well-resolved concentration-time profiles at both sites. Experiment 1 was, in effect, a trial run designed to discover the general solute transport features of the Murray Burn for the reach between Sites 1 and 4. As such it was successful, but due to only a small amount of tracer being used and to initial estimates of the travel time being incorrect, the concentration-time profiles were rather noisy and incomplete. Experiment 2 was much more successful, but the final parts of the profiles were not completely captured. Experiments 3–6 were successful in giving complete profiles at both observation sites. Experiments 4–6 yielded well-resolved profiles, but the profiles from experiment 3 were not quite so well resolved. In the following sections, experiments 3–5 are used to illustrate some concepts, and experiments 2 and 6 are used to validate an enhanced version of the method of moments.

3 Similarity of Tracer Profiles

The idea that tracer data might exhibit similarity was first proposed by Day and Wood (1976). When they non-dimensionalised observed tracer concentration-time profiles they found that profiles from different experiments had similar shapes. Concentration is non-dimensionalised by dividing all observed concentrations by the maximum observed concentration, or peak value (c_{p}): time is non-dimensionalised by choosing two points on the profile as reference times. The choice is arbitrary, but Day and Wood chose the two points on the profile when the concentration was half of the peak value. One of these occurs on the rising limb or leading edge of a profile, at time t_{L}, the other occurs on the falling limb or trailing edge of the profile, at time t_{T}. Day and Wood's approach was used in this work. Thus, non-dimensional profiles were constructed using the following two equations, where c and t are the observed concentration and time, and where C and τ are the non-dimensional concentration and time.

$$C = \frac{c}{c_{\mathrm{p}}} \tag{1}$$

$$\tau = \frac{t - t_{\mathrm{L}}}{t_{\mathrm{T}} - t_{\mathrm{L}}} \tag{2}$$

Since the times t_{L} and t_{T} did not correspond exactly with a time at which a sample was collected, they were estimated by linear interpolation.

Figure 2 shows non-dimensional tracer concentration-time profiles from both sites for experiments 3–5. It is clear that all six profiles have a similar shape, showing that the concept of similarity applies not only over a range of flows at an individual site but also at different sites. The small deviation (on the leading edge) between the non-dimensional profile at Site 3 for experiment 3 and the other profiles is caused by that profile not being as well resolved as the others.

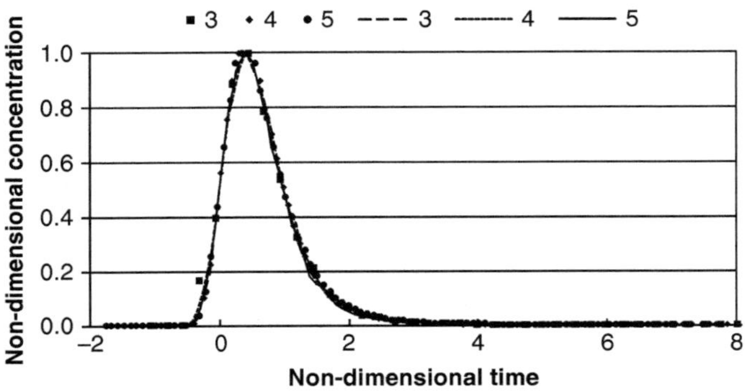

Fig. 2 Non-dimensional concentration profiles for experiments 3–5: *symbols*, Site 3; *lines*, Site 4

4 Method of Moments

The method of moments was used to derive a reach-average velocity and a reach-average dispersion coefficient for each experiment. Although this is a simple procedure, the results are not necessarily reliable. For example, concentration profiles must be complete because the trailing edges of the profiles contain a lot of relevant information. In contrast to the traditional method of moments, which uses the dimensional profiles, the recently proposed idea of using the non-dimensional profiles (Wallis and Manson 2010) was adopted. Formally, the area (α_τ), the centroid time (μ_τ) and the variance (σ_τ^2) of a non-dimensional profile are given by the following equations:

$$\alpha_\tau = \int_0^\infty C(\tau)\mathrm{d}\tau \tag{3}$$

$$\mu_\tau = \frac{\left[\int_0^\infty \tau C(\tau)\mathrm{d}\tau\right]}{\alpha_\tau} \tag{4}$$

$$\sigma_\tau^2 = \frac{\int_0^\infty \tau^2 C(\tau)\mathrm{d}\tau t}{\alpha_\tau} - \mu_\tau^2 \tag{5}$$

In practice, the integrals were approximated using numerical integration, the limits of the integration were defined by the first and last water samples collected and the results were observed in cumulative form. Thus, the development of the profile properties with increasing non-dimensional time could be observed.

Figures 3 and 4 show the cumulative development of the centroid and variance of the non-dimensional profiles from both sites for experiments 3–5. The similarity of the

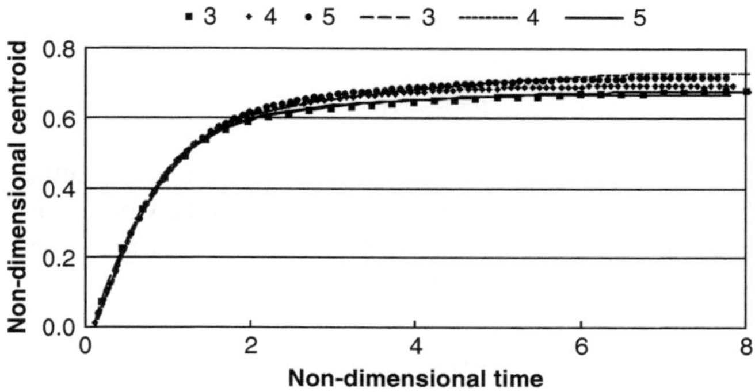

Fig. 3 Cumulative development of the centroid time of non-dimensional profiles for experiments 3–5: *symbols*, Site 3; *lines*, Site 4

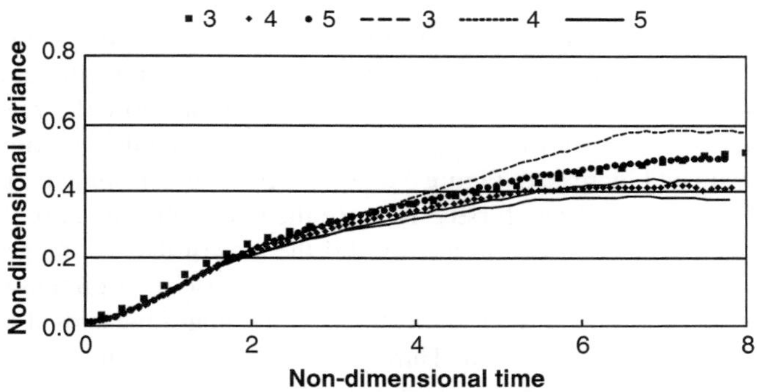

Fig. 4 Cumulative development of the variance of non-dimensional profiles for experiments 3–5: *symbols*, Site 3; *lines*, Site 4

non-dimensional concentration profiles is illustrated by the very close grouping of these cumulative results. Only for non-dimensional times greater than about 2 do the lines start to diverge. The significance of the information in the trailing edges of the profiles is particularly striking for the variance, where its value doubles or even triples as the non-dimensional time increases from 2 to 8, even though less than 5% of the tracer mass is contained in this part of the profiles. Also there are considerable differences between the variances of the non-dimensional profiles for non-dimensional times greater than 4, and the profile from Site 3 for experiment 3 is slightly offset, but again this is only due to it being poorly resolved.

The following equations were used to estimate the area (α), centroid time (μ) and variance (σ^2) of the original profiles from the properties of the non-dimensional ones:

$$\alpha = \alpha_\tau c_{\mathrm{p}}(t_{\mathrm{T}} - t_{\mathrm{L}}) \tag{6}$$

$$\mu = \mu_\tau(t_T - t_L) + t_L \tag{7}$$

$$\sigma^2 = \sigma_\tau^2(t_T - t_L)^2 \tag{8}$$

which reflect that the area is related to the height and width of a profile, the centroid time is related to the timing and width of a profile, and the variance is related only to the width of a profile. Finally, the reach-average velocity and dispersion coefficient were evaluated using the following equations:

$$U = \frac{L}{(\mu_4 - \mu_3)} \tag{9}$$

$$D = 0.5L^2 \frac{(\sigma_4^2 - \sigma_3^2)}{(\mu_4 - \mu_3)^3} \tag{10}$$

where L is the reach length and the subscripts refer to the sampling sites that define the reach boundaries. By using values of μ and σ^2 at the reach boundaries at the same non-dimensional time, (9) and (10) were used to explore how the reach-average velocity and dispersion coefficient changed as the non-dimensional time at which they were evaluated increased. Figure 5 shows some results from experiments 3–5. Three things should be noted. Firstly, because the non-dimensional times at which data is available cannot be controlled a priori the closest data to integer values of non-dimensional times were used. Secondly, because non-dimensional times also vary between sampling sites, an average of the non-dimensional times at the two sites was used to plot the data. Thirdly, the average non-dimensional centroid time and variance (derived from experiments 3–5) were used in (6)–(8). Figure 5

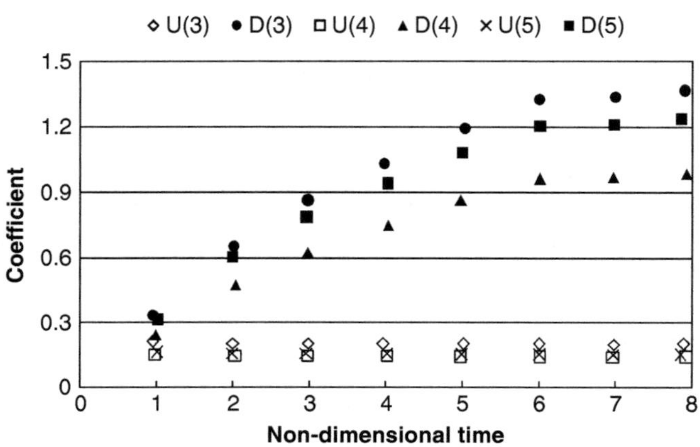

Fig. 5 Cumulative development of velocity, U (m/s), and dispersion coefficient, D (m^2/s), for experiments 3–5

shows that the dispersion coefficient increases significantly with increasing non-dimensional time, approaching an asymptotic value for non-dimensional time > 6. The velocity reduces slightly, and also approaches an asymptotic value. When the individual non-dimensional profile properties for each experiment were used in (6)–(8), the results for non-dimensional time > 6 were very noisy and the asymptotic behaviour was more difficult to discern than when the average profile properties were used.

5 Discussion

It is interesting that the cumulative development of the properties of the non-dimensional concentration profiles shown in Figs. 3 and 4 exposes differences between the profiles that are not apparent from the profiles themselves (shown in Fig. 2). Furthermore, Figs. 3 and 4 suggest that a non-dimensional time of 2 has some significance for the study reach. When $\tau < 2$ the properties of all the profiles are very similar, but when $\tau > 2$ the properties diverge. Interestingly, however, they all fall within an envelope. A possible reason for this behaviour is the effect of transient storage.

Transient storage refers to the temporary trapping of material in dead zones and bed interstices. It usually manifests itself by the presence of extended and elevated trailing edges on concentration-time profiles. Certainly, the trailing edges on the profiles are very long, but they are not significantly elevated. The divergence of the profile properties in Figs. 3 and 4 could be caused by variations in the capacity of the transient storage at different flow rates.

The cumulative development of the properties of the non-dimensional concentration profiles also give useful insight into some of the disadvantages of the method of moments for evaluating dispersion coefficients. In particular, the extremely heavy dependence on the information in the trailing edges of the profiles, which is a well-known phenomenon, is quantified. Figure 5 suggests that the dispersion coefficient can double between non-dimensional times of 2 and 8. Indeed, an interesting question arises, namely, at what non-dimensional time should the dispersion coefficient be evaluated? The answer depends on how the dispersion coefficient is interpreted. For example, in the presence of transient storage, evaluating the dispersion coefficient from the profile properties at large non-dimensional times provides an overall dispersion coefficient, which includes the transient storage effect. But this is not the same as the Taylor/Fischer shear flow dispersion coefficient, which only describes the dispersion created by cross-sectional gradients of (longitudinal) velocity and cross-sectional mixing (Taylor 1954; Fischer 1967). If a non-dimensional time of 2 is significant for identifying the onset of the effects of transient storage, perhaps the Taylor/Fischer shear flow dispersion coefficient should be evaluated at this time.

The evaluation of the velocity is much more robust because it only depends on the centroid times, which are more closely grouped than the variances, and shows only about a 2% reduction between non-dimensional times of 2 and 8. A velocity reducing in this way is consistent with it accounting for the presence of transient storage at large non-dimensional times because transient storage delays the passage of the tracer through the reach.

Finally, the similarity of the non-dimensional concentration-time profiles offers a means of predicting the velocity and the dispersion coefficient from tracer experiments that, at first glance, might seem to have been unsatisfactory. For example, if the trailing edge of a concentration-time profile is not observed it is clear that the moments of that incomplete profile are of little use. However, because all non-dimensional profiles have similar shapes, it is possible to use non-dimensional properties derived from one or more other profiles instead, provided that these profiles were completely observed. This method is termed the enhanced method of moments (Wallis and Manson 2010), and is illustrated here by estimating U and D for experiments 2 and 6, but using severely truncated profiles. Provided that the profiles capture the peak and the 50% peak concentration reference points the method can be used. For example, Fig. 6 shows the truncated profiles used for experiment 2. If this data were analysed with the method of moments the results would be very unreliable. However, using (7)–(10) with average values of μ_τ and σ_τ^2 at each site (derived from experiments 3–5 at a non-dimensional time of 2) and values of t_L and t_T derived from Fig. 6, a velocity of 0.066 m/s and a dispersion coefficient of 0.245 m^2/s were found. These compare very favourably with values obtained using the complete profiles for experiment 2 (0.066 m/s and 0.273 m^2/s, respectively). Figures 7 and 8 show the benefits of being able to use poor-quality tracer data in this way, where the range of flow over which velocity and dispersion can be estimated is significantly extended by including the analysis of truncated profiles from experiments 2 and 6, compared to only using experiments 3–5. The flow rates were estimated using dilution gauging with the complete profiles at Site 4.

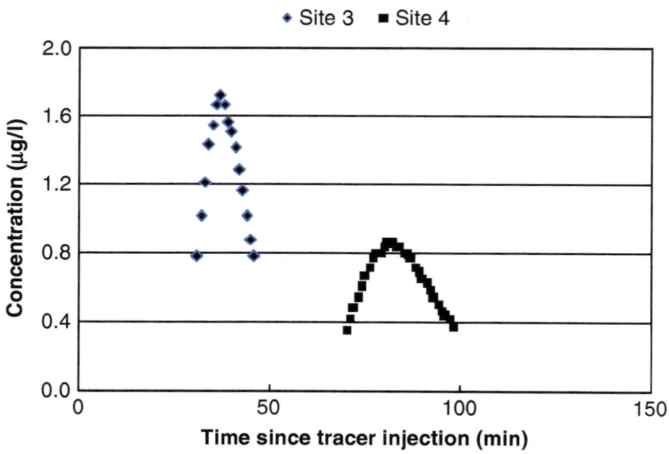

Fig. 6 Truncated concentration-time profiles for experiment 2

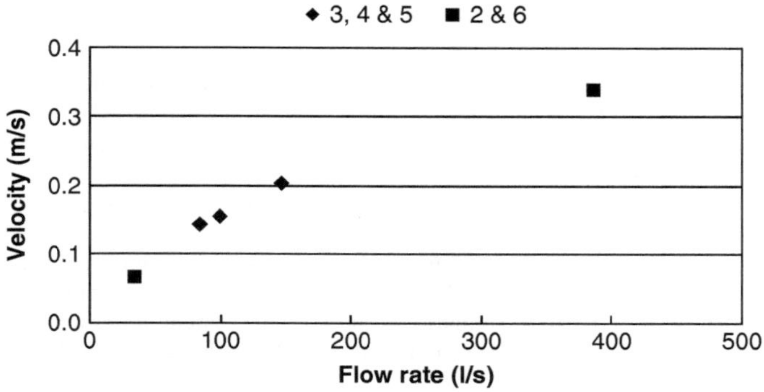

Fig. 7 Variation of velocity with flow rate in the study reach

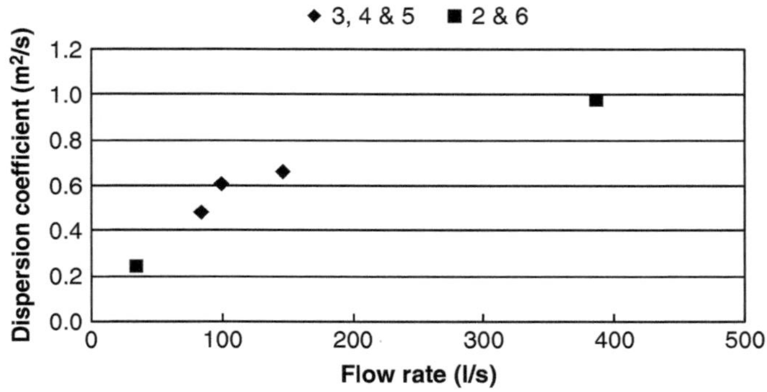

Fig. 8 Variation of dispersion coefficient with flow rate in the study reach

6 Conclusions

The chapter has described the collection and analysis of concentration-time profiles from a recent series of tracer experiments conducted in a small stream in Edinburgh, UK. Once converted to a non-dimensional form it was found that all profiles from both measurement sites had similar shapes. An analysis of the cumulative development of the centroid time and variance of the non-dimensional profiles revealed that the profiles were very similar for non-dimensional times <2, but became less similar at greater non-dimensional times. The same type of analysis applied to the reach-average velocity and dispersion coefficient, which were calculated by applying the method of moments to the non-dimensional profiles, illustrated the sensitivity of these parameters to the length of the trailing edges of the profiles used in the analysis. Some insight into the interpretation of the parameters obtained in this way was achieved. Finally, using an enhanced form of the method of moments that exploited the similarity of the non-dimensional profiles, it was shown that even severely truncated profiles could yield valuable information on velocity and dispersion.

References

Day TJ, Wood IR (1976) Similarity of the mean motion of fluid particles dispersing in a natural channel. Water Resour Res 12:655–666

Fischer HB (1967) The mechanics of dispersion in natural streams. J Hydraul Div Proc Am Soc Civ Eng 93:187–216

Taylor GI (1954) The dispersion of matter in turbulent flow through a pipe. Proc R Soc Lond A 233:446–468

Wallis SG (2005) Experimental study of travel times in a small stream. In: Czernuszenko W, Rowinksi P (eds) Water quality hazards and dispersion of pollutants. Springer, New York

Wallis SG, Manson JR (2010) A similarity inspired enhancement for estimating dispersion coefficients in rivers. In: Proceedings of HydroPredict2010, Prague, Sept 2010

Index

P. Rowinski (ed.), *Experimental Methods in Hydraulic Research*,
Geoplanet: Earth and Planetary Sciences, DOI 10.1007/978-3-642-17475-9,

MIX
Papier aus verantwortungsvollen Quellen
Paper from responsible sources
FSC® C105338

If you have any concerns about our products,
you can contact us on
ProductSafety@springernature.com

In case Publisher is established outside the EU,
the EU authorized representative is:
Springer Nature Customer Service Center GmbH
Europaplatz 3, 69115 Heidelberg, Germany

Printed by Libri Plureos GmbH
in Hamburg, Germany